THE
WAGES
OF GLOBALISM

The Wages of Globalism

Lyndon Johnson and the Limits of American Power

H. W. Brands

New York   Oxford
Oxford University Press
1995

Oxford University Press

Oxford   New York   Toronto
Delhi   Bombay   Calcutta   Madras   Karachi
Kuala Lumpur   Singapore   Hong Kong   Tokyo
Nairobi   Dar es Salaam   Cape Town
Melbourne Auckland

and associated companies in
Berlin Ibadan

Published by Oxford University Press, Inc.,
200 Madison Avenue, New York, New York 10016

Oxford is a registered trademark of Oxford University Press

Library of Congress Cataloging-in-Publication Data
Brands, H. W.
   The wages of globalism: Lyndon Johnson and the limits of American power /
H. W. Brands.
   p. cm.   Includes bibliographical references and index.
   ISBN 0-19-507888-8
   1. United States—Foreign relations—1963–1969.
   2. Johnson, Lyndon B. (Lyndon Baines), 1908–1973. I. Title.
E846.B65   1994
327.73—dc20   93-40140

987654321

Printed in the United States of America
on acid-free paper

Preface

It is impossible to consider American foreign policy during the presidency of Lyndon Johnson without thinking immediately of the Vietnam War. The reasons for this require no elaboration. But for all its prominence, the war in Vietnam was essentially a manifestation of a much broader phenomenon confronting the United States during the mid-1960s. Twenty years after the end of the Second World War, America was running up against limits inherent in the globalist policies American leaders had pursued since the late 1940s. For nearly a generation the United States had bestridden the world, pledging to defend territories in every time zone and at almost every latitude against radical challenges to the status quo. For most of this period, American leaders had managed to cover most of their commitments. But by the middle 1960s the debts were falling due faster than Washington could refinance or retire them. While Vietnam witnessed the most spectacular default, the problem of oversubscription was a general one.

Lyndon Johnson faced a whole world of troubles during his five years in the White House. In western Europe, a presumably friendly region, Charles de Gaulle effectively declared the era of American preponderance finished when he threw American troops out of France and withdrew France from NATO in 1966. Next door, West Germany, whose prosperity and even existence as a sovereign state owed very much to

American actions, did not treat Americans so rudely as de Gaulle did, but Bonn's growing economic strength gave rise to increasing political assertiveness, and in attempting to hold NATO together in the face of de Gaulle's challenge, Johnson and other Americans had to learn to deal with the Germans as equals—in some respects more than equals.

To the south and east, where Europe meets Asia, two American allies nearly went to war twice in 1964 and not quite so nearly once in 1967. The Cold War had begun in earnest for the United States in 1947 when the Truman administration announced a policy of backing Greece and Turkey against Soviet bullying and communist encroachment. The policy had succeeded well enough that the Greeks and Turks now felt free to indulge their age-old enmity for each other. Or almost free: when Johnson threatened to fold the American nuclear umbrella, the latter-day Trojan warriors stood down. Even so, the fact that Athens and Ankara worried more about each other than about Moscow demonstrated the degree to which things had changed since the early postwar period.

Farther east still, the Arabs and Israelis were at each other's throats as usual. But the usual acquired unusually portentous overtones in 1967 when what started as simply the third round of fighting over Palestine resulted in the occupation by Israel of large portions of Israel's neighbors' territory. In 1956 the United States had stopped a previous war in the Middle East, one including Britain and France as well as Israel, with a crack of its economic whip. After that war Washington had forced the invaders to withdraw from territory captured. In 1967 the United States could not keep even Israel alone from attacking, nor could it get the Israelis to relinquish the most important portions of the land its soldiers seized.

In South Asia, an American balancing act begun early in the Cold War toppled over during Johnson's tenure. The United States had leaned toward Pakistan during the early 1950s, later toward India. In 1965 the two countries went to war, leaving Johnson to choose or not choose between them. Johnson, like Melville's Bartleby, preferred not. Unlike Bartleby's employer, neither India nor Pakistan respected Johnson's preference, leaving the president to reassemble the pieces of a broken policy.

In Southeast Asia, Vietnam commanded American attention, partly because certain other parts of the region appeared a lost cause. The largest and strategically most significant country of Southeast Asia, Indonesia, through most of 1965 seemed headed straight for the communist camp. At one time Indonesian president Sukarno had sufficiently valued America's goodwill to steer no more leftward than neutralism between the superpower blocs. But during Johnson's first two years in office Sukarno dropped any pretense of evenhandedness and threw American support back in America's face. Washington, painfully aware of its lack of leverage with Sukarno, resigned itself to whatever fate might befall Indonesia. In the event, the Indonesian army elbowed Sukarno aside and reversed Indonesia's course. Yet the turnaround was a matter of luck for the United States rather than a matter of power, as officials of the Johnson administration recognized.

The loosening of the American grip affected even the Western Hemisphere, that traditionally safe sphere of American influence. Fidel Castro had pried American fingers off Cuba a few years before Johnson entered the White House, and during Johnson's time Castro sought to extend the prying to other parts of the Caribbean. In 1965 a revolt broke out in the Dominican Republic against a regime favored by Washington. The revolt resulted less from Castro's inspiration than American officials believed, but having seen one American neighbor go communist, they determined that no more should be allowed to follow. As leaders of relatively declining powers often do, Johnson overreacted. He reversed three decades of military nonintervention and poured more than twenty thousand American troops into Santo Domingo. The invasion squelched the revolt in the Dominican Republic but corroborated Castro's claim that the Yankee imperialists were running scared.

It was Lyndon Johnson's peculiar bad luck to preside over American foreign policy at the moment when the scales of world power were tipping away from the United States. To be sure, American military, economic, and political resources still outweighed those of any other single country. Nor, absolutely, were American resources declining. Yet whereas in the aftermath of the Second World War the United States had possessed projectable power matching that of the rest of the world combined, by the twentieth anniversary of the war's end other countries were rapidly catching up with the United States. When Truman announced the Truman doctrine in 1947, the idea that the United States could guarantee a friendly status quo in the far corners of the globe appeared ambitious but not inconceivable. During Johnson's half-decade as president, this idea shattered on the hard reality of a new international order.

Though Johnson handled the war in Vietnam disastrously, with the fault lying partly in his stars and partly in himself, he did better managing other aspects of the transition to a world no longer dominated by the United States. He dealt prudently and gracefully with de Gaulle. He negotiated a satisfactory solution to economic and political problems with West Germany. He prevented Turkey and Greece from going to war. He helped keep the third Arab-Israeli war short and localized. He held back from the second India-Pakistan war and did a fair job sweeping up the shards of his predecessors' policies afterward. He got out of trouble's way in Indonesia and positioned the United States to capitalize on the bad luck that befell Sukarno and the Indonesian left. He contained Castro in the Caribbean, if only by blunderbuss means.

Lyndon Johnson will never be judged a master of the diplomatic arts.

Vietnam precludes any such verdict. But Vietnam was not as important as Johnson made it out to be—which was the fundamental misperception that produced most of the agony—and as the Vietnam War recedes into the past, the opportunity arises to examine and evaluate Johnson and his foreign policy in wider terms. What follows is a first step in that direction.

It is only a first step. Anything approaching a comprehensive account of Johnson's foreign policy remains years off. The Johnson Library in Austin has opened many of its files, but much high-level material is still in the classified vaults. Available records of the State and Defense Departments are even thinner. If the CIA *ever* releases the bulk of its working papers for the 1960s, many researchers, including this one, will be surprised.

But one has to start somewhere. As the notes at the back of this book indicate, I have relied principally on documents at the Johnson Library. Most of these documents are part of what the archivists there call the National Security File. I have also made use of the large and extremely informative collection of interviews conducted by the Johnson Library's oral history program. The majority of these interviews took place within a few years after the events they describe; as a result, the memories of the interviewees were sharp and relatively unclouded by subsequent developments. I have supplemented these interviews with interviews and correspondence of my own with several surviving members of the Johnson administration. But because most of what I would have asked was answered already, more satisfactorily than would be possible a quarter-century later, I have not attempted any large-scale interviewing project. And because it would be presumptuous to expect individuals to recall details of what they were doing during a very busy time of their lives many years ago, I have used recent recollections chiefly for general background. For their kind assistance, I would like to thank George Ball, Robert Barnett, George Benson, McGeorge Bundy, William Bundy, Chester Cooper, Michael Forrestal, Marshall Green, Robert Komer, Thomas Mann, Edward Masters, Robert McNamara, Walt Rostow, and Dean Rusk.

I would also like to thank the staff of the Johnson Library. Linda Hanson managed the morning shift with professionalism and good humor. David Humphrey, who knows all of the Johnson administration's secrets, told what he could (before moving on to the State Department).

Robert Divine and Douglas Little read an early version of the manuscript and offered characteristically cogent comments. Thanks to them.

Austin   H. W. B.
February 1994

Contents

THE
WAGES
OF GLOBALISM

One | **Great Expectations**

THE OVERCOMMITTING began when Truman pledged the United States to support "free peoples who are resisting attempted subjugation by armed minorities or by outside pressures." Truman was thinking particularly of the peoples of Greece and Turkey—neither country quite a paragon of individual freedom, but close enough for the purposes of the hour—yet the loose language of Truman's doctrine rendered it potentially applicable almost anywhere. The Marshall Plan, announced several weeks later, was more modestly conceived, while more lavishly financed. American strategists had learned enough from the unfortunate aftermath of the First World War to guess that gifts would probably be less expensive in terms of political fellow-feeling and perhaps even monetary payback than loans. They guessed correctly. The Marshall Plan proved the cheapest antiwar and antidepression insurance the United States ever purchased.

William Jennings Bryan had opposed shipping American money to Europe in 1915 on grounds that where money went, much more would follow; his ghost might have issued a similar warning in 1948, with equal prescience. Truman, having requested and received approval to invest billions of dollars in the future of the continent, in 1949 requested and received senatorial authorization to send American troops to protect the investment in the event of war. In 1951 he followed up the North Atlantic treaty by requesting and receiving the troops themselves. Off they went, becoming hostages who guaranteed that any European war would immediately escalate into a third world war.

Dwight Eisenhower and John Foster Dulles elaborated the globalism implicit in the Truman doctrine. The Republican president and secretary of state piled pact after pact atop NATO, eventually completing a chain of allies and clients that girdled the earth. From Canada the chain ran across the Atlantic via Iceland to northern and western Europe, thence south and east through Greece and Turkey to Iran and Pakistan, through Southeast Asia to Taiwan, up to South Korea and Japan and back to North America, with lateral branches arcing down through the Philippines to Australia and New Zealand and through Central America and the Caribbean into South America. The world had never seen the like of this sphere of influence—known to friends as the "free world." The "free" in the label did not imply freedom to defect from the American sphere, as the Mossadeq government in Iran discovered in 1953 and the Arbenz government in Guatemala in 1954. Nor did it imply the right to indulge in policies of which Washington disapproved strongly, as the Eden government of Britain and the Mollet government of France discovered during the Suez affair of 1956. Freedom for individuals within member countries was optional, as dissidents in Greece, Turkey, Iran, Pakistan, South Vietnam, Taiwan, South Korea, Nicaragua, Guatemala, and Cuba—to cite the more egregious instances of unfreedom—discovered.

American commitments did not diminish during John Kennedy's brief tenure. If anything, Johnson's immediate predecessor expanded the scope of American activity with his vow to pay any price and bear any burden to assure the survival of liberty. In certain parts of the world—Latin America, for instance—Kennedy placed greater emphasis than Eisenhower on combatting communism by economic development. But elsewhere—notably Southeast Asia—he showed himself readier to spill American blood than the general-president had been.

Johnson had no desire to extend the American sphere further. Johnson set his heart instead on remaking the empire's core: on regenerating America. But much was expected of Lyndon Johnson. What others had created he must defend. By the early 1960s the United States was the greatest power in world history, and like other great powers it aimed to hold what it had. Because it had so much, the holding was an enormous job. Harry Truman, on learning of Franklin Roosevelt's death, had said he felt as though the sun, moon, and stars had fallen on him. Johnson, possessed of a larger ego than Truman, never would have uttered such words. But if he had said he felt that half a world had fallen on him, he would not have missed the truth by much.

II

Along with a foreign policy, Johnson inherited a team of foreign-policy advisers. The policy of a presidential administration is never better than the people who make it. Often it is worse. Johnson's was neither, reflecting fairly accurately the strengths and weaknesses of the men—there were no women in responsible foreign-policy positions in Johnson's administration—who made it.

As is supposed to be the case, constitutionally and by custom, the president's right-hand diplomat was the secretary of state. Dean Rusk brought to the head of the State Department a great deal of experience. Following service with General Joseph Stilwell in the China-Burma-India theater during the Second World War, Rusk had joined the War Department's policy staff. At the beginning of 1946 he switched to the State Department. In the bureau responsible for American activities in the United Nations, he displayed a tendency to look to international law for solutions to world problems—a tendency that placed him somewhat outside the realist mainstream of George Kennan, Paul Nitze, and others of the dry-powder school. Yet the legalists and the realists concurred regarding the importance of collective security in averting a third world war, and by the time the United States sponsored the Atlantic treaty of 1949 Rusk was drawing a harder line against communist expansion than even Kennan. (No one drew a harder line than Nitze, the lead author of the apocalyptic NSC-68, which called for permament American mobilization against the global communist conspiracy.)

Events in Asia proved the most formative for Rusk's future. After a brief stint as Dean Acheson's deputy "for substance," as Acheson denominated him, Rusk accepted the thankless and dangerous—in career terms—position of assistant secretary of state for the Far East. With China fallen and Korea falling, the Republicans wanted scalps. They preferred the scalp of "Red Dean" Acheson but in a pinch would settle for the other Dean's. Fortunately for Rusk, though Joseph McCarthy's snoops investigated him, other matters distracted the Wisconsin demagogue, and Rusk emerged relatively unscathed. All the same, Rusk witnessed the damage raging rightists could do to American foreign policy and the costs of appearing soft on communism. He would remember.[1]

Rusk also learned and remembered the perils of underestimating China. He had helped select the thirty-eighth parallel as the line dividing Korea into zones for the surrender of Japanese forces. He appreciated the artificial character of Korea's division, and when Douglas MacArthur's stunning reversal of the Korean War's momentum opened the way to Korea's reunification, the assistant secretary pushed hard for letting the general chase the communists clear to the Yalu. Truman, hot to put the communists in their place and the Republicans in *theirs*, accepted the idea. Neither Rusk nor Truman credited Beijing's warnings that an American approach to China's border would bring the Chinese into the war. After the roof caved in at the end of November 1950, weighted by three hundred thousand Chinese troops, Rusk realized that on matters touching their security and especially their frontiers the Chinese were no paper tigers.

Rusk soon left the State Department for the quieter confines of the Rockefeller Foundation. During the eight years of Republican management of American world affairs he kept out of controversy's reach, doling out John D.'s millions and writing an occasional piece for *Foreign Affairs* and other staidly establishment publications.

With Kennedy he crossed the New Frontier back into government. Rusk was a safe choice for secretary of state, which suited Kennedy, especially since the Democratic president intended to direct foreign policy from the White House. But as Rusk did with most people he worked for, he grew on Kennedy. He won Kennedy's respect when he gave no public hint of having repeatedly told the president the Bay of Pigs operation was the screwball scheme it turned out to be. Rusk had first apprenticed under George Marshall, and from Marshall he acquired the conviction that a secretary of state answers only to the president and God, in that order. Not for Rusk the intellectual free-for-alls that passed for policy-making under Kennedy. He saved his thoughts for the Oval Office and gained a reputation, among those who didn't really count, for dullness and lack of imagination. Yet his influence with the one who really did steadily increased.

Upon Kennedy's assassination, Lyndon Johnson, like others who had been in the new president's awkward position, pledged his administration to continuity. At least until he won election on his own, Johnson would not consider letting any of Kennedy's top people resign. He most wanted to keep Rusk. During Johnson's vice presidency, Rusk, perhaps remembering how Roosevelt's death had thrown the totally unprepared Truman into the hurricane of international affairs, had taken pains to brief Johnson on what America was up to around the world. Johnson appreciated the attention. As he told his brother Sam Houston Johnson, Rusk was not like some of the "bastards" in the Kennedy administration. "He's a damned good man. Hard-working, bright, and loyal as a beagle. You'll never catch him working at cross purposes with his President. He's just the kind of man I'd want in my Cabinet if I were President."[2]

Rusk remained as tight-lipped under Johnson as under Kennedy. Referring to Johnson, Rusk later said, "I never let any blue sky show between his point of view and my point of view." Rusk added, "It's not good for a president and a cabinet officer to debate each other in the presence of other people." Johnson agreed. Johnson concurred equally in another of Rusk's sentiments. "Dean Acheson once said that in a relation between a president and a secretary of state it is important that both understand which is president," Rusk recalled. "President Johnson never had any doubt about who was president. Nor did I."[3]

In terms of qualities that endeared Rusk to Johnson, the secretary's barefoot boyhood ran a close second to his loyalty. Cherokee County, Georgia, conceded nothing to the Texas hill country in gully-rutted living. Conditions on the Rusk farm got so bad by Dean's fourth year that the family gave up and moved to Atlanta. Rusk bookstrapped his way to Davidson College in North Carolina and from Davidson to England's Oxford, the latter venture courtesy of Cecil Rhodes. Further finishing in Berlin might have erased completely the memory of Georgia's red clay, but it didn't. Even after heading the Rockefeller Foundation, which could have bought half the Ivy League, Rusk remained more comfortable swapping lies with Johnson about growing up poor and southern—Johnson was the better liar—than mixing with the eastern and transatlantic elite into which he had won membership.

Rusk was conscientious to a fault. He kept longer hours than anyone in the administration (except Johnson, who was notorious for squeezing two work days into twenty-four hours by breaking for an afternoon nap—pajamas and all—and rising to a second round of meetings, telephone calls, and state functions). Rusk observed weekends by coming to the office sans tie. He rarely took holidays and then usually on orders from the president. Occasionally the strain told. In June 1964 he mentioned to McGeorge Bundy that he had been having nightmares. Bundy relayed the message to Johnson, who insisted that the secretary break his breaking routine.[4]

Seconding Rusk at the State Department was George Ball. Ball entered life in 1909, ten months after Rusk and sixteen months after Johnson—an eventful period that also included, as Ball liked to point out, the final tottering of the Manchu dynasty in China, Japan's absorption of Korea, and a variety of assassinations, deposings, and minor wars that presaged the lively decades he and his generation would inhabit. An Iowan, Ball went to Washington with the New Deal, arriving a bit before Johnson. In Washington, Ball met Dean Acheson, then acting treasury secretary. Acheson showed Ball a thing or two on the handball courts of the Hotel Ambassador, but when a policy dispute split the Treasury Department from the White House, Acheson taught Ball a more valuable lesson, one that would come in handy for Ball and Johnson both three decades later. Acheson taught Ball how to resign like a gentleman.

In 1935 Ball decided to make some money. He practiced law for several years in Chicago, where he met Adlai Stevenson. Pearl Harbor pulled Ball back to Washington to work on Lend-Lease. In 1944 he received an assignment that proved crucial to the development of his thinking on the efficacy of military—especially air—power in achieving political objectives. With Paul Nitze and John Kenneth Galbraith, Ball directed a survey of Allied strategic bombing, the aim being to determine whether the air offensive against German cities was significant in shortening the war. The survey's conclusion, which Ball would recite during the bombing campaign against North Vietnam, was that any shortening was insignificant.

Ball kept close touch with Stevenson during the decade and a half after the war, and in the intraparty fence-mending that followed Kennedy's 1960 defeat of Stevenson and then Nixon, Ball got the job of undersecretary of state for economic affairs. When principal undersecretary Chester Bowles took only months to demonstrate that his many gifts did not include modesty, an essential quality in undersecretaries, especially under Dean Rusk, Ball inherited the number-two post in the State Department. He held that job when Johnson became president.

Ball combined intellectual integrity with good-soldiership. Though—and because—Ball consistently challenged conventional wisdom within the Johnson administration, the president insisted on hearing Ball out. Johnson also relied on Ball for delicate missions abroad. In the spring of 1964 Johnson sent Ball to Paris to talk with de Gaulle about Vietnam. To some observers Johnson's choice of Ball for the assignment appeared surprising since Ball had discreetly but vigorously voiced skepticism about the president's approach to both Vietnam and the French general. Yet Johnson knew his man. "George," he said, "you're like the school teacher looking for a job with a small school district in Texas. When asked by the school board whether he believed that the world was flat or round, he replied, 'Oh, I can teach it either way.' That's you. You can argue like hell with me against a position, but I know outside this room you're going to support me. You can teach it flat or round."[5]

Ball taught it flat for three years. He faithfully served president and country, perhaps most significantly by helping prevent Cyprus's endemic mayhem from spreading to Greece and Turkey. "George is a powerhouse," remarked David Bruce, ambassador to Britain during the Kennedy and Johnson years, in explaining Ball's capacity for adding special assignments to his regular job as undersecretary. Ball possessed "phenomenal energy" and "great ability," Bruce said. He was "an activist who got things done." Lincoln Gordon, Johnson's ambassador to Brazil who subsequently assumed the post just below Ball in the State Department, did not disagree with Bruce. But Gordon added that Ball sometimes let administrative matters slide. While undersecretaries customarily keep an eye on the State Department's running, Ball conceived himself a policy man. "He really didn't give a damn about the internal organization," Gordon said.[6]

Yet eventually Ball decided that his and Johnson's views on Vietnam diverged too greatly for him to provide the president meaningful advice. He followed Acheson's example of thirty years before and left like a gentleman.

Together Rusk and Ball managed to hold the State Department's own against the Defense Department. For most of the period from 1947, when the War and Navy Departments had merged to form the Department of Defense, to 1961, the Pentagon had played a relatively small role in formulating foreign policy. During the early phase of the Korean War, the generals and their muftied associates gained influence, but George Marshall, secretary of defense in the crucial months from September 1950 through the following spring, considered the subordination of the military—especially the subordination of General MacArthur—to the civil authorities a matter almost sacred. John Foster Dulles dominated Charles Wilson during the Eisenhower years, and Eisenhower completed the pacification of the Pentagon by acting in important respects as his own secretary of defense.

Kennedy knew nothing about running the country's largest bureaucracy, and cared less; for the job he hired Robert McNamara, a brilliant, driven man who immediately made his mark on Washington. Clark Clifford, McNamara's successor at the Defense Department and no lightweight himself, later commented, "No one ever held the capital in greater sway than Robert S. McNamara did from 1961 until the end of 1967." McNamara cracked the Defense Department into shape, and with energy to spare he began encroaching on the turf of others, notably Rusk. The low-key Rusk seemed no match for McNamara. In some ways he wasn't. McNamara understood how America's defense posture in the Cold War could mold its diplomacy, and used this understanding to great effect. Rusk, for his part, recalling the bickering between Dean Acheson and George Marshall's predecessor as defense secretary, Louis Johnson, sought accommodation with McNamara. Rusk did not question McNamara's prerogatives in the military area, an area that grew with growing involvement in Vietnam. Where accommodation did not challenge traditional State Department prerogatives, Rusk was willing to give ground. Yet the secretary of state sometimes stood fast. Insisting that the State Department retain primary responsibility for the activities of American officials at the embassy in Saigon, Rusk told McNamara, "If you want to direct our embassy in Vietnam, then give me the marines."[7]

Absent Vietnam, McNamara probably would have slipped relative to Rusk after Johnson assumed the presidency. McNamara's sharp style was much more in tune with Kennedy's preferences than with Johnson's, and though Johnson respected McNamara's formidable gifts, the rapport between Johnson and McNamara never approached that between Johnson and Rusk. The escalation of the Vietnam War, however, required the president to lean heavily on the defense secretary.

The third corner of the administration's foreign-policy triangle was occupied by the national security adviser, Johnson's in–White House expert on international affairs. McGeorge Bundy created the modern office of the national security adviser—a dubious achievement in light of what some of Bundy's successors made of the post. Truman's and

Eisenhower's special assistants for national security had assisted but not advised. Their circumscribed duties reflected mostly the preferences of the two presidents, and partly the preferences of the strong and highly visible secretaries of state they served with. Kennedy, convinced that the foreign-policy bureaucracy had contributed materially to the drift he perceived during Eisenhower's last years, aimed to streamline the process and concentrate power in the White House. In appointing Bundy, Kennedy shot better than he knew.

Bundy had entered the arena of international affairs working for Henry Stimson at the War Department during the 1940s. After helping the venerable statesman write his memoirs, Bundy returned to academic life, scorching a trail across Harvard Yard to become dean of the college at the age of thirty-four. Such a record naturally attracted the attention of Kennedy, who kept a close eye on his alma mater. Assuming that the prima donnas of government could not prove significantly more fractious than the prima donnas of academia, Kennedy set Bundy to coordinating White House affairs relating to national security. Bundy soon rewrote the job description. Between Rusk's reticence and his own capacity for work, not to mention a temperament and style that meshed wonderfully with Kennedy's, Bundy transformed a coordinating role into a policy-making one. He enlarged his staff by adding area experts until it grew to rival the State Department. As long as Bundy held the security adviser's position, the rivalry remained muted. Snide memos about lethargy and lack of imagination at the State Department sometimes circulated in the Bundy shop. But Bundy respected Rusk, and he left the serious guerrilla warfare to the Kissingers and Brzezinskis who came after. Rusk returned the respect and appreciated the good work that came out of Bundy's office. "They had some extraordinarily competent people on that staff," the secretary remarked later. "You were glad to get their help most of the time because they had judgments to contribute." George Ball found no cause for complaint against the White House staff. "Bundy was extremely fair in all of his dealings," the undersecretary said. "We used to have a lot of arguments, but we never had any problems between us."[8]

Those who worked for Bundy found the experience stimulating to the point of fright. James Thomson, Bundy's assistant for East Asia, described his boss as "the most interesting, exciting employer I ever worked for." Bundy had no tolerance for the slow and not much for the reflective. Bundy was "a super-pragmatist," Thomson said. Thomson remembered a typical Bundyism: "The fact of the matter is: We're here, and what do we do tomorrow?" Thomson continued, "If you begin to suggest reconsidering the whole, unraveling the entire ball of yarn, you're in danger of being viewed as a sorehead and a long-winded fool." The high-pressure atmosphere did not encourage humility. Thomson explained: "One of the delights and horrors of working for Mac Bundy is that no one ever tells you anything. You come onto the job, and there's no briefing as to what you're supposed to do. You're just supposed to be a quick learner and fall in fast." If ignorance intruded, one did not admit it. Thomson recalled an instance when a visitor from the Pentagon arrived with a sheaf of maps and diagrams, requesting approval for what he described as "armed reconnaissance." Thomson recognized the country in question as Laos but did not know what armed reconnaissance entailed. Not wishing to appear uninformed, he declined to ask. When he relayed the request to Bundy, the national security adviser told him to exercise his own judgment. Thomson approved the request. "I only sort of much later realized that what I'd been signing off on was rather heavy bombing of an entire country."[9]

Although Johnson got on tolerably with Bundy, the latter was too much Kennedy's man—in reality and in Johnson's opinion—to make a comfortable transition to the new administration. Bundy worked at the relationship. He adjusted himself to Johnson's way of doing business, sending recommendations generally ending in three options:

Approve _____

Disapprove _____

See me _____

Customarily Johnson would check one and Bundy would implement the decision. Occasionally Johnson would scribble some remarks at the bottom elucidating the reasons for a decision.

Bundy grew to appreciate Johnson's gifts, different though they were from Kennedy's. At least he gave that impression. Nine months into Johnson's full term, Bundy sent the president a Superball, a lively toy currently the rage of the nation's playgrounds. In a covering memo Bundy wrote, "This ball is very bouncy—it comes back fast—it gathers speed as it goes along—and it has a pressure and spin which are all its own. So it is like you, and I hope you like it."[10]

But Bundy's heart fell out of his job with Kennedy's killing, and Johnson did not exert himself to keep Bundy once the Johnson administration was fairly launched under its own steam. At the end of 1965 the Ford Foundation offered Bundy its directorship. Bundy accepted early the following year.

Robert Komer followed Bundy briefly, after having served as as Bundy's deputy on the White House staff. If Bundy possessed a reputation for impatience with the slow and ill-prepared, at least his years in academic administration had fostered a veneer of gentility in suffering fools. Komer also came to the Johnson White House from Harvard, but only after several years' seasoning in the CIA, which had sharpened his edges and shortened his fuse. When Bundy recommended Komer to Johnson, the national security adviser described his deputy as "able, energetic, quick and highly knowledgeable." Bundy added, "While he presses his point of view with energy, he is disciplined in the execution of decisions, whether or not they accord with his recommendations." Komer's energy was what impressed his associates, who tagged him "Blowtorch Bob." Komer had little use for most of the career diplomats in the State Department. Of the disdainful comments around the White House about the pettifoggers of Foggy Bottom, more than his share came from Komer. Johnson evidently recognized the trouble that might follow Komer's permanent appointment as national security adviser; consequently the president, after working with Komer in the White House a few months, decided that Komer's gifts could find better employment elsewhere. He assigned Komer to head the administration's Vietnam pacification program, on the view that Komer ought to turn his blowtorch on some real bad guys.[11]

Instead of Komer, the president tapped Walt Rostow to be Bundy's permanent replacement. Like Bundy, Rostow had hitched to Kennedy's star during the 1950s. Yet Rostow's connection was more political and doctrinal than personal, and it did not extend to the rest of the Kennedy family. A native of New York, Rostow attended Yale and Oxford—the latter on a Rhodes scholarship, like Rusk—before joining the Office of Strategic Services during the war. A brief stint with the State Department gave way to teaching posts at Oxford and Cambridge. In 1951 he settled down at the other Cambridge, taking a position in economic history at the Massachusetts Institute of Technology. While working on a capitalistic reply to Marx entitled *The Stages of Economic Growth,* Rostow joined forces with Senator Kennedy to promote aid to Third World countries like India.

No one ever accused Rostow of thinking small (he subtitled his opus on economic growth *A Non-Communist Manifesto*), and after Kennedy's inauguration Rostow assumed direction of the State Department's big thinkers: the Policy Planning Council. The planning post encouraged a tendency toward wordiness that had to be remedied when Rostow succeeded Bundy at the White House in March 1966. "I think Walt's the man to take Bundy's job," Johnson told aide Jack Valenti. "But I want you to talk with him because I think one of Walt's faults is his prolixity, his verbosity. I want you to tell him how I like to be handled." Valenti explained the situation to Rostow: the president wished his memos terse and his briefings straight to the point. Valenti said that Rostow, in opening meetings, should "state the issue and briefly, very briefly, maybe state the pros and cons, and then shut up." Johnson noted substantial improvement, as did Rusk. Rusk remarked that at first Bundy was a better drafter than Rostow, that Rostow ran on too long. But with time Rostow cut his length and became "a very efficient special assistant in all respects."[12]

Johnson once compared Rostow and Bundy in a conversation with a journalist. "Bundy is sharp, humorous, incisive, decisive, scintillating, positive," the president said. "Rostow is philosophical—this is pro, this is con. He doesn't try to shove anything unless it is wanted. Both of them are ideal staff men. Rostow doesn't try to win an argument and lose a sale. He says, 'Here's my judgment,' as an able analyst."

In the same conversation, Johnson reflected on the other two leading members of his foreign policy team: Rusk and McNamara. "If you went in with a cabinet of thirteen men and you asked who is the ablest, wisest in the group to bet the lives of your wife and daughters, Rusk would get twelve of the thirteen votes. The thirteenth would be his." McNamara's abilities lay in a different direction. "If you wanted an organization run right, wanted to make a profit and wanted a thoroughly honest operation, McNamara would get the vote. He is a genius as an organizer." The two worked well as a team since each appreciated the responsibilities of the other. "McNamara has a deep understanding of the diplomatic side, and Rusk was only twenty minutes from being a professional military man, so they understand each other's job."[13]

III

A rough rule of American policy-making during the Cold War was that Republican presidents generally liked the Central Intelligence Agency and Democrats didn't. Harry Truman distrusted the spooks as a proto-Gestapo and agreed to the agency's creation only under duress of the deepening conflict with the Soviet Union. Eisenhower, by contrast, reveled in the freedom the covert operators gave him to pursue his objectives against the communists by means short of open military conflict. Eisenhower authorized operations to topple governments in Iran, Guatemala, and, unsuccessfully, in Indonesia. He countenanced assassination plots against leaders of the Congo and Cuba. He directed preparations for a paramilitary invasion of Cuba. He ordered the penetration of Soviet airspace by American spy planes. He brought Allen Dulles, the CIA director, into his inner circle of policy advisers, where Dulles exercised considerable influence over the shape of American foreign policy. By the time Eisenhower left office, the CIA—still under Allen Dulles, while the State Department had passed to a successor far less influential than brother John Foster Dulles, and while the Defense Department remained in its pre-McNamara doldrums, and while the White House staff had yet to be Bundified—was arguably the predominant branch of the American national security bureaucracy.

Kennedy, burned by the agency's bungling at the Bay of Pigs just three months after inauguration, booted Allen Dulles and lassoed the CIA's operatives abroad. Kennedy installed the concept of the country team, under which all American personnel in a foreign country, including the

CIA station chief, reported to the ambassador. Kennedy did allow efforts to destabilize Castro—mortally if necessary—and he approved continued covert operations in Vietnam, but on the whole the secret warriors found their scope considerably constructed from the Ike age.

In the post-Johnson era, Richard Nixon would turn the CIA loose on enemies abroad, including Chilean president Salvador Allende, and enemies at home, including student protesters and government leakers. Nixon used the CIA in the Watergate coverup, inadvertently creating a climate for exposing the agency's dirty secrets. Following the Ford interregnum, Jimmy Carter attempted to clean up the mess the Senate investigators had laid bare, again curtailing the agency's activities. Ronald Reagan once more reversed the trend, giving agency director William Casey a large voice in policy-making and allowing Casey to mount anticommunist operations in Afghanistan and Angola and a covert war against the government of Nicaragua—the last financed by secret weapons sales to Iran.

Johnson fit the pattern for Democrats. Johnson shut down the anti-Castro operation, although not at once. He increased the agency's involvement in Indochina, but since this was part of the much larger military effort there, the involvement did not quite qualify as peacetime meddling in other countries' business. Equally important, Johnson downgraded the CIA director to the status of one who spoke when spoken to, and spoke then to convey information rather than advice. The president met with his intelligence chief only when he wanted specific facts and figures, and he conspicuously declined to include the CIA in matters of policy-making. Richard Helms, the CIA director who served Johnson longest, remembered, "On no occasion in all the meetings I attended with him did he ever ask me to give my opinion about what policy ought to be. Even during the most intense decisions and most difficult decisions over Vietnam and other problems he never deviated from this." The extremely narrow manner in which Johnson defined the role of the director of central intelligence encouraged his first director, John McCone, to quit. Johnson substituted Admiral William Raborn, the father of the American missile-submarine force, but a man somewhat out of his depth in intelligence matters. Helms took over in June 1966.

Helms was an intelligence professional, and he did not mourn the CIA's lost policy influence. To some degree Johnson's compartmentalization of intelligence made Helms's job easier. Helms said of the president, "I do not believe that he really had the faintest idea of how the Central Intelligence Agency was organized or how the intelligence community was organized." Helms went on, "He expected me to produce the goods. When I produced them, which I think I did with regularity, he never asked any questions about where they came from." Because the intelligence agency was essentially divorced from policy-making, it had few positions to defend. "I helped to keep the game honest," Helms recalled.

The CIA director found Johnson open to reasoned dissent. The dissenter, however, must defend himself and his position. "It behooved one to go right back at him," Helms said, "and go back at him forcefully and without any undue amount of kid-glove work so that he clearly understood your side of the case. President Johnson did not like being opposed, but I think he liked far less any individual for whom he didn't have any regard and who he didn't think could stand up for his side."[14]

IV

In treating intelligence gathering as a black box, leaving the inner workings to professionals like Helms, Johnson exhibited a trait he displayed toward the foreign-policy process generally. Unlike Kennedy, whose calls to desk officers drove their superiors crazy, Johnson only occasionally jumped the chain of command. He insisted on approving bombing targets in Vietnam, and he got deeply involved in the details of the Dominican crisis of 1965, but for the most part he contented himself to work from the top down.

If Johnson did not usually look below his principal appointees for foreign-policy advice, he did look beyond them. Like Franklin Roosevelt, against whom he measured himself in many areas, Johnson sought the counsel of amateurs and outsiders. Johnson's frequent consulting with Abe Fortas, the Supreme Court associate justice, raised constitutional questions that helped scuttle Fortas's nomination for the chief justiceship. Johnson chose to suffer the complaints rather than forgo Fortas's advice. Fortas had more to tell the president about domestic politics than about international affairs, but in addition to possessing common sense, useful in any realm, Fortas enjoyed influential contacts abroad. During the Dominican crisis, Johnson sent Fortas to Puerto Rico, whose principal figures Fortas knew personally. Through his Puerto Rican friends Fortas met with Juan Bosch, the deposed president of the Dominican Republic and the focus of much of the agitation in Santo Domingo. Fortas's negative impression of Bosch confirmed Johnson's inclination to seek elsewhere for a solution to the Dominican troubles.

Clark Clifford played a role for Johnson similar to that of Fortas. The three often huddled to discuss particularly vexatious matters. Clifford, the consummate Washington insider, served as a self-described "general utility man" before succeeding Robert McNamara at the Pentagon in 1968. Clifford's political sensibilities matched Johnson's in acuity, while his unofficial status afforded him a perspective the president lacked. Although Johnson overrode Clifford's warnings against Americanizing the Vietnam War, the president took Clifford's objections seriously and demanded of the hawks in the administration answers to the questions

Clifford raised. In addition Clifford provided a link to previous Democratic administrations, a matter of great importance to Johnson, especially when things went wrong.[15]

Johnson also solicited the views of an advisory panel of senior foreign-policy experts, the "wise men." George Ball did not think them so wise. After one meeting on Vietnam, the undersecretary went up to Dean Acheson and Arthur Dean and burst out, "You goddamned old bastards! You remind me of nothing so much as a bunch of old buzzards sitting on a fence and letting the young men die. You don't know a goddamned thing about what you're talking about. You just sit there and say these irresponsible things." Pointing at Acheson, Ball demanded, "Would you have ever put up with this if you had been secretary of state?" (As he related the story later, Ball chuckled over Acheson's reaction. "He said afterwards that I shook the hell out of him. He said he was very upset by it.")[16]

Yet the irresponsibility that enraged Ball provided detachment Johnson could not get from persons within the administration. Clifford, Acheson, Arthur Dean, John McCloy, Robert Murphy, and occasional others, having relatively little riding on current policies, had little to gain or lose by defending or attacking them. At times like the one Ball described, the sages seemed simply to ratify administration actions. This was hardly surprising, in light of Johnson's efforts to continue the policies Acheson and the rest had set forth during their turns at responsibility. But they did not always nod assent. If a single factor caused Johnson to change his mind about the winnability of the war in Vietnam, it was the 1968 shift of the collective mind of this advisory group. Meanwhile Johnson got double duty from some members. He persuaded Acheson to mediate between Greece and Turkey in 1964 and John McCloy to negotiate with the British and Germans in 1966 and 1967.

One outside consultant held a unique position. During the 1950s Dwight Eisenhower and Lyndon Johnson had forged a fairly effective working relationship, the former as head of the executive branch and spokesman for the Republican Party, the latter as majority leader in the Senate and the most powerful Democrat in the country. The relationship worked most smoothly in international affairs, on which, after the Republicans dropped their containment-as-defeatism campaign rhetoric, there existed the closest thing the United States has enjoyed to a foreign policy consensus during the twentieth century. Eisenhower's administration cultivated the consensus assiduously. Cultivating the consensus required cultivating Johnson. Gerard Smith, a State Department official under John Foster Dulles, described his boss's attitude toward the majority leader. "I can just hear Dulles saying time and again, 'What does Lyndon Johnson say?,' whenever there was a crisis or some problem. It was quite a refrain. 'What does Lyndon Johnson say?' " Usually Lyndon Johnson said yes, not least because Eisenhower and Dulles took care to avoid anything that derogated from the prestige of Congress. Eisenhower refused to commit American planes or troops to Indochina in 1954 without congressional approval, which the legislature declined to give. In 1955 Eisenhower submitted America's secondhand dispute with China regarding the Taiwan Strait to congressional consideration, receiving in response the Formosa resolution authorizing the use of American military force in the area. In 1957 Eisenhower asked Congress to grant similar authority for the Middle East, which Congress did. In each case Johnson played an important part in determining the outcome.[17]

Whether Johnson at the time expected Eisenhower to return the favors should Johnson become president, only the majority leader knew. As president, Johnson certainly expected help from Eisenhower, though not so much from back-scratching considerations as from a well-founded conviction that he was pursuing the same objectives Eisenhower had. McGeorge Bundy, writing to General Andrew Goodpaster, an assistant in the Eisenhower White House and Johnson's liaison to the former president, said in 1965, "President Johnson's belief is that what he is trying to do now is what all American Presidents have been trying to do since 1947. He takes great encouragement from the belief that General Eisenhower and he see eye to eye on these fundamental principles."[18]

Johnson repeatedly picked Eisenhower's brain. In the aftermath of the Kennedy assassination, Johnson immediately turned to Eisenhower for advice on his new responsibilities. In February 1965, when Johnson was pondering a sustained air war against North Vietnam, the president brought Eisenhower to the White House for a lengthy interview. At this meeting Johnson told Eisenhower how he had modeled the Tonkin Gulf resolution of the previous summer on Eisenhower's Formosa resolution. Eisenhower encouraged Johnson to stand firm in Vietnam. "We should be sure that the enemy does not lack an appreciation of our stamina and determination," the former president said. "When we say we will help other countries, we must then be staunch." During the next several months, as Johnson weighed the arguments for and against increasing American troop levels in Vietnam, Johnson regularly sought Eisenhower's views. Eisenhower did not hesitate to give them, admonishing Johnson to see the venture through to the end. "When you once appeal to force in an international situation involving military help for a nation, you have to go all out," Eisenhower said. "Do what you have to do."[19]

V

As a Democratic president and a distinguished alumnus of Congress, Johnson should have been able to count on the support of the Democratic congressional leadership. He got it only in part. The speaker of the House of Representatives, John McCormack, took no particular interest in foreign affairs and was usually content to follow the president's direction and example. Until January 1965 Johnson made a habit of inviting McCormack, next in line to the presidency, to meetings of the National Security Council. In the Senate, the majority leader, Mike Mansfield, was a longtime student of international affairs, especially affairs touching East Asia. Johnson benefited from Mansfield's temperate and useful advice on a variety of topics. Mansfield not infrequently told Johnson things the president did not want to hear. Mansfield judged Americanization of the Vietnam War a mistake, and he thought the United States should reduce its military presence in Europe. Yet Johnson's dealings with Mansfield remained constructive for both and valuable to the country.

Johnson's dealings with J. William Fulbright were another story. As chairman of the Senate's Committee on Foreign Relations, the Arkansas Democrat held the most influential post in Congress on matters of American international policy. While Johnson finished Kennedy's term and for the first few months of Johnson's own, Fulbright hewed to the party line. But during the Dominican intervention of 1965, a rift developed between the two southerners. Although Fulbright originally backed the president's decision to send troops, the senator gradually grew convinced that Johnson had deceived him regarding the reason for the decision. In September 1965 Fulbright went into opposition with a widely noted speech attacking the president's handling of the Dominican matter. Before long Fulbright was publicly questioning Johnson on Vietnam as well. Fulbright repented of having guided the Tonkin resolution through his committee with administration-pleasing swiftness, and eventually he came to challenge nearly everything about Johnson's conduct of the war.

The Johnson-Fulbright schism had as much to do with personalities as with policies. George Ball, a longtime Fulbright friend, remembered that Fulbright had taken it hard after the 1952 election when Johnson brashly and successfully put himself forward for Democratic leader in the Senate. Fulbright, who had wanted someone more liberal for the post, was "deeply upset," Ball explained. Ball went on to describe Johnson and Fulbright as "incompatible." Only half in jest, Ball asserted, "One of Bill Fulbright's great disabilities was that he read books. Nobody in the Senate should read books." Johnson hadn't read many books while in the Senate and hadn't changed habits since. "They weren't the same kind of people," Ball declared. "Fulbright thought of himself as an intellectual. He was a man who was interested in ideas. He wasn't interested in politics per se." Johnson, of course, ate, slept, and dreamed politics. "So they were incompatible." Johnson himself put the dispute simply—too simply: "Fulbright has never found any president who didn't appoint him secretary of state to be satisfactory."[20]

The dispute between the two men made for some tense congressional briefings. Occasionally Johnson let his dissatisfaction show. During a session with the legislative leadership in the summer of 1967, the president told Fulbright to put up or shut up on the war. "If you want me to get out of Vietnam, then you have the prerogative of taking back the resolution under which we are out there now. You can repeal it tomorrow. You can tell the troops to come home." Eventually Fulbright and other skeptics did just what Johnson suggested, but by the time they repealed the Tonkin resolution, the Vietnam War was Richard Nixon's responsibility.[21]

VI

Johnson had as little patience with the news media as with Fulbright. Much of the nation had laughed at Eisenhower when the Republican president pretzeled the language in press conferences. Yet Eisenhower had long been used to the spotlight the media lavish on war heroes, presidents, and serial killers, and if his sentences sometimes failed to parse they served his purposes—which more than occasionally included obfuscation. The media loved Kennedy, who invented the sound bite (or employed speechwriters who did) and, in contrast to Nixon, did not sweat. The camera presence Ronald Reagan spent years learning in Hollywood came to Kennedy naturally.

While anyone would have had trouble following Kennedy, Johnson brought special disabilities to his relations with the media. Johnson's political style was intensely personal. Through decades of practice he had learned that his most effective method of persuasion was to put himself in the place of the person he was trying to persuade. America has never seen Johnson's equal in one-on-one jawboning. But the method failed when applied to large numbers of people, especially to people on the far side of a television camera. How could Johnson put himself in the places of millions of people at once, people he did not know and could not see?

To worsen matters, many of the correspondents for the leading news organizations in the country—the *New York Times,* the *Washington Post,* the television networks—had the same kind of well-heeled background that discomforted Johnson among the polished easterners of Kennedy's crowd. That the media's job was to make the president look bad, or at least to discover discrepancies between presidential profession and presidential practice, discomforted him still more.

For some journalists Johnson developed a special loathing. David Halberstam of the *New York Times*—whose perceived lack of objectivity regarding Vietnam had infuriated Kennedy as well, provoking the president to suggest to the *Times*'s management that Halberstam be pulled off the Saigon beat—seemed to Johnson the epitome of the self-important, smart-alecky types who plagued his administration. Johnson once remarked that Frederick Nolting, ambassador to South Vietnam during the early 1960s, was a good officer who ought to have been allowed to stay longer at that sensitive post. "But he wasn't a Charles River man and he wasn't a fellow who would give up his conviction to satisfy Halberstam," the president said bitterly. Unlike Richard Nixon, Johnson developed no formal enemies list. If he had, Halberstam would have been near the top, along with fellow *Times* correspondent Tad Szulc and Dan Kurzman of the *Washington Post*. Szulc's and Kurzman's original sin against Johnson was to report searchingly about the Dominican intervention. From this crime—which doubly damned them by abetting Fulbright's defection—they proceeded to more generic forms of lese majesty.[22]

Johnson had scarcely more liking but rather more respect for Walter Lippmann. Lippmann had been writing about American presidents and American foreign policy from before Johnson had learned to read, and Johnson could not easily dismiss his criticism. On the contrary, Johnson tried hard to bring Lippmann around to the administration's point of view. White House personnel, particularly McGeorge Bundy, spent long sessions apprising Lippmann of administration goals and attempting to answer Lippmann's objections to administration policies. The sessions were largely unavailing, and the president tried other methods of dealing with Lippmann. In 1966 Johnson authorized a "Walter Lippmann project." The idea was to neutralize the veteran journalist's opposition to the administration's Vietnam policy by showing Lippmann to be a perennial naysayer who had never met a president he didn't dislike.[23]

Johnson's distrust of the media caused him to order his aides to file regular reports on their contacts with journalists. Every week Bundy and later Rostow gave detailed accounting of whom they had seen and what they had talked about. If an interview threatened to produce a less than favorable story, the president received advance warning. If a slip of the tongue cast the administration in an unflattering light, Bundy, in particular, quickly provided explanations to avert the president's anger.

VII

Johnson's worries over the reporting on his administration contributed to the development of one of the distinctive institutions of his presidency: the "Tuesday lunch." In 1947 Congress had created the National Security Council by the same omnibus act that folded the War and Navy Departments into a single Defense Department and established the CIA. The National Security Council's assignment was to ensure that the president receive the accurate and timely advice needed to cope with the dangers of the Cold War. Truman, already set in his ways, never made full use of the National Security Council. Eisenhower, by contrast, found it congenial and valuable, and records of council meetings during the

1950s show the Republican administration thrashing out matters of high policy in lengthy gatherings. Kennedy, distancing himself from Eisenhower in this area as elsewhere, deemed the National Security Council too clumsy for his ambitious purposes. Kennedy essentially placed the council on a shelf.

Johnson came down between Eisenhower and Kennedy regarding use of the National Security Council, landing nearer Kennedy. Although Johnson convened the council regularly, he did so primarily as a political gesture. The president often brought legislative leaders to meetings to inform them about current administration policies. He almost never made decisions at council meetings. "Those meetings were held because certain people expected them to be held," Clark Clifford stated later. Beyond politicking, Clifford said, nothing much happened at the formal hands of the council. "Subjects would be assigned to the national security council. They would be taken up. The subjects would be briefed. There would be some desultory discussion of the points. Then everybody would pick up and leave."[24]

Johnson felt uncomfortable with the National Security Council as a decision-making forum chiefly because he could not control attendance. By the 1947 statute, the council comprised several individuals who could and invariably did insist on attending meetings. When the principals could not attend, they sent deputies. Other persons, though not statutory members of the council, over time came to function as de facto members. Eisenhower, for example, included the treasury secretary and the budget director. As a result, council meetings were often attended by more than a dozen people, some of whom the president knew hardly at all.

If Johnson could have trimmed the attendance list without raising a fuss, he might have done so. Johnson instinctively shunned large groups in favor of small. To a certain degree, his preference for intimate gatherings may have indicated insecurity in the presence of strangers, but to a greater extent it reflected years of practice in the politics of congressional caucuses and cloakrooms. Johnson's decision-making style never lost the collegial character of his congressional days. Senators and representatives would have short careers if their public expressions were required to match their private actions—or else the machinery of government would freeze up. Johnson knew better than to expect candor in front of audiences. Eisenhower, accustomed to giving orders and resigned to the inefficiences of large bureaucracies, had few reservations about airing differences before full-dress conclaves of the National Security Council. Johnson, the veteran logroller, kept his counsel more closely. Johnson saw no reason to tip his hand to undersecretaries, particularly ones he had inherited from Kennedy.

Johnson's affinity for small groups also revealed a preoccupation with preventing leaks. "More than any other president I have known," William Bundy, assistant secretary of state for East Asia, said, "he was obsessed by this, and that he must keep it up his sleeve if he was thinking of doing anything new." Johnson himself explained his feelings on persons speaking out of school at a 1967 meeting, following unauthorized remarks by an indiscreet admiral. "If the Johnson administration goes down next year," the president declared, "I want it to go down on *my* words and *my* policies and not on what some goddamned admiral says."[25]

To minimize leakage and facilitate free discussion of foreign policy, Johnson inaugurated the Tuesday lunch. During his Senate years, Johnson had met frequently with the members of the Democratic policy committee to bat around ideas on legislation. Tuesday at lunch often proved a convenient day and time. As president, Johnson continued the tradition of Tuesday working lunches. On Tuesday, February 4, 1964, he gathered Dean Rusk, Robert McNamara, and McGeorge Bundy for a casual discussion of foreign policy. The same group met again two weeks later and soon fell into a more or less regular habit, with time off for elections, presidential trips, and other unavoidable distractions.

Over the subsequent months, the Tuesday group expanded to include the chairman of the Joint Chiefs of Staff, Earle Wheeler; CIA director Raborn (and later Richard Helms); and Johnson's press secretary William Moyers (subsequently George Christian). Initially the meetings included no designated note taker, and such records as were kept were haphazard and incomplete. Johnson liked things this way. Eventually, though, largely from the lunchers' desire to make sure they followed up on the president's directions, Johnson consented to let deputy press secretary Tom Johnson attend and jot down major points.

The Tuesday meetings covered a wide range of topics. Vietnam claimed the greatest amount of attention, but Johnson and the others also discussed the protracted crisis in Cyprus in 1964, the Dominican revolution and intervention of 1965, the India-Pakistan war of 1965, the Middle East war of 1967, the North Korean capture of the American intelligence ship *Pueblo* in 1968, recurrent difficulties regarding NATO, arms-control matters relating to the Soviet Union, and an assortment of additional issues.

In the confines of the president's dining room, the members of the Tuesday group could speak without fear of loosing secrets for the consumption of the whole country. "They were invaluable sessions," Rusk recalled, "because we all could be confident that everyone around the table would keep his mouth shut and wouldn't be running off to Georgetown cocktail parties and talking about it." The secretary continued, "We'd have a full discussion, and it was in a relaxed fashion. We could debate with each other, we could expose different points of view, we could look at all the alternatives, we could talk about the attitude of

other personalities and individuals such as senators or leading congressmen. It was a most valuable institution and made a great difference to
the ease of working relationships among those who were carrying the
top responsibility."[26]

While the Tuesday lunch naturally appealed to the reticent Rusk,
others in the foreign-policy chain of command—especially those not
invited but charged with acting on the decisions handed down by the
lunchers—found it less commendable. "The Tuesday lunch was an
abomination," William Bundy said. "It was *so* unstructured, so without
any opportunity to know what might be discussed." The assistant secretary added, "There was no preparation. And there was almost no read-
out."[27]

Bundy exaggerated. The fact that *he* didn't know what was happening
simply indicated that he was not in Johnson's closest circle. (And Bundy
lost an important source of sidestream information when brother Mac
left the administration.) Rusk's special assistant Benjamin Read described the "great deal of staffing" that went into preparations for the
meetings. McGeorge Bundy and Rostow faithfully assembled agendas
and not infrequently provided the president briefing papers for items
on the menu. After Tom Johnson began recording the meetings, such
problems of "read-out" as had existed diminished greatly.[28]

Yet William Bundy's complaint had certain merit. Even when officials
at the middle and lower levels of the administration began receiving
regular reports of the decisions announced at the Tuesday meetings,
they often lacked the context of the decisions: the reasoning behind
approvals, the perceived deficiencies of the arguments rejected, the positions of the competing factions of the bureaucracy. They operated not
entirely in the dark but in the half-light of incomplete knowledge. They
had little basis for guessing which direction the president would move
next, and, unable to anticipate, they were unable to smooth the way—
for themselves as well as for the president. Some people like being surprised, but those who do don't become bureaucrats.

For all its undeniable importance, the Tuesday lunch was sometimes
overrated, often by those not there. It was not a decision-making body,
even in the Lincolnian sense in which the president's ballot outweighed
those of the rest of his counselors combined. Though it became such a
calendar fixture that Rostow felt compelled to chide McNamara for
scheduling a Tuesday speech out of town—"Tuesdays must be held for
lunch," Rostow said—to Johnson it remained chiefly a forum for receiving information and opinions. Sometimes he announced decisions
at the Tuesday lunch. More often he took the information and opinions
back to his private quarters, where he compared them with intelligence
obtained from his night reading and from his telephoning to Fortas,
Clifford, and who knew who else, and only then gave his verdict.[29]

VIII

Robert Anderson, a Texas oilman and former treasury secretary who was among the president's who else, described Johnson's as a "personalized presidency." Whatever the role of the Tuesday lunch, of the panel of "wise men," of the National Security Council, and of the other institutional structures Johnson utilized in making policy, the foreign policy of his administration ultimately depended on him.[30]

When Kennedy alluded in his inaugural address to the fact that he was the first American president born in the twentieth century, Johnson might have reflected that he too was a child of the present century, only he had seen a good deal more of it than Kennedy. What Johnson saw most distinctly—so distinctly that the view forever colored his outlook on the world—was the decade of the 1930s. Johnson experienced this most formative period in modern American history in a way Kennedy only read about. The 1930s molded Johnson's understanding of both domestic and foreign affairs. As a junior congressman from a district wrung by the depression, Johnson learned to appreciate the potential of government for ameliorating poverty and restoring hope to communities suffering from too much of the former and too little of the latter. He witnessed the depth and force of the American desire to be left alone by the world, to tend America's garden unmolested. He observed the failure of the Western democracies to stand up to aggression, especially at Munich in 1938. Once Johnson became president, his domestic program evinced his desire to complete the New Deal and extend the comforting hand of government to those parts of society still suffering. His foreign policy demonstrated his determination to avoid the mistakes of the 1930s and spare humanity a repetition of the horrors of the subsequent world war.

The postwar period did nothing to alter Johnson's fundamental opinions on world affairs. His rapid rise in the Senate placed him among those regularly briefed on significant issues, first by Truman, then by Eisenhower. Though Johnson paid more attention to domestic developments than to international events, he learned much about global affairs from the front-row seat his position in the Democratic leadership afforded him. John Foster Dulles's frequent consultations privied Johnson to the diplomatic thinking of the Eisenhower administration, which Johnson broadly shared. With Dulles and Eisenhower, Johnson accounted communism close kin to fascism—fascism with a red face. As fascism had battened on the irresolution of the democracies, so would communism if given a chance. What alone would keep the Cold War from becoming another world war was the fortitude of the Free World.

While vice president, Johnson waved the flag where required and undertook occasional missions of somewhat greater substance, as to Southeast Asia in the spring of 1961 and to Berlin shortly after the erection of the wall later the same year. Dean Rusk kept Johnson closely informed regarding breaking events. During the Cuban missile crisis, Johnson attended all the important meetings of the top-level Executive Committee of the National Security Council. The Cuban case confirmed his conviction of the need to stand firm against the communist challenge.

Beyond reminding him of what he already knew, Johnson's exposure to the American policy-making process as majority leader and vice president increased his knowledge of world affairs. Johnson never became a student of international relations the way Richard Nixon, for example, did. But Johnson, like Nixon, early developed designs on the presidency, and he understood that anyone who intended to be president needed to keep an eye on happenings abroad. Johnson listened carefully to Dulles and Rusk, and he noted what he saw when he traveled.

Besides, a man as smart as Johnson, in his position, would have learned a great deal about the world without even trying. Johnson inspired a variety of feelings in those who knew him, yet all agreed that he possessed daunting mental firepower. Robert Kennedy remarked, "I can't stand the bastard, but he's the most formidable human being I've ever met." Eric Goldman, a Princeton historian who worked in the Johnson White House, commented that he had never encountered anyone with more raw intellectual ability than Johnson. Richard Helms remarked upon the president's "enormously intelligent mind" and his "great capacity to grasp facts." When Johnson believed an issue merited his attention, he blotted it up. "He mastered the details down to the last riffle," Helms said. John McCloy depicted Johnson as "much more exacting and penetrating" in his efforts to get at the root of questions than Kennedy had been. "Mr. Johnson always gave me the feeling that he knew a great deal about his subject," McCloy remembered. "I was always impressed by the depth of his penetration." Dean Rusk was too. The secretary of state said of Johnson, "He was a man of great intellectual capacity and had an ability to understand the issues that were in front of him clearly and in great depth."[31]

Allied with Johnson's intellect was his overwhelming personal presence. David Bruce described the Johnson force field. "I'm not frightened of him," the ambassador said, "but I must say that when he entered a room, particularly if you were going to be the only person in it, somehow the room seemed to contract—this huge thing, it's almost like releasing a djinn from one of those Arabian Nights' bottles. The personality sort of fills the room. Extraordinary thing." John McCloy related a narrow—and rare (and temporary)—escape from the Johnson treatment. In 1964 the president attempted to persuade the former high commissioner to Germany to replace Henry Cabot Lodge as ambassador in South Vietnam. "Talk about twisting your arm!" McCloy said.

I will never forget it. It was not in the big oval room but in that little room at the side there. And he was quite insistent. He's a pretty tall man, and he leaned over me and he said, "We're organizing for victory there, McCloy, and I want you to go out there and help in the organization." He rang all the changes. He went from appealing to my patriotism and shaming me with my lack of it, or lack of willingness to take on a tough job. . . . He said to me, "I want you to go out there, McCloy, because you're the finest"—or greatest or something, I forget what the adjective was but the indication was that I was a pretty successful proconsul, having in mind my German experience. And he said to me—these may not have been his exact words but they were close to it—"You're the greatest proconsul the Republic has ever had." I saw myself with a Roman toga and a laurel wreath around my head. I must say, he almost got me at that point. . . . I came out of there limp and feeling a bit ashamed of myself because I hadn't agreed to it.

(Johnson later nailed McCloy to serve as negotiator in talks with Britain and Germany over troop levels and problems regarding America's balance of payments.)[32]

Outsiders often assumed that the Johnson treatment did not work with foreigners. Unmodified, it didn't. But Johnson was clever enough to adapt his approach to changing needs. Eugene Rostow, undersecretary of state for political affairs and Walt Rostow's brother, found the president "extraordinarily sensitive and adept" in diplomatic conversations. "Simply superb," the undersecretary summarized. David Bruce conceded that many in Britain considered Johnson a "picturesque character," yet the ambassador noted that Johnson got on admirably with British prime minister Harold Wilson. George Ball explained that Johnson's political background stood him well in dealing with that most difficult of foreign leaders, Charles de Gaulle. "I think the president respected de Gaulle as a brilliantly effective politician. He had a sort of high professional respect for him, and at the same time totally distrusted him." Benjamin Read described how Johnson managed visiting dignitaries at Kennedy's funeral. Read and others at the State Department prepared index cards for Johnson identifying the scores of guests and suggesting pertinent topics of conversation. Johnson palmed the cards and worked the reception line as if he had known the visitors since birth. "He handled it just extraordinarily skillfully," Read recounted. "It left us with the greatest feeling of admiration."[33]

IX

Congressional politics is the art of coalition building. Occasionally one party or faction enjoys sufficient cohesion or numerical superiority to flout the desires of the rest of the legislature, but sustained success requires seeking common ground. Not surprisingly in light of his experience, Johnson brought to the presidency an ingrained tendency to fashion coalitions and seek consensus. From the first, Johnson made plain that he did not wish to have to resolve disputes between executive departments and offices. "He was impatient about the inability or the unwillingness of senior colleagues to agree among themselves," Rusk observed. "He disliked the role of refereeing among senior colleagues." Richard Helms detected the same dislike. "It was very clear over the years that President Johnson did not like to make decisions when he had to decide between one or the other of his cabinet officers," Helms said. "He always liked a consensus, and he liked his cabinet officers to agree about these things."[34]

In any such refereeing, the president would have to side with one of his secretaries against others. The siding likely would create hard feelings, which would distract from the important business of governing. Rancorous disputes would foster the pernicious practice of creating paper trails of objections in case chosen policies turned sour. Quarrels would increase the tendency to leak information to the press. All this Johnson could do without.

Yet while he valued consensus, Johnson did not attempt to stifle differences of opinion. At critical moments in the development of policy on Vietnam, to cite the most conspicuous and telling instance, he actively sought dissenting views. He urged George Ball and Clark Clifford to speak their minds, and he listened carefully to what they said. From the records of meetings of February and July 1965, when the president approved a bombing campaign against North Vietnam and subsequently ordered a major increase in American troop strength, one gets the impression that Johnson wished for others to join Ball's and Clifford's challenge.

At the same time, though, Johnson desired to keep differences in the family. This desire lay behind his preference for the Tuesday lunch over the National Security Council. It accounted for his preoccupation with leaks and his monitoring of the media contacts of Bundy and Walt Rostow. And it reflected the first of two basic principles that characterized Johnson's foreign policy.

This first principle was that all politics was local politics, even international politics. Johnson would not have admitted that he tailored decisions regarding the nation's security and external interests to fit political circumstances at home. The leader of the Free World is supposed to be above such things. But a professional vote counter like Johnson knew that no administration could sustain a foreign policy without the support or at least acquiescence of majorities in Congress and among the American people. Johnson also knew that any president commands a limited reserve of political capital, even a president elected as overwhelmingly as he was in 1964. Seized by a vision of reshaping American society at home, Johnson preferred to expend his capital on civil rights, on education, and on medical care for the elderly rather than on an undiminished supply of troops in Europe, on aid to Indonesia, or on a declaration of war for Vietnam. An able and energetic president can change the direction of American politics and American society to a limited degree on a limited number of issues. Johnson sought, with singular success, to change America's direction in matters touching minorities, the poor, and the otherwise disadvantaged. He made almost no effort to change America's direction in international affairs, even when change was necessary.

Johnson's concentration on domestic affairs—hardly an unusual trait in presidents, but still one worth noting—exerted a profound influence on the manner in which he conducted American foreign policy. William Bundy described how the influence appeared to the specialists who attempted to make decisions based on the diplomatic merits of cases. "It was as if you were studying a planet's motion in astronomy and you realized it was forty-five degrees off the course you'd expected, and there must be some very attractive body in the area that was pulling it off course." Johnson's perception of what politics in the United States would allow continually pulled American foreign policy off the course it would have followed under different circumstances. In some areas— relations with the NATO allies, for example—the deviation was minor. In others—relations with the Middle East and Indochina—the deviation was considerably greater.[35]

Johnson's domestic orientation contributed to his wish to present a common front on foreign affairs. Especially during his first two years in the White House—during the fight for the 1964 civil rights act and the launching of the Great Society—Johnson feared that a debate on foreign policy would disperse the coalition he was constructing for domestic reform. Having dealt with legislators for decades, he questioned the capacity of Congress to deal with more than one contentious issue at a time. If his coalition started scattering on policy toward Europe or Asia, he might never get it back together on policy toward Appalachia and the inner cities.

The second principle of Johnson's foreign policy likewise reflected an aversion to boat rocking, although for a different reason. This second principle was a strong inclination toward the international status quo. Such an inclination is again nothing unusual in American presidents, especially in the period after the Second World War. The status quo has suited the United States better than any other country since 1945, and attempts by American leaders to maintain it occasion no surprise. Yet where a Truman might embrace such novel schemes as the Truman doctrine, the Marshall Plan, and NATO to bolster things-as-they-were, and a Nixon might embrace novel bedfellows such as the Chinese, Johnson attempted to hold the line by the tested practices and allies of his predecessors. Johnson could foment a revolution in American domestic affairs, but on international issues he stuck to the traditional verities of the Cold War. With dogged lack of imagination, he strove to deliver on the communist-containing pledges of Truman, Eisenhower, and Kennedy.

Changing circumstances since the onset of the Cold War delayed delivery in some areas of American policy and prevented it in others. Where the status quo was not beyond salvage—where American power, even if relatively diminished, remained substantial—Johnson's unimaginative orthodoxy allowed his administration to muddle through. Without addressing the underlying sources of unrest in Latin America, he held a lid on troubles in Panama, Brazil, and the Dominican Republic. Without discovering a solution to the Cyprus dispute, he kept the Greeks and Turks from outright warfare. After India and Pakistan put their guns down, he moved to restore useful if not especially warm relations with the two South Asian countries. He waited out de Gaulle and patched up trade and troop troubles with West Germany. His patience regarding Indonesia eventually bore anti-Sukarnoist fruit. Only where the status quo had rotted away entirely did Johnson's conservatism fall flat. At great cost, Johnson attempted to defend the status quo in South Vietnam, not realizing, or not admitting, that by the mid-1960s there was essentially nothing left to defend.

| # Who Lost Cuba?

FROM THE BEGINNING, Johnson demonstrated a sensitiv-
ity to possible domestic repercussions of international setbacks. Having
lived through the scurrilous debates that followed the communist conquest of China and terminated, for the time being at least, the creative
period of American postwar policy, Johnson had no desire to suffer
through a similar experience as president. His sensitivity in this area
would figure in his assessment of his options in Southeast Asia, but it
first surfaced closer to home in Latin America.

Better than Ho Chi Minh (too grandfatherly), better than Mao Ze-
dong (too inscrutable), Fidel Castro fit the American stereotype of a
radical revolutionary. His furious beard, his ubiquitous battle fatigues
(How many sets did he have? Or did he never change?), his volcanic
speeches, and his excessive charisma marked him in the American imagination as the preeminent foe of hemispheric peace and a leading threat
to American security. Castro had reason for wishing the United States
ill. Since his accession to power in 1959, the American government had
deployed an arsenal of coercive instruments to bring him down. Eisenhower suspended American aid and eliminated the sugar quota on
which the island's economy had depended since the nineteenth century.
Eisenhower escalated to an embargo of nearly all American products,
which with the sugar blockade amounted to a declaration of economic

war. When Castro still survived, the Republican president ordered the CIA to develop assassination and covert invasion plans. Kennedy initially tried the invasion, which produced the Bay of Pigs fiasco. Then he attempted assassination, which likewise failed but at least remained quiet until after Kennedy died.

Johnson may or may not have credited reports of a Cuban connection in the murder of Kennedy. There is no evidence Johnson felt physically threatened by Castro. Yet he certainly felt politically threatened. The Bay of Pigs, besides casting the United States in the role of hemispheric thug and international liar, had converted Cuba from a merely Republican defeat to a bipartisan sore spot. During his five years as president, Johnson pursued one objective in Latin America above all others: to prevent another Cuba. Strategic factors entered into the president's calculations: a second Soviet ally in the region would hardly enhance American security, and the Panama Canal remained vital to American welfare and vulnerable to sabotage or other disruption. But more than anything else, the fear of political fallout from another successful communist revolution in the Americas motivated Johnson to take whatever measures were required to prevent it.

II

Johnson's fidelphobia became apparent only weeks after he entered the White House. At the beginning of January 1964, anti-American riots broke out in Panama. The excuse for the violence was the hoisting of an American flag in front of Balboa High School in the Canal Zone, an act prohibited by an agreement between the United States and Panama narrowly specifying where the flags of the two nations should fly and in what company. The deeper cause was Panamanian resentment at the semicolonial relationship Panama simultaneously benefited from and suffered under with respect to the United States. While the canal and the American presence provided employment and revenue, the Canal Zone sliced Panama in two, and the American troops stationed in the zone often acted like the occupying force they were.

After American students raised the flag above the high school, a group of Panamanian students responded with a protest. Officials at the American embassy in Panama City doubted that some of the Panamanian demonstrators had seen the inside of a classroom in years. The embassy's Wallace Stuart described many of the demonstrators as "somewhat elderly" for students. Stuart and others had no doubt that communists in cahoots with Castro were behind—and in some cases at the front of—the violence. Stuart identified Victor Avilo, a "known Communist," as a leader of the rock throwers. The assistant secretary of state for Latin America, Thomas Mann, warned of "Castro agents" busily undermining Panama. Intelligence officers of the American Southern

Command in the Canal Zone blamed much of the fighting on the Vanguardia Accion Nacional, "a pro-Castro, violently anti-U.S. revolutionary group" that was fomenting "a Castro-type revolution in Panama." The intelligencers went on to warn of future attacks by "terrorist groups composed of pro-Castro personalities, some of whom have received guerrilla-type training in Cuba."[1]

As with many American descriptions of Third World leftists, these characterizations of Castro sympathizers were at once accurate and misleading. A Panamanian nationalist in the mid-1960s was almost of political necessity pro-Castro. Castro had booted the American imperialists out of Cuba, and Panamanians who desired to regain control of their country from the Americans naturally emulated him. Castro indeed provided help to would-be revolutionaries in the Caribbean basin, from an instinct for self-preservation—since more revolutions would keep the Americans from focusing on him—as much as from socialist solidarity. Yet American officials sometimes misled themselves in thinking that Castro was directing revolutionary activities beyond Cuba or that Castro's sympathizers were Castro's agents. Unrest in Panama, Brazil, the Dominican Republic, and elsewhere in Latin America arose principally from conditions in those countries. Castro might try to capitalize on the unrest, but people in most of the region did not need Castro's guidance to find cause for complaint against the status quo.

Whatever its relation to reality, the American fixation on Castro served an important psychological purpose. Identifying Castro as the culprit throughout the hemisphere simplified the conceptual process of dealing with the complexities and confusions of Latin America. It was easier to think of the region's troubles as being chiefly the work of a malevolent individual than to untangle the social, political, and economic cords that bound the peoples of the area to inequality and repression—and bound the inequality and repression to the United States. American officials, by and large, were not fools. Most understood that if Castro disappeared, Latin America's troubles would persist. Yet even when they saw the systemic sources of turmoil in Panama and elsewhere, they found it easy to exaggerate the influence of the bad boy of Havana. In planning the Bay of Pigs debacle, American intelligence had grossly underestimated popular support in Cuba for Castro. Stung once, American analysts subsequently erred on the side of caution in assessing Castro's appeal.

Even more important, the Castro-as-devil theme played well in domestic American politics. American voters did not have the patience or interest to seek the socioeconomic causes of turbulence in the continent-and-a-half to their south. Nor had they been prepared by nearly twenty years of Cold War–mongering on the part of the American government to do so. The crisis-charged and oppressively conforming atmosphere of the late 1940s and early 1950s had crushed most of what subtlety had existed in American interpretations of the world. Trained to spot communists wherever they existed—and in many places where they did not exist—Americans easily latched onto Castro as the *eminence rouge* of the Antilles.

American political leaders abetted the attachment, and it abetted them. Thumping denunciations of communism usually sufficed to satisfy voters regarding a candidate's correctness on foreign affairs. Personifying communism in a bombastic bombthrower like Castro gave the denunciations a bite they had often lacked since Stalin's heirs started talking peaceful coexistence. Searchers for subtext could find in the rhetoric of the Cold War an effort to vindicate American virility in the face of the Marxist challenge. Much of John Kennedy's appeal had rested on his manly youthfulness and vigor—a point underlined by the football games on the White House lawn, in contrast to Eisenhower's mere golfing. (Ike had been a real football player, of course, but long ago.) Castro, exuding sweaty revolution and menacing machismo, threatened the self-image of the New Frontiersmen. He defied them to go mano a mano with him. They could not back down. Neither could Johnson.

On a more mundane level, solving Latin America's problems would take many years and lots of money if the problems had fundamentally systemic causes. Kennedy initially tried to tackle the troubles of the region at the systemic level. The Alliance for Progress conditioned large-scale American aid on land reform and other measures designed to reduce the inequality and poverty that rendered unrest inevitable. But Congress gutted the program by forbidding the use of American funds for land redistribution (too socialistic), and the executive branch could not bring itself to force the oligarchs, most of whom swore undying opposition to communism, to share the wealth. Rather than blame themselves for the alliance's failure to spread prosperity and peace across the region, American leaders preferred to blame the communists, particularly Castro.

For the current troubles in Panama, they also blamed Roberto Chiari, the Panamanian president. Amid fighting between demonstrators and American troops in the Canal Zone that left twenty Panamanians and four Americans dead, the Panamanian president tossed fuel on the fire by vowing that "the blood of the martyrs who perished today will not have been shed in vain." Chiari's foreign minister, Galileo Solis, contributed comments castigating Washington for "ruthless aggression" against Panama's "defenseless civilian population."[2]

Johnson believed Chiari's first obligation was the restoration of order, and he thought Chiari was cynically aggravating the situation—at the cost of American and Panamanian lives—for his own political purposes. Primary among these purposes was revision of Panama's treaties with the United States governing operation of the canal and control of the Canal Zone. Revision might or might not benefit Panamanians generally, but it certainly would make a hero of Chiari. Thomas Mann, dispatched by Johnson to survey the ground, reported back: "While the possibility that Chiari is simply a weak, vacillating and uncertain president cannot be entirely eliminated, we have a growing feeling that the strategy of the Panamanian Government is to apply pressure on the U.S. in order to obtain a commitment for structural revision of the existing treaties." Mann noted that Panamanian police had made little attempt to subdue rioters and that news reports disseminated from government offices contributed to "the highly inflammatory, distorted and exaggerated accounts" passed as truth by the Panamanian press. When Chiari announced a break in diplomatic relations with the United States, the announcement—especially the "sensational way" in which the president himself made it—confirmed Mann in his thinking that Chiari was playing to the street. Either that or Chiari had lost his senses. "We fail to see how any rational Panamanian could conclude that a break in relations at this time would serve Panamanian interests."[3]

Mann lacked imagination. Chiari's actions were not irrational. Panamanians for some time had been trying to persuade Washington to renegotiate the treaties. Washington invariably had other things on its mind. Repeatedly the Americans put the Panamanians off. Now Chiari had grabbed America's attention, and he intended to hold it.

Chiari certainly got Johnson's attention. After consulting with Rusk, McNamara, and Bundy, the president talked to Chiari personally. Johnson telephoned Chiari and said he greatly regretted the violence that had claimed the lives of Panamanians and Americans. He expressed hope that their governments could work together to restore peace and stability, suggesting that elements "unfriendly to both of us" might try to exploit the situation. Chiari pointed out that the present circumstances reflected his country's frustration at Washington's "indifference" to the legitimate demands of the Panamanian people. Chiari reminded Johnson that the Panamanian government had raised the issue of treaty revision with President Kennedy in 1961. Washington had ignored the matter. Johnson replied that both countries must look to the future instead of the past. Without committing himself to anything of substance, Johnson indicated a willingness to discuss matters of mutual concern. But first the violence must end.[4]

Chiari elaborated on Panama's views in meetings with Mann. Chiari demanded "structural revision" of the 1903, 1936, and 1955 treaties, which he criticized as unfair and anachronistic. Unless President Johnson agreed in advance to such revision, Panama would not resume relations with the United States. The two countries had nothing to talk about.

Johnson refused to give such a commitment. "You should tell the President," Johnson cabled Mann, "that we cannot negotiate under pressure of violence and breach of relations and that therefore his demand for agreement to structural revision of treaties is unacceptable." Johnson did, however, reiterate his willingness to discuss matters of mutual concern once peace was restored.[5]

In Mann's meetings with Chiari, the assistant secretary repeatedly stressed the need for restoring peace and order. Although Mann eventually changed his mind about Chiari's irrationality, he feared that Chiari had unleashed forces that would get out of control. Mann worried that communists within Chiari's administration were plotting against the Panamanian president—and against the United States, of course. On the third day of his visit, Mann asked Chiari to send his aides other than Foreign Minister Solis out of the room. (Fluent in Spanish, Mann dispensed with a translator throughout his sessions with Chiari.) Mann proceeded to tell Chiari that American intelligence sources believed Castroite communists had penetrated the highest levels of his government. Havana would soon be trying to smuggle weapons into the country. Asserting that communist-inspired violence posed as great a threat to Panamanian interests as to American, Mann added that the United States would react swiftly to a recurrence of trouble. "There should be no mistake. We would have to defend ourselves, including women and children in the Zone, if mobs should again force their way into the Zone. The casualties would be heavy."[6]

During the following week, Panama City bubbled with rumors of coups against Chiari. No one knew when the blow would fall or whether it would come from leftists hoping to seize the government or from rightists aiming to preempt the left. Preparing for all eventualities, the White House directed General Andrew O'Meara, head of the Southern Command, to make contact with Colonel Bolivar Vallarino, the Panamanian first commandant, and to "tell Vallarino that in the event of a communist coup he can count on our assistance to prevent it." At the same time, Washington warned O'Meara to keep his eye on Vallarino and ensure that the Panamanian general not fabricate a communist coup to vault himself into power. "In the event that he subsequently calls on you for such assistance, alleging a communist coup, you will grant assistance if you have reason to believe, based on your knowledge of the situation, that it is a communist coup." Should Vallarino come into conflict with Chiari, O'Meara should consult the White House. "Don't hesitate to call at any time of the night."[7]

Despite annoyance at what it deemed Chiari's irresponsibility, the Johnson administration decided that it could do worse than keep Chiari in power. Chiari, for all his faults, was a known quantity. Mann explained: "We would prefer to see Chiari continue in office because of the inevitable risks for us inherent in any political upheaval and the probability that the United States will be blamed for causing Chiari's downfall." On the other hand, the American government should not alienate any of the important conservative elements in Panama. "They may come to power no matter what we do or say. If they do we will have to deal with them." Mann advocated a policy of nonintervention with regard to anticommunist groups and of intervention, if necessary, with regard to communist groups. "Our main concern at the moment is to prevent the growth of commie influence and especially any commie takeover."[8]

Mann had serious doubts about Chiari's future. The rightists thought Chiari was too easy on the left, while the left thought he was too soft on the Americans. Mann suggested that the right was holding back only to see whether Chiari would succeed in squeezing concessions from Washington—which rendered firmness in administration policy all the more vital. "If this estimate is correct," Mann advised Rusk, "then it would make good sense to disabuse all Panamanians, and indeed all Latin Americans, of any ideas that, in the end, we are going to save Chiari by agreeing to his preconditions. Only then will the anti-communists adjust to reality and begin to organize and plan." Mann conceded the perils of such a policy, which dangled Chiari precariously between left and right. "But we gain by giving anti-communist elements the time and the opportunity to organize an alternative to a communist-infiltrated or communist-controlled government."[9]

Johnson was willing to play for time, which allowed him to touch base with congressional leaders. At a White House meeting at the end of January 1964, Johnson had Mann, back from Panama City, brief the legislators. Mann summarized the crisis to date, described Chiari's position, and identified the sticking issue between the United States and Panama: whether or not Washington would agree in advance of resumption of diplomatic relations to negotiate a new treaty. Johnson asked the assembled legislators for their thoughts. The opinions divided into two categories. Liberals Hubert Humphrey and Wayne Morse, joined by William Fulbright, contended that while the president should not negotiate under the gun of violence in Panama, the Panamanians had legitimate grievances. Inequity and high-handedness had often marked American relations with Panama. The president could afford to be generous. He ought to be able to find some path to the negotiating table without compromising American honor. Conservatives Richard Russell and Everett Dirksen dissented. Russell held that a treaty was a treaty. The canal treaties granted the United States certain rights in Panama. The United States should stand on those rights. Russell allowed that the administration might budget more aid for Panama as a means of calming the situation, but it must not yield on the central question of treaty revision. Everett Dirksen argued that capitulation to Chiari would call into doubt American credibility around the world. Other countries would learn that the way to get the American government to negotiate about grievances was to assault Americans and sever relations with Washington. The Senate minority leader said American prestige and security could not survive the spread of such notions.[10]

Johnson listened carefully to Dirksen and Russell, more carefully than to Humphrey and Morse and Fulbright. Never a liberal himself in foreign affairs, Johnson recognized that while liberalism on domestic matters often plays well in the United States, liberalism on international issues rarely does. Americans can accept the idea that Washington should bestow favors on Americans, but they have substantially more difficulty accepting that Washington should bestow favors on foreigners. Johnson knew that the Marshall Plan, to cite the most conspicuous example of American largess, owed more to American fear of Stalin than to American compassion for suffering Britain and France. (As soon as the war had ended, Harry Truman slammed the door on Lend-Lease assistance, and when the British came calling for help in 1946, Washington negotiated a loan the most careful collateralist could hardly have faulted.) The Alliance for Progress, to take an example pertinent to Panama, was largely a response to Castro's injection of communism into the Western Hemisphere. Moreover, Johnson understood from watching conservatives drive Harry Truman to distraction and back to Independence that while a Democrat might safely appear too tough on foreign affairs, he must not risk appearing weak. Johnson had participated in the campaign of 1960 in which Kennedy had argued the historically unlikely and factually unsupportable—but not politically unsuccessful—case that Eisenhower had let America's defenses lag.

The commentary the Panama affair provoked in Congress warned Johnson against anything hinting of capitulation to Chiari and the Panama rioters. The Senate, which would have to ratify treaty revisions, rang with charges that the violence in Panama was the work of Castro and his henchmen. New York Democrat Jacob Javits blamed the riots in the Canal Zone on "Castroites and other Communist leaders in Panama." Javits's fellow New Yorker, Republican Kenneth Keating, declared that "Castro-Communist agents" had played "a very large role" in the turmoil in Panama. Keating wondered when the administration was going to wake up to the need for a tougher line against the menace that stalked the Caribbean. "How long will we continue to face the Cuba problem inadequately and ineffectively? How long will we continue to wait until a major crisis occurs?" Everett Dirksen accounted the Panamanians irresponsible and ungrateful. The Illinois Republican recited the history of American good works in the isthmus, from the eradication of yellow fever and malaria to the building of the canal. The canal provided employment to Panamanians and revenue to the Panamanian goverment—revenue, Dirksen reminded, that the United States had repeatedly increased. "I think we have been pretty generous." Now the Panamanians were biting the hand that fed them. The fact that some Americans suggested giving in to Panama's demands boggled Dirksen's mind. "We are in the amazing position of having a country with one-third the population of Chicago kick us around." Dirksen simulta-

neously commended and admonished Johnson: "Thank goodness the
President has stood up to it. I hope he will not retreat from his position."[11]

Johnson did not intend to retreat. But he did not rule out moving
sideways. While the United States might by main force keep Panama
quiet, Johnson was not Theodore Roosevelt, and the 1960s were not the
turn of the century. Bullying Latin American nations came harder politically and diplomatically than in the bully days of the Rough Rider.
Besides, Johnson already had one war on his hands in Southeast Asia.
Far better to meet the Panamanians part way, perhaps halfway, so long
as no basic principles were surrendered, than to stonewall and risk
greater turmoil down the line.

Johnson chose to follow the advice of Mike Mansfield, though not all
of it and not at once. On January 31 the Senate majority leader, who had
missed the earlier White House meeting, recommended a firm but conciliatory policy. "We have only one fundamental interest to protect in
the present situation," Mansfield wrote Johnson. "We have got to insure
untroubled and adequate water-passage through Central America." The
president should keep this interest foremost in mind and not get mired
in details. The Panamanians sought an American commitment to "negotiate" a new treaty. Administration officials had said they would only
"discuss" a new treaty—the latter term connoting less of a commitment
to revision. The president of the United States, Mansfield declared,
should not quibble over terminology. Johnson should remember that
unrest in Panama, as in most of Latin America, was not an import from
abroad. "The pressure comes primarily from the inside, from the decay
and antiquation of the social structures of various Latin American countries. Even if we desired to do so, we could not, as a practical matter,
stop the pressure for change. But we may have something constructive
to contribute to the form and pace of the change if we play our cards
carefully and wisely."

Mansfield suggested involving the Organization of American States
closely in attempts to resolve the dispute. Panama would find a settlement endorsed by the OAS far easier to accept than one presented solely
by the United States. The administration might make noises about building a new canal through Mexico or Nicaragua. A new canal would be
needed eventually. More to the immediate point, the thought that Panama might not be indispensable to the United States would sober Chiari
and his more unrealistic supporters. The administration should avoid
leaning too far in the direction of the American residents of the Canal
Zone, whose interests did not coincide with those of the United States
as a whole. Finally, tempting as it might be, the president and the Democratic leadership should "avoid boxing ourselves in at home against
change through the fanning of our own emotions by crediting Castro
and Communism too heavily for a difficulty which existed long before

either had any significance in this Hemisphere and which will undoubtedly continue to plague us after both cease to have much meaning." To
blame Castro would play into the hands of hard-line anticommunists
who might render any reasonable adjustment of American policies in
Latin America impossible.[12]

In this case, if not in others, Johnson managed to avoid the Castro
box. And while the president ultimately got to the Mansfield position
on nonquibbling, he started out closer to the Russell-Dirksen view. The
president spoke softly but made clear he had no intention of meeting
Chiari's demands. "We have stated our willingness to engage without
limitation or delay in a full and frank review and reconsideration of all
issues between our two countries," Johnson announced in a televised
address. "We have set no precondition to the resumption of peaceful
discussions. We are bound by no preconceptions of what they will produce." Since preconditions and "negotiations," rather than "discussions," were what Chiari wanted, Johnson's declaration represented a
rejection.[13]

To encourage Chiari to reconsider, the Johnson administration applied a moderate amount of economic pressure. While Washington continued to process various loan applications Panama had filed with the
United States Agency for International Development, Rusk told the embassy in Panama City to relay word to Chiari that approval of the loans
would require assurances of security for the AID officials who would mon-
itor spending of the money. Additionally the State Department insisted
on an item-by-item review of AID projects—a procedure whose efficiency
presumably would be a function of Panamanian cooperativeness. Mann
summarized the administration's thinking on the use of the assistance
lever. "We do not want to convict ourselves of Panama's charges of economic aggression," Mann explained. At the same time, though, "We do
not wish to give the appearance of rewarding irresponsibility."[14]

Johnson was fortified in his firm approach by the public response to
the Panama affair. McGeorge Bundy informed the president that tele-
grams to the White House ran ten or fifteen to one in favor of a staunch
position. Bundy's staff passed up to the president the results of a *Washington Post* poll indicating that of those persons following the story, an
overwhelming majority—some 85 percent—thought the United States
should not give in to Panama's demands for greater control over the
canal.[15]

Meanwhile Johnson followed Mansfield's recommendation to draw
on the services of the OAS. Both at OAS headquarters in Washington and
in various more confidential locations around the capital, Mann and
Ellsworth Bunker, Johnson's representative to the OAS and general-pur-
pose fixer, held meetings with Latin American diplomats. Further, Johnson called on the Inter-American Peace Committee of the OAS to look
into the U.S.-Panamanian troubles.

Progress came slowly. After several weeks Mann thought he had an agreement, mediated by Mexico's ambassador to the OAS, for resumption of diplomatic relations and initiation of talks on the treaties. To finesse the issue of "discussions" versus "negotiations," the arrangement specified that the United States and Panama would appoint representatives "with sufficient powers to carry out discussions and negotiations."

Johnson rejected the plan as soon as he read the word "negotiations." As before, the president refused to commit in advance to any new treaty, which by now "negotiate" plainly implied. He told Mann to go back and try again. Mann did, but not in time to prevent the OAS from releasing news of the purported agreement. Johnson's immediate disavowal of the agreement knocked the process several squares backward.[16]

Yet having made his point, Johnson proceeded to soften his position. On March 21 the president surprised reporters and his new press secretary, George Reedy, by walking into a briefing and personally giving the assembled journalists an advance tip. In a few hours, Johnson said, he would deliver to the OAS a statement clarifying his position on the Panama question. In a sympathetic tone, he said the differences that had arisen between the United States and Panama greatly pained him. He recalled how Panama had rallied to the side of the United States during the Second World War, with Panama's congress declaring war against Japan even before the American legislature did. Regarding Panama's position in the present dispute, the president said, "We are well aware that the claims of the Government of Panama, and of the majority of the Panamanian people, do not spring from malice or hatred of America. They are based on a deeply felt sense of the honest and fair needs of Panama. It is, therefore, our obligation as allies and partners to review these claims and to meet them, when meeting them is both just and possible." Pausing for emphasis, Johnson continued, "We are ready to do this. We are prepared to review every issue which now divides us, and every problem which the Panama Government wishes to raise."

A reporter at once asked whether the president meant he was willing to renegotiate the canal treaties. Johnson did not say no. He said he meant just what he had said. Though he did not use the word *negotiate,* he referred to a solution "subject to the appropriate constitutional processes of both our Governments," which definitely suggested something along the lines of a new treaty.[17]

By this time, Chiari was looking for an honorable way to end the impasse. His party faced an election in a few months, and its prospects seemed to be suffering from the deadlock. The Panamanian economy, not particularly robust before, was weakened by the cloud hanging over the country's future.

Chiari made some last, half-hearted efforts to get Johnson to utter the word *negotiate,* but the American president refused. All the same,

Johnson's spokesman, Bunker, dropped another hint that Washington was not ruling out treaty revision. Meeting with his Panamanian counterpart, Bunker said that President Johnson wanted an accord for resuming relations and starting talks that avoided "complicated language which might stand in the way of getting an agreement through the Senate at some later time." Talking about the Senate was tantamount to talking about a new treaty.[18]

Johnson closed the deal by a call to Chiari. He told Chiari he was appointing a top man—Robert Anderson—as his special representative to settle differences between Panama and the United States. Chiari said this was the right way to handle the situation.[19]

Minutes later, Johnson announced the settlement to a meeting of the National Security Council, to which he had invited key congressional leaders. The president explained that although he had not committed to any particular conclusions in the Panama talks, he had told Chiari he was willing to review all problems between the two governments. Mike Mansfield registered approval of the arrangement, as did William Fulbright and Wayne Morse.[20]

III

The Panama talks begun by Johnson would not culminate until the presidency of Jimmy Carter, but the April 1964 agreement allowed Johnson to turn to other matters. Even while he sought to prevent Castro's isthmian supporters from gaining a foothold on the Panama Canal, the president found himself face to face with Fidel nearer Cuba. At the beginning of February 1964, the American Coast Guard discovered four Cuban fishing boats in American waters in the Florida keys. The guard escorted the boats to Key West, where the crews were arrested.

The Cuban government denounced the American action as arrogant and illegal. Cuban foreign minister Raul Roa filed a complaint with the United Nations Security Council, asserting that the boats had been seized in international waters and charging the Yankee cutthroats with "intolerable aggression." Castro blasted the Americans. "What have the imperialists come to?" he demanded. "They have come to being pirates, captors of fishing vessels. The symptom of a decaying system, this only disgraces them more." Ever supportive, at least rhetorically, of its socialist client in the Western Hemisphere, Moscow added its quota of contempt. After the missile crisis of 1962, Nikita Khrushchev no longer rattled Russia's rockets so noisily as he had, but *Pravda* intoned ominously that "the anti-Cuban policy which the rabid reactionaries are trying to impose on the United States may lead to disastrous consequences." The counselor of the embassy of Czechoslovakia in Washington, which handled Cuban affairs in the absence of diplomatic relations between the United States and Cuba, registered a formal protest with the State Department.[21]

To underscore its upset, the Cuban government announced a shutoff of water supplies to the American naval base at Guantanamo, still operating under lease terms to run until 2002 The shutoff carried no real threat to the base since the American navy had long anticipated such an action. The only puzzle was why Castro had not acted sooner. Part of the answer rested with the two thousand Cubans employed on the base, whose layoff would add to Cuba's economic difficulties. When Castro closed the water valve, Washington responded by handing the employees their pink slips.

Johnson ordered the navy to implement its contingency plan for making the base self-sufficient in water. Tankers brought water from Florida while sailors skipped showers and ate from paper plates. Engineers expedited work on a plant to desalinate seawater. To demonstrate American unconcern at Castro's troublemaking, the base commander directed that the water main leading from the Cuban side of the fence be conspicuously severed and the pipe removed.

The brouhaha subsided within a few weeks. Florida courts released the fishermen, who went home, with the exception of two men who preferred bright Miami to now-dull Havana. Castro offered to turn the water back on, but Washington told him not to bother.

Johnson interpreted the affair as a probing action. "I had no doubt about Castro's purpose," Johnson recalled afterward. "He had decided, perhaps with Soviet encouragement"—Castro recently had returned from a trip to Moscow—"to take the measure of the new President of the United States, to push me a little and see what my response would be."[22]

IV

The Guantanamo affair had just ended, and the Panama negotiations or discussions or whatever were about to begin, when developments farther south caught Johnson's attention. Although the Monroe Doctrine described the entire Western Hemisphere in terms of a special U.S. sphere of influence, in practice American leaders have generally contented themselves with pushing around the countries of the Caribbean basin. The greater proximity of Mexico, Central America, the northern coast of South America, and the islands of the Caribbean has made them particularly susceptible to American attention and influence. So also has the existence of important trade routes through the Caribbean and across Central America—the most important, of course, being the Panama Canal and its approaches. Occasionally the United States has looked beyond Colombia and Venezuela to the rest of South America. In the 1890s the Cleveland administration exchanged hot words with the Chilean government over a perceived affront to the dignity of the United States Navy. Argentina's pro-German tendencies worried Washington during subsequent decades. Radicalism in Bolivia in the 1950s put the CIA on alert. In the early 1970s the Nixon administration helped destabilize the government of Chilean president Allende. But compared to the constant and often heavy-handed involvement of the United States in the affairs of Caribbean region, most of South America has been an oasis of U.S. nonintervention.

Johnson bent this rule slightly in March 1964. Three years earlier, Brazilian Vice-President Joao Goulart had succeeded to the presidency of his country on the unexpected resignation of his predecessor. At that time the Brazilian military had looked askance at the elevation of the populist Goulart, judging him—not without justification—guilty of collaborating with communists and other shakers of the status quo. Although the generals and admirals failed to keep Goulart from taking office, they did not hide their dissatisfaction. A few months after Goulart's swearing-in, one admiral told the American ambassador to Brazil, Lincoln Gordon, that the navy, the army, and a variety of civilians were planning a coup. The admiral said that although the date of the coup had not been set and might be some way off, he and his friends wanted to keep the ambassador informed. "One of these days we will act," he said. "And I hope when that happens the United States will not be unsympathetic."[23]

Whether the admiral's sails were full of hot air or the coup simply took longer than expected to organize, Gordon never determined. Goulart remained in power through 1962 and into Johnson's administration. During that time Goulart's policies slipped increasingly to the left. In 1962 he announced the expropriation of a Brazilian subsidiary of the American-owned International Telephone and Telegraph company. He allowed communist influence in labor unions to grow. His associates in government made a habit of verbally attacking the United States for neocolonialism and related crimes. Goulart's actions did nothing good for the Brazilian economy, and in October 1963, amid intensifying economic and political unrest, he requested the Brazilian legislature to grant him martial law powers. Although he subsequently withdrew the request, critics read it as evidence of latent dictatorial tendencies. At about the same time, Washington received reports of Soviet offers to provide planes to the Brazilian air force.

Johnson took note of the Brazilian situation as soon as he became president. Goulart sent a letter conveying condolences on Kennedy's death and indicating a desire for American help in easing cash-flow problems his government was experiencing. Johnson replied with a pro forma affirmation of "the bonds of natural affection that exist between our two peoples," but with no encouragement on the money question. The president pointed out that the United States held only a small part of Brazil's foreign debt. He said his administration would be willing to facilitate talks between Brazil and Brazil's major external creditors, but he conspicuously declined to offer substantive aid to see Brazil through its straits.[24]

Johnson had no wish to prop up a regime he perceived as combining irresponsibility with incompetence. The degree of Goulart's dependence on the radical factions in Brazil occasioned debate in Washington, as did Goulart's political and economic intentions. But few American officials believed that Brazil would revive with Goulart at the helm. George Ball wrote to Lincoln Gordon in December 1963: "Goulart is bound to be harassed by a progressively worsening inflation, increased political turbulence, and the cacophony of external pressures for drastic action." Gordon had advocated efforts by the new administration in Washington to restrain Goulart's excesses. Ball was not sure this was worth trying. The United States might better look to the person who would come after Goulart, either by election in 1965 or by extra-electoral means. In an addendum to Johnson's letter to Goulart, the undersecretary commented, "While we share your view that every effort must be made (as through the above letter) to develop and maintain a constructive and moderating impact on Goulart wherever possible, unless or until Goulart is clearly to be replaced by a more acceptable leadership, the basic situation nonetheless has many bleak aspects."[25]

During the first several weeks of 1964, Goulart succeeded in keeping his balance. By the beginning of March, however, the Brazilian president's position was starting to totter. On March 4 Gordon cabled that the military was growing more restive by the day. The ambassador judged this a positive development. The military had traditionally been a force for stability in Brazilian politics, he said. So the officers remained, more than ever. "They are an essential factor now in a strategy for restraining the left-wing excesses of the Goulart government and maintaining prospects for a fair election in 1965 and the installation of a successor in 1966." Gordon deemed Brazilian military leaders different from the military in other Latin American countries. The Portuguese influence had something to do with the difference, the ambassador explained. "Unlike many Spanish-American officer corps, they are not an aristocratic caste separate from the general public. The basic orientation of the great majority is moderately nationalist but not anti-U.S., anti-communist but not fascist, and pro-democratic constitutionalist. The military not only have the capability of suppressing possible internal disorders but also serve as moderators on Brazilian political affairs directed at keeping them within constitutional and legal limits."

During the next four weeks, Gordon would make much of this idea of the Brazilian military as guardian of Brazilian democracy. Meanwhile the ambassador pushed for more American military aid to Brazil. "U.S. military assistance does not increase the danger of a fascist-type military takeover," he asserted. Despite a few bad eggs in the Brazilian services, "the great majority are still basically faithful to constitutional and democratic tradition and are predominantly anti-communist and pro-U.S." Gordon thought the chances of protracted rule by military leaders, should a coup occur, no more than slight. "If a reaction to a left-wing coup attempt led to a temporary military takeover, they would be quick to restore constitutional institutions and return power to civilian hands." The United States ought to throw its backing to the Brazilian military, "the most powerful single force of resistance against the inroads of anti-democratic (as well as anti-American) elements in Brazil."[26]

Gordon did little to disguise his sympathies from the Brazilians, doubtless encouraging the planning against Goulart. That planning accelerated on March 13 when Goulart's supporters held a huge rally in Rio de Janeiro. A close associate of the Brazilian president publicly urged Goulart to overturn the constitution and dismiss the legislature. The crowd roared approval. When Goulart appeared, he flourished a decree nationalizing the last remaining independent oil refineries, completing the monopoly held by Petrobras, the state-owned petroleum concern. The crowd roared more.

The Rio rally stirred the fears of Brazilian conservatives about the direction Goulart was leading the country. Anti-Goulart demonstrators marched in Sao Paulo, carrying banners reading "Resignation or Impeachment" and "Down with Red Imperialism." The army chief of staff, General Humberto Castelo Branco, remarked to an American embassy official that the only signs he had seen at the Rio rally were hammers and sickles. Castelo Branco added that he now thought Goulart would not agree to leave office when his term expired.[27]

Gordon interpreted the expropriation of the oil refineries as particularly significant. The move would strengthen the power of the leftist labor unions that controlled the work force of Petrobras, and likely would push Goulart yet further to the left. Gordon relayed to Washington rumors circulating in Rio of government designs on the country's oil distribution companies. Should Goulart take over the distributors, the government's grip on the oil industry would be complete. Gordon said he did not believe Goulart was ready for such a move at this stage, but he added, "We cannot rule out this as a possibility."[28]

Rumors did not emanate only from Brazil. Some came from Washington. After the March 13 rally, Gordon flew home for a conference of American diplomats assigned to Latin America. At this conference Thomas Mann described an important shift in American policy—or so the story was reported. According to the *New York Times*, Mann said the United States no longer would seek to punish military juntas for toppling elected governments. John Kennedy had tried to establish a principle of strong preference for democratic governments and aversion to military regimes. Under Lyndon Johnson, Mann was reported to have said, opposition to communism would matter more than strict adherence to constitutional norms.[29]

Whether or not Mann put the administration's policy in precisely such terms was less important than that the administration declined to disavow the reports, and what quickly became known as the Mann doctrine encouraged the Brazilian military leaders to go ahead with their designs to push Goulart aside. On March 30 the CIA predicted that a conspiracy in the east-central state of Minas Gerais was about to hatch. The agency's source said, "A revolution by anti-Goulart forces will definitely get under way this week, probably within the next few days."[30]

A few hours was more like it. Later that very day, the governor of Minas Gerais publicly denounced Goulart and called on the military to vindicate the constitution against the usurper. Shortly thereafter the commander of the military district headquartered in Minas Gerais began moving troops toward Rio. Soon other generals, including Castelo Branco, joined the revolt.

Gordon applauded the action. Shortly before the revolt began, the ambassador had wired Washington that Goulart intended to grab total power. "My considered conclusion is that Goulart is now definitely engaged on a campaign to seize dictatorial power, accepting the active collaboration of the Brazilian Communist party and of other radical left revolutionaries to this end." The probable consequences were forbidding. "If he were to succeed it is more than likely that Brazil would fall under full Communist control, even though Goulart might hope to turn against his Communist supporters on the Peronist model, which I believe he personally prefers." Chances of stopping Goulart were dwindling. "A desperate lunge for totalitarian power might be made at any time." Gordon reiterated his confidence in the Brazilian military, especially Castelo Branco, whom he described as "a highly competent, discreet, honest and deeply respected officer who has strong loyalty to legal and constitutional principles." When the anti-Goulart uprising in Minas Gerais began, Gordon significantly—and mistakenly—labeled it a "democratic revolt."[31]

On receiving word of the military's moves, Dean Rusk ordered all stations in Brazil onto twenty-four-hour alert status. During the next three days, Gordon filed dozens of reports on the progress of the revolt. Several times Gordon communicated with Washington via teletype, the State Department's equivalent of the hot line.[32]

The breaking events confronted the Johnson administration with the question of what it could do to facilitate a favorable outcome. Like Gordon, administration officials in Washington hoped the Brazilian military would succeed in preventing Goulart from delivering Brazil to the communists intentionally or by inadvertence. But the administration hesitated to back a movement that might fail. Johnson had learned by watching Kennedy in the Bay of Pigs affair, and Rusk and Bundy and

McNamara and Ball had learned from personal experience, the costs of careless cooperation with unready rebels. No one wanted to go that route again. As Ball commented to Johnson, "We don't want to get ourselves committed before we know how the thing is going to turn out."[33]

On the other hand, neither did administration officials wish to see Goulart take Brazil the way Castro had taken Cuba. The loss to communism of Brazil, the fifth-largest country in the world, possessing enormous natural resources and dominating the Atlantic coast of South America, would have a strategic impact far greater than the loss of South Vietnam. Ball and Thomas Mann warned that the administration must take care "not to let an opportunity pass that may not recur." Johnson said he wanted to "take every step we can" so that it wouldn't.[34]

Some days earlier, Lincoln Gordon had suggested a two-pronged contingency plan in the event of what now was happening. Gordon recognized the critical role supplies of petroleum products would play in a military revolt. He also appreciated the capacity for supply choking that Goulart enjoyed as a result of the government's near-monopoly of the oil industry. And what Goulart did not cut off, the radicals in the oil unions would, either during a revolt or after, by seizure of refineries or by sabotage. A shortfall during a revolt might kill the revolt at once. A shortfall afterward might make the country ungovernable for the military. Consequently Gordon recommended that the United States prepare to deliver petroleum supplies to the Brazilian military in case of need.

Gordon also advocated a show of American naval force. An aircraft carrier task group should be sent to the area of Rio or Sao Paulo or wherever its presence might be most effective. If the president desired to offer an explanation for the movement, he could say the ships were being readied to evacuate American citizens. But the ships' presence would send an unmistakable signal of support to the generals. This message alone might discourage any diehard Goulartists from taking to the hills and carrying on a guerrilla war. Should the president wish to provide more substantial backing, in the form of air cover or the landing of American marines, the option would be readily available.[35]

When the revolt commenced, Johnson approved the plan Gordon recommended. The Joint Chiefs of Staff ordered the loading of four oil tankers with automotive and aviation fuel. A carrier task group headed by the U.S.S. *Forrestal* called all sailors back from shore leave in Norfolk and steamed south. In addition the president authorized the stockpiling of a modest amount of military aid—mostly small arms and ammunition—for air delivery to Brazil if developments required and allowed.[36]

In the event, the military situation in Brazil moved swiftly enough in favor of the rebel generals that American military support became unnecessary. On April 1 Goulart fled Rio for Brasilia, thence to Rio Grande do Sul. On the same day the president of the Brazilian supreme court, acting independently of his associates, swore in an interim Brazilian president, Paschoal Ranieri Mazzilli.

The pressing question for the Johnson administration then shifted to whether and when to recognize the new government. On April 2 Johnson called a White House meeting to consider his options. J. C. King of the CIA gave a summary of the latest news from Brazil, then went into some detail regarding the status of Mazzilli's presidency. King said that because the Brazilian constitution had no provision for ousting a chief executive, Mazzilli occupied legal limbo. President Johnson should bear this in mind in deciding on recognition. "While we do not wish to cast doubt on the legitimacy of Mazzilli as President, we do not wish formally to accept a government which the Brazilian courts may later decide is illegal."

Johnson asked whether the Brazilian legislature, which appeared happy enough to see Goulart gone, could name Mazzilli president pro tem. Dean Rusk said Gordon was suggesting just such a course to the appropriate persons. Johnson inquired about remaining pro-Goulart resistance. Earle Wheeler of the Joint Chiefs replied that one loyalist regiment and possibly one cavalry unit had not given up. But Wheeler anticipated that the rebels would have little difficulty cleaning out these pockets in short order.

Rusk urged recognition of Mazzilli, which would show American support of events thus far, yet the secretary of state said that beyond this the administration should "sit and wait." Things were moving in a favorable direction on their own. President Johnson could let the *Forrestal* keep going toward Brazil since he had plenty of time to turn it back if its presence proved superfluous. Robert McNamara agreed, saying that the task group had just passed Antigua and was still several days away from the scene of the action in Brazil.[37]

Johnson took Rusk's advice and recognized Mazzilli. The president sent Mazzilli a letter conveying "warmest good wishes on your installation." Johnson went on to say, "The American people have watched with anxiety the political and economic difficulties through which your great nation has been passing, and have admired the resolute will of the Brazilian community to resolve these difficulties within a framework of constitutional democracy and without civil strife."[38]

Johnson expected little difficulty from Congress on his handling of the Brazilian affair. Brazil was not a priority in American politics, and Johnson had not placed America's prestige on the line. But as usual the president sought to keep key legislators apprised of his thinking. On April 3 he summoned the leaders of the two parties for a briefing. Rusk explained that prior to the military revolt Goulart had been heading toward the creation of "an authoritarian regime politically far to the left." The secretary assured the senators and representatives that the recent events were the work of the Brazilians. "The United States did not engineer the revolt. It was entirely an indigenous effort."

Wayne Morse said he approved "thoroughly" of the president's response to events in Brazil. Everett Dirksen wondered what the new government's attitude toward private property would be. Would Mazzilli and the generals return property seized by Goulart? Rusk responded that this was one of the first questions the administration would be looking into.

Johnson urged the legislators to assist the new leaders of Brazil in stabilizing the Brazilian economy. The administration had already begun the task. "We are hard at work with our allies to provide the economic help which the new Brazilian government will need," the president said. "We are doing everything possible to get on top of the problem of helping the new government."[39]

During the first week of April, the Johnson administration congratulated itself on its good luck and good judgment in managing the Brazilian situation. The CIA reported to the president that the turnaround had occurred none too soon. "The revolt probably spared Brazil from falling under control of the extreme left," the intelligence agency asserted. The CIA was only slightly less confident in its evaluation of Brazil's new president. "Mazzilli has occasionally flirted with the left but he is basically a representative of the vested interests, economically conservative and seemingly pro-Western."[40]

Before long, though, American officials were beginning to question whether the restoration of Brazilian democracy would come as quickly as Gordon had predicted. On April 2 Gordon Chase of the White House staff had remarked with satisfaction that the revolt was proceeding on legally legitimate lines. "We have the constitutional base—no military junta business," Chase declared. But by April 10 the State Department's Thomas Hughes was noting with concern that the generals seemed in no hurry to relinquish power. Citing recent statements by the new government, Hughes commented that the language of the statements was "hardly compatible with constitutional forms."[41]

Even so, the delay did not prevent Washington from sending American aid. The State Department cobbled together an emergency package of $50 million, and Johnson approved it.

The money did nothing for Brazilian democracy, which seemed to disappear into the future as the weeks—then months, then years—passed. Undiscouraged, Johnson supplemented the first aid package with nearly a half billion dollars more by the end of his presidency.

The American aid signified Johnson's relative satisfaction with the outcome of the situation in Brazil. All things being equal, Johnson and his associates would have liked for the Brazilian government to guarantee civil liberties and political participation to the Brazilian people. Military-controlled juntas were not Washington's first choice for collaborators. But convinced that the choice in Brazil was rapidly reducing to either the communists or the generals, Washington backed the generals. That the generals kept the communists out of power without conspicuous intervention by the United States made the outcome of the situation there the more satisfying.

V

Johnson's next crisis in Latin America involved intervention considerably more conspicuous. Since the assassination of Rafael Trujillo in 1961, the Dominican Republic had suffered through an extended bout of social, economic, and political instability. The Kennedy administration had hoped the removal of the longtime dictator would open the door to reform—some thought the hope might have prompted Washington to wink to the assassins—but Trujillo's elected successor, Juan Bosch, proved a distinct disappointment. John Bartlow Martin, the American ambassador in Santo Domingo, believed the United States could not dismiss the possibility Bosch was a "deep-cover Communist"—a comment that said more about Washington's obsession with Castro than about Bosch's left-liberal politics. George Ball did not think Bosch possessed such subtlety. "I had lunch with Bosch when he came to Washington just before he took office, and I have rarely met a man so unrealistic, arrogant, and erratic," Ball wrote. "I thought him incapable of running even a small social club, much less a country in turmoil. He did not seem to me a Communist, as some were asserting, but merely a muddle-headed, anti-American pedant committed to unattainable social reforms."[42]

American dissatisfaction with Bosch doubtless encouraged the Dominican military to topple the president in 1963. Ambassador Martin's replacement, W. Tapley Bennett, quickly warmed up to Bosch's replacement, Donald Reid Cabral. Reid was rather less popular with many of the Dominican people. Constitutionalists demanded the restoration of Bosch, while radicals agitated for an entirely new order. When Reid attempted to clean up corruption in the army, he succeeded only in adding uniforms to the ranks of those who wished him ill. Early in 1965 reports circulated that he would cancel or fix elections slated for the fall. The news moved the country closer to a major upheaval. Bennett judged the left best positioned to gain. "Little foxes, some of them Red, are chewing at the grapes," the ambassador warned Washington.[43]

In April the chewing grew louder, until on April 24 the foxes tried to gobble the government at a bite—although precisely which foxes and what their prospects of success the American embassy had difficulty discerning. "We believe the Government of the Dominican Republic is in serious straits," the chargé d'affaires, William Connett, cabled on the afternoon of the twenty-fourth. "But we have no solid evidence it is in fact overthrown." Early the next morning Connett clarified slightly: "The situation here is rapidly deteriorating as rebel army troops are entering the capital in large numbers from outlying camps."[44]

By now the outlines of what was happening were taking shape. Reid, learning of a conspiracy by pro-Bosch military officers, had tried to fire several of the plotters. The plotters instead arrested the general sent to relieve them and announced the overthrow of Reid's regime. The announcement, quite premature, split the Dominican military, with some elements backing Reid and others calling for the return of Bosch.

Through the rest of April 25, Connett, filling in for Bennett, who was in Georgia visiting family, described the progress of the fighting. The rebels had seized army headquarters west of downtown Santo Domingo, but pro-Reid forces were beginning to react. The air force chief of staff, Juan de los Santos Cespedes, told the air attaché of the American embassy that the air force would stick with the government. The navy appeared inclined to do likewise. The extent of disaffection among the army remained unclear. A key figure, General Elias Wessin y Wessin, was keeping his intentions to himself. On balance, Connett saw reason for optimism. "We are gratified by signs the military is stiffening and talking of pulling together to prevent the collapse of public order and the return of Bosch."[45]

The optimism dissipated quickly. The next twelve hours demonstrated that opposition to Reid ran deeper than Connett had perceived. The rebellion spread until two-thirds of the army units in the Santo Domingo district had joined the uprising. A march on the downtown area from camps in the suburbs seemed imminent. Much of the navy had defected as well. By the morning of April 26, Generals Wessin and de los Santos were calling for American troops to keep the government in power and the Boschists out. At ten o'clock that morning Connett cabled Washington, "Probably nothing short of major U.S. involvement could prevent Bosch's return at this time."

Connett nonetheless advised against sending troops just yet. Along with officials in Washington, Connett had no desire to see Bosch back, but military intervention, he predicted, would have "extremely serious implications" for American foreign relations generally and with regard to Latin America particularly. "We would be cast in the role of an interventionist power opposing a popular revolution of democratic elements overthrowing an unpopular unconstitutional regime." The chargé d'affaires thought the United States would have difficulty justifying intervention. "We do not believe we could effectively make the case that this is a communist-controlled movement at the present time, although we all know that communists are deeply involved in the rebel movement." Connett advocated a strong diplomatic effort to keep Bosch from returning and to persuade the government to schedule elections. Washington must move swiftly. "We believe there is a serious threat of a communist takeover in this country and that very little time remains in which to act." Connett did not rule out the use of American military force. Indeed he saw a "distinct possibility" that the United States would have to send marines to prevent a leftist takeover. But he wanted to hold the big stick in reserve for the present.[46]

The CIA seconded Connett's concern over Dominican communism. From the beginning of the uprising, the agency's contacts in Santo Domingo kept close watch on the pro-Moscow Partido Socialista Popular and related radical groups. Hours after the revolt began, the CIA station in Santo Domingo reported a source saying that PSP members were "making contact with party members and other anti-government young people in order that they take advantage of the unstable situation in the country." The PSP evidently had not known in advance of the revolt. The CIA said the party was "completely unaware of the attempted coup." But the communists were recovering quickly from their surprise. "They are now enthusiastic and spirited at the possibility of a successful coup."[47]

Before long the CIA was describing the revolt as falling increasingly under the influence of the radicals. On the afternoon of April 25 the Santo Domingo station asserted, "The military men who planned the coup were completely outmaneuvered by the Communists." The next morning the agency quoted a source close to the revolt saying, "The revolutionary movement is being controlled by the Communists." Later on April 26 the CIA asserted that the radicals were hoping to do considerably more than depose Reid and restore Bosch. Speaking of the Maoist Movimiento Popular Dominicano, the agency declared, "The MPD is very pleased with the way the insurrection is going, and party leaders are now considering continuing the revolution towards Marxism-Leninism."[48]

It was significant, in light of Johnson's subsequent justification for landing American troops, that both Connett and the CIA station in Santo Domingo registered serious concern regarding the possibility of a communist takeover days before they began worrying much about the danger to Americans from the revolt. As soon as the fighting started, the embassy laid plans for evacuating Americans from Santo Domingo, but there seemed no particular hurry until April 27.

On that day Bennett arrived back at the embassy. In midafternoon the ambassador described a "despicable action this morning by a rabble of about 300 rebels" who had surrounded a hotel where some American evacuees had gathered. The rebel rabble, Bennett said, had "terrorized our evacuees by pointing tommy-guns at them and shooting up the lobby."[49]

By next morning Bennett was pronouncing Americans "seriously endangered" by the insurrection. Still angry over the confrontation at the hotel, the ambassador said, "Tactics used by the rebels have reached the limits of human decency." The violence was closing in on the embassy as well. "Our garden is growing bullets this spring," he remarked.[50]

The increasing turbulence reflected a turn for the worse in what was quickly becoming a civil war. The rebels captured the presidential palace and announced the imminent return of Bosch. Generals Wessin and de los Santos responded by ordering the Dominican air force to strafe the palace. Dominicans were not accustomed to this level of domestic violence, which polarized the situation more than before. The rebels seized members of the families of the air force pilots and held them as insurance against repeated attacks. What inclination had existed among the various military factions toward compromise vanished.[51]

With it vanished hopes for Connett's proposed diplomatic solution to the conflict. Besides, Bennett preferred the military option. After the confrontation at the hotel, Bennett believed American lives were in danger. Americans might be killed, or the rebels might take hostages to forestall American intervention, which, with the government inviting American troops, seemed likely. Bennett equally feared a communist takeover. In the early afternoon of April 28 he asked Washington to provide some communications equipment to the government forces. Bennett cited a dual reason for the request. "I do not wish to be over-dramatic," he told the State Department, "but if we deny simple communications equipment and the opposition to a leftist takeover here loses for lack of heart or otherwise, we may very well be asking in the near future for the landing of marines to protect U.S. citizens and possibly for other purposes." In the current context, the "other purposes" clearly included putting down the leftists. Bennett asked, "Which would Washington prefer?"[52]

Within hours Bennett decided walkie-talkies would not suffice. In the late afternoon of April 28 the ambassador sent a highest priority cable to Washington. "The situation is deteriorating rapidly," he said. Government troops were tired and discouraged. The army leadership was disorganized. The loyalist generals who now ran the government had just delivered a formal request for American troops, saying that if the United States did not respond favorably and at once they would have to quit. The threat to Americans continued to increase. "American lives are in danger," Bennett declared. "The country team is unanimously of the opinion that, now that we have a request from the military junta for assistance, the time has come to land the marines." Bennett closed, "I recommend immediate landing."[53]

The State Department quickly passed Bennett's message to the White House, where Johnson was discussing Vietnam with Rusk, Bundy, McNamara, and Ball. Bennett's account of the situation left Johnson little room for choice. With the man on the spot declaring American lives in danger—and with Bennett, Connett, and the CIA having expressed concern regarding the possibility of a communist power grab—the decision was preordained. As Ball remembered, "There was nothing to do but react quickly. Though none of us wanted to repeat history by stationing troops in the Dominican Republic, as America had done from 1916 to 1924, we had no option." Johnson gave the order, and five hundred marines landed from ships assigned to the area.[54]

Had the five hundred simply evacuated the Americans and left, their visit would have occasioned little controversy, and that little would have dissipated with their departure. But though evacuation dealt with the Americans-in-danger theme of the intervention, it did nothing for the reds-on-the-rampage aspect. The Johnson administration was determined to prevent a leftist victory. On April 27 Rusk listed the administration's priorities in the Dominican Republic. "Our primary objectives," Rusk wired Bennett, "are restoration of law and order, prevention of a possible Communist takeover, and protection of American lives." Bennett, nearly as soon as he asked for the marines, urged that the president expand their role beyond safeguarding Americans. "I recommend that serious thought be given in Washington to armed intervention which would go beyond the mere protection of Americans and seek to establish order in this strife-ridden country," the ambassader said. "All indications point to the fact that if present efforts of forces loyal to the government fail, power will be assumed by groups clearly identified with the Communist party." Should the junta slip further, as appeared probable, the United States must act at once. "My own recommendation and that of the country team is that we should intervene to prevent another Cuba from arising out of the ashes of this uncontrollable situation."[55]

Almost as Bennett was wiring this recommendation, Johnson met with congressional leaders at the White House. The pressure of events in Santo Domingo had precluded prior consultation, but the president wanted the legislators to learn of his decision before he announced the landing to the American people. Rusk told the group how the rebels had begun the revolt and how the antirebel forces, after apparently containing the uprising, had lost control of the situation. "A number of civilians in the city of Santo Domingo had been armed by rebels, including some Castro-trained Communist supporters," Rusk said. "Law and order had broken down as a result." In the wake of the breakdown, the military leadership had informed Ambassador Bennett that Dominican security forces could not guarantee the safety of Americans in Santo Domingo. Bennett had requested protection for the Americans, in the form of American troops.

Other administration officials augmented Rusk's remarks. CIA director William Raborn said his agency had "positive identification of three ring-leaders of the rebels as Castro-trained agents." Robert McNamara explained that the American aircraft carrier *Boxer* had moved into position to land two thousand marines in a matter of hours. The defense secretary added that two divisions of paratroopers had been placed on alert at Fort Bragg, North Carolina. The administration was prepared to deal with all contingencies.

Johnson reiterated that Ambassador Bennett had advised Washington that the authorities in the Dominican Republic no longer could provide protection for American lives. The embassy, the president said, had tried to arrange a cease-fire to allow evacuation. But the fighting had grown worse. Johnson said he would announce his decision to send marines ashore—"to protect and escort American citizens to safety"—within the hour. He emphasized that he had made his decision "on the basis of unanimous decisions of our country team in the Dominican Republic, the recommendation of the ambassador, and on the recommendation of the secretary of state, the secretary of defense, and the joint chiefs of staff."

To highlight his point, Johnson read the group the cables Bennett had sent from Santo Domingo during that day. Once more he stressed that he had "no alternative" to sending troops, in view of "the unanimous recommendations received from all responsible officials." Johnson declared, "We can't waste one moment in taking action."

The legislators at the meeting raised no objections to the president's action. If anything, they encouraged Johnson to go further. Speaker McCormack said it was "obvious" that conditions in the Dominican Republic were not consistent with the American national interest. McCormack asked, "Can we afford another Castro situation of this sort?" Everett Dirksen said it was "imperative that the United States go to the heart of the matter." Dirksen said he had heard from friends in Miami that Castro was making "a concerted effort to take over the Dominican Republic." The administration, Dirksen asserted, must "take into account the factor of Castro." He expressed his hope that the president's actions would be "vigorous and adequate." The Illinois Republican concluded, "I will stand up for you."

Democratic senator George Smathers of Florida asked whether the president planned to bring the marines out as soon as the Americans were evacuated. Johnson answered cautiously, "We haven't crossed that bridge yet." Dean Rusk was a bit more forthcoming. "This is not a twenty-four hour operation," the secretary of state said.

William Fulbright and Mike Mansfield thought the president should engage the OAS in the operation. Johnson expressed approval of the good idea. The president directed that the OAS be written into the speech he was about to give. As he had once before in the meeting, Johnson asked whether there were any objections to the move he was taking or criticisms of his handling of the affair. There were none.[56]

Johnson left the cabinet room at 8:25 P.M. At 8:40 he was on the air in the White House theater telling the nation he was sending American troops into Santo Domingo. The action, he said, was "necessary in this situation in order to protect American lives." He cited the request for troops by the Dominican authorities, who had informed the American embassy that "American lives are in danger." He said he had ordered the secretary of defense to put the necessary troops ashore "in order to give protection to hundreds of Americans who are still in the Dominican Republic and to escort them safely back to this country." He promised that the OAS would be kept fully informed of American actions.[57]

VI

Johnson's repeated emphasis that the mission of the marines was to protect American lives was natural enough, given the time compulsion under which he was operating. In midafternoon on April 28 Bennett had said American troops were not necessary. By the evening he had changed his mind. Before nine o'clock Johnson had to explain why he was sending American troops into the thick of a foreign civil war. The narrow justification—the need to protect American citizens—was the obvious one and the one most easily defended.

Unfortunately, the narrow justification got Johnson into trouble when it became apparent, as Rusk had predicted to the legislative leaders, that the intervention in the Dominican Republic was not a twenty-four-hour operation. Nor did Johnson help his case by progressively exaggerating the violence and the threat to the Americans. On May 2 the president said Bennett had declared Americans would "die in the streets" absent American intervention. On May 3 he had the ambassador declaring, "You must land troops or blood will run in the streets, American blood will run in the streets." On May 4 he had Bennett frantically speaking to Washington above the din of bullets shattering the windows of the American embassy. By June, Bennett was under a desk while bullets were zinging overhead. On June 17 the carnage of late April became particularly appalling. "Some fifteen hundred innocent people were murdered and shot and their heads cut off," Johnson asserted.[58]

Meanwhile the American troops set about preventing a leftist victory. On the day after the first marines landed, Rusk cabled Bennett saying Washington was considering interposing American troops between the insurgents and the junta forces. The purpose of the interposition would be to stop the insurgency while the junta held on. "This is consistent with our primary purpose, which is to protect American lives, and with our general policy of opposing the spread of communist controlled governments in this hemisphere," Rusk explained.[59]

The interposition began soon after. The shift in emphasis from evacuation to pacification required a great increase in the size of the American force. From five hundred troops the operation grew to five thousand, then to ten thousand and twenty thousand before topping out at more than twenty-two thousand.

Justifying the American role in pacifying the Dominican Republic was more difficult than justifying intervention for the purpose of evacuation. Not only did the larger operation place American lives in jeopardy and cost lots of American money—at a time when Johnson was escalating in Vietnam and spending as fast as he could on the Great Society—it also flew in the face of the public American commitment to nonintervention in the internal affairs of the countries of the Western Hemisphere.

Johnson did his best. On April 30 the president again went before the American people to explain his actions in the Dominican Republic. He was pleased to report that American forces had successfully completed the evacuation of more than twenty-four hundred Americans and nationals of other countries. He reminded his listeners that he had ordered American troops to effect the evacuation only upon the express invitation of Dominican authorities. However, he continued, the danger in the Dominican Republic had not ended. Violence and disorder had increased. Now foreign elements were trying to manipulate the instability, to their own selfish purposes. "There are signs that people trained outside the Dominican Republic are seeking to gain control." Johnson did not utter either the name "Castro" or the word *communism* in this brief speech, but his message was plain. "The legitimate aspirations of the Dominican people and most of their leaders for progress, democracy, and social justice are threatened and so are the principles of the inter-American system." The United States was cooperating with the OAS to secure peace and the return of constitutional processes. Until these objectives were achieved, the United States would remain at the side of the Dominican people. "The United States will give its full support to the work of the OAS and will never depart from its commitment to the preservation of the right of all the free people of this hemisphere to choose their own course without falling prey to an international conspiracy from any quarter."[60]

Having conjured the specter of "outside elements" and "international conspiracy," Johnson had to supply the specter with substance. On May 2 he brought the congressional leaders back to the White House for a progress report. He explained how the situation in the Dominican Republic had grown more threatening. A truce previously agreed to by both sides had broken down. The American embassy had been under "constant fire." The Dominican airwaves were filled with "Castro anti-American oratory."

Robert McNamara described the American buildup in Santo Domingo. At the moment the United States had approximately nine thousand men on the ground. More were on their way. "These may sound like large numbers," the defense secretary said, "but to do the security job we need these men." Earle Wheeler estimated the size of the "hard-core rebel" force at twenty-five hundred. The Joint Chiefs chairman added that the rebel leadership included "several hard Communist-guerrilla types."

Johnson acknowledged that the intervention would not make him popular in Latin America. Yet he said he could not allow American safety to be jeopardized by fear of criticism. The American people would not stand for it. Johnson summarized his dilemma: "If I send in the marines, I can't live in the hemisphere. If I don't, I can't live at home."

The meeting ended with a comment by William Fulbright indicating that Johnson had successfully made his case, for now at least. The foreign-relations chairman remarked, "This has been the most informative meeting we have ever had. I feel much better informed. I support you fully."[61]

Fulbright would not be so supportive when he discovered that the evidence of an imminent second Cuba was considerably thinner than Johnson had indicated. In a meeting on April 30—without congressional leaders present—Rusk pointed out that it was impossible to prove that *all* the rebels would turn out to be communists. Johnson agreed, but he was not going to take any chances. "Not all Cubans were communist," the president remarked.[62]

Short on evidence, Johnson tried to stretch what he possessed. Just after his May 2 meeting with the legislators, the president once more went on television. This time he spoke directly about the danger of communism in the Dominican Republic. Announcing further increases in American troop strength, Johnson said changing circumstances required the escalation. The uprising had originated in an honest desire for reform but had gone bad. "The revolutionary movement took a tragic turn. Communist leaders, many of them trained in Cuba, seeing a chance to increase disorder, to gain a foothold, joined the revolution. They took increasing control. And what began as a popular democratic revolution, committed to democracy and social justice, very shortly moved and was taken over and really seized and placed into the hands of a band of Communist conspirators." (Johnson repeated this sentence nearly verbatim a few moments later, either because he liked it particularly or because he got his notes mixed up.) Resurrecting John Kennedy, Johnson reminded his listeners of Kennedy's commitment, pledged "less than a week before his death," that the United States would use "every resource at our command to prevent the establishment of another Cuba in this hemisphere."[63]

In support of the Castro-communist-conspiracy theory of the Dominican revolution, the CIA filed dutifully dire reports. The agency asserted that during the several days after the commencement of the revolt "a modest number of hard-core Communist leaders in Santo Domingo managed by superior training and tactics to win for themselves a position of considerable influence." By April 28, the CIA explained, the communists had outmaneuvered the moderates. "There appeared to be no organization within the rebel camp capable of denying them full control of the rebellion within a very few days." Putting the issue succinctly, the CIA declared, "The prospect at the time of U.S. intervention clearly was one in which a movement increasingly under the influence of Castro and other Communists was threatening to gain the ascendancy in the Dominican Republic."[64]

Johnson did not make this report public, but he did allow the release of a list of fifty-three, then fifty-eight, then seventy-seven alleged communists and fellow travelers allegedly involved in the Dominican revolt. The list, which at least in its growth rate recalled the days of Joseph McCarthy, proved a public-relations disaster. Some names were duplicated, betraying a hasty effort at padding. American reporters in the Dominican Republic who attempted to verify the identities and activities of the listees discovered that certain persons said to have taken part in the revolt had been in prison during the whole affair. Six had been out of the country. One individual fingered as a PSP sympathizer had served on the Dominican supreme court for twenty-six years—which showed either the ludicrousness of the list or tameness of Dominican radicalism. Washington soon let the matter drop.[65]

VII

The skepticism provoked by the leftist list and by Johnson's overwrought explanations of the need for American intervention turned out to have more profound consequences—for American foreign policy if not for life in the Dominican Republic—than the intervention itself. With some difficulty, the American troops, augmented by far smaller forces from several OAS states, succeeded in stopping the fighting in Santo Domingo. Negotiating a settlement between the opposing sides required a longer period and three tries by the Johnson administration. Johnson's first try was a mission by former ambassador John Bartlow Martin that began two days after the American troops landed. The second try sent McGeorge Bundy, Thomas Mann, the Pentagon's Cyrus Vance, and the State Department's Jack Vaughn to Santo Domingo in mid-May. The third try engaged the services of Ellsworth Bunker, who worked in conjunction with the OAS representatives of Brazil and El Salvador. Between June and August, Bunker and his OAS colleagues drafted and redrafted one peace proposal after another. In the second week of August they presented a document calling for the creation of a provisional government under Hector Garcia Godoy, a respected senior member of Bosch's cabinet, and for elections in the spring of 1966. The previously warring but now exhausted parties accepted the formula with relief. In the elections of June 1966, former president Joaquin Balaguer defeated Bosch, to Washington's satisfaction.[66]

Yet the Dominican affair lived on in American politics. Johnson did not like the beating his policy took in the American press. Reporters and editors of the *New York Times* and the *Washington Post*, to cite the two most visible critical papers, challenged the need for sending twenty-two thousand troops and questioned the reasons the administration adduced for the sending. Tad Szulc of the *Times* and Dan Kurzman of the *Post* piled injury on insult by hurrying home from Santo Domingo to revise their dispatches into books that made a splash in the reviews and to a lesser extent in the bookstores.

But Johnson particularly objected to criticism by William Fulbright. The foreign-relations chairman had sufficient good taste and political acumen not to attack the president while American troops were still under threat in the Dominican Republic. In addition, Fulbright wanted to conduct committee hearings on the Dominican business before stating his conclusions. In September 1965, however, just after the establishment of the provisional Garcia Godoy government, Fulbright gave a long speech in the Senate blasting Johnson's actions. Fulbright essentially accused Johnson and his advisers of lying—of making statements marked "by a lack of candor and by misinformation"—when they cited the danger to American lives as the primary reason for landing troops. Declaring the danger "more a pretext than a reason for the massive U.S. intervention," Fulbright asserted, "The United States intervened in the Dominican Republic for the purpose of preventing the victory of a revolutionary force which was judged to be Communist dominated."

And wrongly judged at that. Fulbright said the administration seriously overestimated the influence of communists among the rebels. "In their apprehension lest the Dominican Republic become another Cuba, some of our officials seem to have forgotten that virtually all reform movements attract some Communist support." Further, and ironically, by assuming that all reformers in Latin America were either closet communists, dupes of communists, or potential dupes, the administration encouraged the radicalism it aimed to prevent. "The tragedy of Santo Domingo is that a policy that purported to defeat communism in the short run is more likely to have the effect of promoting it in the long run. Intervention in the Dominican Republic has alienated—temporarily or permanently, depending on our future policies—our real friends in Latin America." By dispatching troops to prop up an illegal military regime, the Johnson administration had cut ground out from under Latin American moderates. "We have lent credence to the idea that the United States is the enemy of social revolution in Latin America and that the only choice Latin Americans have is between communism and reaction." If this was the administration's policy, the consequence was predictable. "There is no doubt of the choice that honest and patriotic Latin Americans will make: they will choose communism, not because they want it but because U.S. policy will have foreclosed all other avenues of social revolution and, indeed, all other possibilities except the perpetuation of rule by military juntas and economic oligarchies."[67]

The alienation of Fulbright may have been the most damaging consequence to Johnson of the Dominican affair. Once the chairman of the Senate Foreign Relations Committee began publicly to question the honesty and prudence of the administration, Johnson found his foreign policy burdens enormously increased. The man who should have been Johnson's strong right arm on Capitol Hill now lent his strength to the administration's opponents. Johnson was usually too busy looking ahead and too shrewd a politician to hold grudges. But he never forgave Fulbright.

Three | # The Trojan War (cont.)

IF THE CARIBBEAN basin proved vexing to Lyndon Johnson, the Middle East proved downright dangerous. Johnson's most important crisis in the Middle East would be the June War of 1967, but his first crisis in that general area had nothing to do with the fight between Arabs and Israelis. The conflict over Palestine and Israel, even if dated from the battles between Moses and the Egyptian pharaoh, or between David's crowd and the proto-Palestinians, was a late starter compared with the struggle for control of the islands and peninsulas of the eastern Mediterranean. The current manifestation of the Homeric contest pitted the descendants of Hector and Aeneas in Turkey against the heirs of Achilles and Odysseus in Greece, and the present prize was Cyprus. Cyprus, at any rate, was the ostensible casus near-belli. Thirty centuries of score settling figured as well.

Johnson never faced a more excruciating dilemma than that posed by the Cyprus problem. With the exception of a noisy and mobilized group of Greek-Americans, almost no one in the United States cared anything for Cyprus. Few more found much to distinguish among the relative claims of Greeks, Turks, Greek Cypriots, and Turkish Cypriots. Lacking visible moral grounds for choosing among the parties at dispute, Johnson sought to decide the matter according to strategic considerations. (Whether morality, had such been identifiable to administration officials, would have overridden strategy is an open question.) The trouble was that strategy gave as little guidance as morality. The Cyprus affair arrayed two American allies against each other. Having declared cold war in 1947 largely in response to communist efforts to topple the government of Greece, and now benefiting from the use of military facilities in that country, the United States government could hardly ignore the demands of Athens. Yet the American commitment to Turkey was of equally long standing, and from the early 1950s the Turks had cooperated in allowing Americans to eavesdrop on the Soviet Union from Turkey, to launch spy flights over the Soviet Union, and to erect missile installations hard against the Soviet border. Further, the Turkish straits constituted a choke point against the Soviets worth half the American Mediterranean fleet in time of war.

Unable to decide between the two allies, the Johnson administration followed a path that predictably antagonized both. Johnson recognized the hazards of evenhandedness but calculated that antagonism on the part of the two countries toward the United States, so long as it did not get out of hand, was preferable to a military showdown between Greeks and Turks. Should Athens and Ankara have at it, the southeastern flank of NATO would collapse, affording the Soviets unfettered entree to the Mediterranean and the opportunity for untold mischief in the Middle East and northern Africa. To a degree matched only in Vietnam, Johnson learned amid the Cyprus troubles the thankless character of a superpower's role—with the difference that in Vietnam he was making war while in Cyprus he was preventing it.

II

In important respects, the Cyprus crisis of 1964 replayed the events in Greece and Turkey that had produced the Truman doctrine seventeen years earlier. For the purposes of gaining Johnson's attention, this was a blessing because as always the president placed great emphasis on the historical basis of American commitments. Both sets of developments began with Britain responsible for maintaining a pro-Western balance of power in the territory in question, and each ended with the United States having assumed Britain's task. But where the 1947 troubles eased not long after Washington took over from London, the Cyprus sore festered and oozed for a decade after the United States got involved. Johnson managed only a holding action, and he left office in 1969 with the fundamental issues as unresolved as ever.

Witnesses at the birth of the republic of Cyprus in 1960 required unwarranted optimism to forecast a happy youth for the infant. The travail forcing the birth, in particular the widespread terror against the British colonial government, shifted in target but did not cease. The three-quarters of the Cypriot population of Greek descent included many who wished for *enosis,* or union with Greece. The Turkish community on the island naturally and vigorously opposed the idea. Lacking the numbers to credibly advocate union with Turkey—though they pointed to the geographic logic of such a move—the Turkish Cypriots divided into two persuasions: the sanguine, who sought ironclad guarantees of equal treatment in an independent Cyprus, and the realistic, who believed they would not be safe until their portion of the island was attached to Turkey.

The president of Cyprus, Greek Orthodox archbishop Makarios III, predictably frowned on both enosis and partition. The latter would steal half his domain, the former the whole thing. (The latter could steal the whole thing too, since Greece might well respond to a Turkish takeover of Turkish northeastern Cyprus with forcible enosis for the rest.) Yet as leader of the Greek Cypriot community, and with a history of Turk-baiting, Makarios seemed to the Turkish Cypriots anything but an honest broker between the two communities. Before the British handed over power in Nicosia, they attempted to enforce honesty on Makarios by an appropriately byzantine five-party arrangement known as the London-Zurich accords, which committed Greek Cypriots and Turkish Cypriots to share power in Cyprus and pledged Britain, Greece, and Turkey to ensure observance of the power-sharing formula. Greece and Turkey were allowed to station troops on the island to protect the London-Zurich formula, which in practice meant that Greek troops protected Greek Cypriots, and Turkish troops Turkish Cypriots. Britain was allowed to station troops on the island to keep the Greeks and Turks apart.

Partly as an inducement not to wash its hands of the frustrating business peacekeeping rapidly turned out to be, Britain was permitted to retain two military bases on the island. The bases were less important than in the days when Cyprus guarded the approach to British Palestine and to British India via Suez, but they still possessed a nostalgia value for holdover imperialists. (The nostalgists tried to forget that the Cyprus bases had served as the staging ground for the 1956 Suez humiliation.) Cyprus also provided an important communications link between Britain and the positions Britain still held east of Suez. Since the British shared information and facility rights with the United States, Cyprus counted in American strategy as well.

As the British (who had more experience decolonizing territory than anyone else, principally because they had more territory to decolonize) might have guessed, the Greek Cypriot majority accepted the London-Zurich accords only long enough to gain Cypriot independence. Immediately the Greek Cypriots began attacking the guarantees as an infringement of Cypriot sovereignty, and they denounced the Greek, Turkish, and British troops as occupying forces. On both counts they were correct—which gave Turkish Cypriots all the more reason to seek the protection of their cousins on the mainland. The seeking fanned fears of fifth-columnism and increased the hostility of the Greek Cypriots for the Turkish Cypriots. Makarios may or may not have held a personal grudge against Turkish Cypriots, but he deemed it politic not to try very hard to suppress the violence against them.

The violence escalated significantly just as Johnson entered the White House. In November 1963 Makarios put forward thirteen constitutional amendments challenging the London-Zurich accords and diminishing the role of Turkish Cypriots in the island's government. The Turkish Cypriots objected militantly. As the year's end approached, the country trembled on the brink of civil war. Greek and Turkish troops in Cyprus mobilized, as did reinforcements in Greece and Turkey.

Washington had been expecting a crisis for some time. In the summer of 1961, Kennedy responded to intelligence reports of heightened tension in Cyprus with the worried comment: "It seems to me that if the situation is as desperate as we hear it is, we cannot continue to rely upon our policy of hoping that the guarantor powers will shoulder the principal share of the Western burden. Shouldn't we have this more carefully reviewed?" The review succeeded chiefly in demonstrating that a war over Cyprus would not conduce to American security. It produced no new ideas on solving the conflict, which continued to deepen. When Makarios visited the United States in 1962, Kennedy urged him to avoid agitating matters between Greek Cypriots and Turkish Cypriots and offered American economic aid as an incentive to good behavior. Makarios took the aid but ignored the advice, leading White House assistant Robert Komer to remark in June 1963, "We're in for real trouble if we don't dampen this communal squabbling down soon."[1]

When the real trouble came six months later, Johnson tried diplomatically to dampen it. He sent a Christmas message to Makarios and the Turkish Cypriot vice president, Fazil Kutchuk, expressing the sadness of the American people that Cypriots were killing one another, particularly during this special season. "I will not presume to judge the root causes or rights and wrongs as between Cypriots of the two communities," Johnson said, adding, "I cannot believe that you and your fellow Cypriots will spare any efforts, any sacrifice, to end this terrible fraternal strife."[2]

The two communities made a half-hearted effort to curb the violence. At Britain's invitation, they holstered weapons briefly and sent representatives to a conference in London in January 1964. But nothing good happened there. The Greek Cypriots demanded cancellation of the portion of the London-Zurich accords allowing intervention by the guarantor powers and insisted on a government based on majority rule— which implied Greek Cypriot rule. The Turkish Cypriots, despairing of peace or fair treatment in a democratic Cypriot republic, advocated partition and refused to reconsider Turkey's continued right of intervention. Neither side would budge, killing the London conference and restarting the killing of Cypriots.

At wits' and patience's end, the British turned to the United States, much as they had done during the Greek civil war in 1947. In the final week of January, British ambassador David Ormsby-Gore paid a visit to George Ball, the acting secretary of state during a brief absence of Dean Rusk. Ormsby-Gore explained that Britain had dealt with Cyprus as long as it could. Greece and Turkey, which were supposed to help preserve peace, instead were goading the opposing parties to violence. The British government recommended a multilateral peacekeeping force from the NATO countries. The peace force would quell the violence while a mediator sought a solution to the island's troubles. The British government hoped the United States would contribute troops to the peace force. Should the force not materialize, Britain would turn peacekeeping in Cyprus over to the United Nations.[3]

The British demarche produced double deja vu in the State Department. In addition to the obvious analogy to the Greek civil war, the prospect of seeing Cyprus handed to the United Nations revived memories regarding Britain's decision, also of 1947, to place Palestine before the same forum. The Greek comparison was unpropitious enough, for though Johnson believed that the Truman doctrine had been necessary and wise, he had no desire to undertake anything similar toward Cyprus. Even had Johnson possessed such a desire, American politics would have prevented it. Mike Mansfield warned the president off. "The crisis in Cyprus is of greater concern to our European allies—notably to Britain—than it is to us," Mansfield told Johnson. "We ought to do everything that we can to keep it that way. There is nothing for us to gain by involving ourselves in replacement of Britain as the number one outside power, after Greece and Turkey, in this situation. There is every danger that if we do so we will reap a harvest of cost, blood and resentment and little more."[4]

If the 1947 Greek example afforded scant comfort, the Palestine parallel was even worse. The United Nations, that final dumping ground for unresolvable conflicts, had failed to resolve the Palestine dispute, as two subsequent wars and innumerable skirmishes and terrorist raids attested. There was no reason to think the United Nations would succeed any better with Cyprus. Taking self-determination as its shibboleth, the international body doubtless would back Makarios's denunciation of the London-Zurich right of foreign intervention. This would relieve the British of their obligation to keep the island's peace (it might also relieve the British of their bases, but London thought something might be worked out with Makarios), yet at the same time it would allow Greek Cypriots to murder Turkish Cypriots, and Turkish Cypriots to murder back, and it would probably provoke even greater foreign intervention—

by Greece and Turkey in perhaps war-level force—than it ostensibly sought to avert.

The Kremlin could be counted on to capitalize on the discomfiture of the NATO camp. At worst the Soviets would become directly involved in a fight between Greece and Turkey when the losing side, unsatisfied with the support it was getting from Washington, looked elsewhere for help. At best they would seize the chance of United Nations consideration of the Cyprus question to meddle in a region the United States had tried for two decades to keep them out of.

The Johnson administration already had a taste of what the United Nations would do with Cyprus. At the end of December 1963, Makarios had called on the Security Council to investigate the violence on the island. The debate the issue elicited confirmed Washington's worst fears. As W. Averell Harriman, undersecretary of state for political affairs, summarized the discussion: "The results of this meeting were to weaken current efforts to reestablish law and order on Cyprus, to strengthen the hands of extremists in both communities on Cyprus and in Greece and Turkey, to reduce the possibility of a settlement of the basic issues on Cyprus, and to provide an opportunity for Soviet and other anti-Western propaganda initiatives." Condemning "the utter sterility and self-defeating nature of this exhibition," Harriman said the United States and its allies should do everything possible to prevent its repetition.[5]

For all their concern about the consequences of Cyprus falling to the United Nations, American officials were not sufficiently fearful to accept Britain's invitation to join a peace force from the NATO countries. George Ball consulted Johnson, then informed Ormsby-Gore that while the United States appreciated the possibly disruptive effects on the Atlantic alliance of the Cyprus situation, the American government could not add to its already extensive global commitments. The intractable nature of the Cyprus question heightened American reluctance. "We would find ourselves moving into a power vacuum and another complicated political problem with no end in sight," Ball said. The American government would happily furnish logistical support to a peacekeeping unit composed of forces from other NATO countries. The undersecretary suggested that a force without Americans might appeal more readily to certain alliance governments—especially that of Charles de Gaulle— than would one including Americans. Washington would do what it could diplomatically with Athens and Ankara. But at present the United States could not undertake the action Britain proposed. Ball urged the British government to reconsider its decision to withdraw from Cyprus.[6]

As soon as Ormsby-Gore left, Ball called a meeting of top officials of the State and Defense Departments and the White House staff to assess the Cyprus situation. With help from the assistant secretary of state for the Near East, Phillips Talbot, Ball explained that any peacekeeping force sent to the island would require the approval of Makarios, who would be even less inclined to countenance the intrusion of NATO into Cyprus than to accept the present British-Greek-Turkish arrangement. The Greek Cypriots had mastered the techniques of terrorism during the fight against British colonialism. They would not hesitate to employ those techniques against what they would consider a new and equally oppressive occupation force.

Robert McNamara asked why London had chosen this particular moment to bug out. Ball replied that the British experience in Cyprus had been long and unhappy, that the British wanted to have done with empire, certainly as it related to Cyprus, and that the British government faced an election in which Cyprus looked like a loser. McNamara suggested, and Maxwell Taylor of the Joint Chiefs agreed, that the British were bluffing. The British would not really pull out, McNamara said. Ball wished he could concur, but could not do so with any confidence. Though London recognized the perils of leaving Cyprus to the United Nations, the British apparently preferred this to the alternatives. Yet Ball did not exclude the possibility that Britain might be persuaded to reorder its preferences.

To a query from McNamara as to what would happen to Cyprus in the United Nations, Talbot responded that Makarios and the Greeks Cypriots would get their way. Makarios would wrap himself in the mantle of self-determination and nonintervention. Because most members of the United Nations had recently been someone's colonies or otherwise subject to the whims of the great powers, Makarios's position would easily muster a majority. Besides, almost no one liked the Turks—for historical reasons as well as more recent ideological and political ones— and few would shed tears over dead Turkish Cypriots. Ankara would feel isolated and might grow belligerent.

Ball summarized the decision facing the United States at the moment. "The prime question," the undersecretary said, "is one of whether we are prepared to let the U.K. off the hook, or do we keep their feet to the fire?" Despite his doubts regarding its efficacy, Ball favored the latter approach, as did McNamara. The administration, Ball said, should remind the British that both countries had duties to world peace. "We have other responsibilities which we will continue to bear. They must bear theirs."[7]

The British were not in a bearing frame of mind, though, as they made unmistakable to Washington. When they did, the Johnson administration began to soften its opposition to participation in a peacekeeping force. Additional violence in Cyprus, which heightened fears of Turkish intervention and Greek counterintervention, caused a further softening. "It now looks as if the situation has become extremely grave," Ball commented at the end of January, adding that circumstances called for more-positive American action.

The immediate need was for measures to keep the Turkish and Greek soldiers in their barracks. There was a "real danger of a blow-up," Ball told Johnson; to which Johnson responded that the United States government needed to tell the Turks and Greeks: "You just behave yourselves." Ball cabled the American embassies in Ankara and Athens relaying the president's order that American officials devote "maximum efforts" to restraining the parties to the imminent war. The governments of Greece and Turkey should be reminded that their obligations to NATO and world peace came before their interests in the strife on Cyprus. The consequences of a war between two important members of the Atlantic alliance were "almost too appalling to contemplate." The governments of Greece and Turkey nonetheless should be compelled to contemplate them, lest they precipitate them. The Turks should be assured that the United States desired what Turkey desired: the security of the Turkish community on Cyprus. But the United States must have Turkey's cooperation in order to achieve this common goal. In particular Ankara must stop threatening to impose a military solution to the Cyprus problem. As for the Greeks, Athens must reassure Ankara that the rights of the Turkish minority in Cyprus would be respected. In a final word, Ball passed along the president's special reminder of what Athens and Ankara owed the United States and an emphatic message that the two countries must not allow their quarrel over Cyprus to create a breach in the wall against communism. Johnson said he found it impossible to accept "that two U.S. allies, whose forces are largely equipped by the U.S. Government and which are under the U.S. protective umbrella, should lose sight of the overriding need to maintain NATO solidarity against the real enemy."[8]

To highlight his concern, Johnson dispatched General Lyman Lemnitzer, the American commander of NATO, to the eastern Mediterranean. Lemnitzer preached patience while hinting at sanctions in the event patience failed. Meanwhile the president invited the British to send a senior general to Washington to discuss the details of a peacekeeping plan. Johnson also directed the State Department and the Pentagon to project the ramifications of American participation in a peace force in order to circumscribe American liability most narrowly. Finally, after receiving Lemnitzer's report from the front, Johnson authorized Ball to tell the British that the United States would support their peacekeeping proposal.[9]

Two days later, Johnson informed Makarios of his decision. The archbishop-president had already rejected the British-proposed force as a transparent device to ensure NATO access to the bases and communication facilities on the island. Makarios insisted that the United Nations take charge of peacekeeping—which was exactly what Washington was trying to prevent. Johnson urged Makarios to accept the British plan. "It fully protects the position of the Government of Cyprus," Johnson averred. "It reduces the present dangerous polarization. It provides for the appointment of a mediator of irreproachable standing and objectivity whose selection would be subject to the approval of your Government." It afforded to all concerned the best prospect of a fair solution to the tragic problems Cypriots of both communities had faced for too long.[10]

Makarios did not immediately say no to Johnson. Fraser Wilkins, the American ambassador in Nicosia, thought the archbishop was readying a counterproposal. Johnson, wishing to forestall a veto, sent Ball to Cyprus to sell Makarios on the British plan.[11]

En route, the undersecretary stopped in London. There he met the (Greek) Cypriot foreign minister, Spyros Kyprianou. Kyprianou apparently knew more than Makarios had told Wilkins, for the foreign minister indicated that the government of Cyprus would not go along with a peacekeeping force from the NATO countries. Kyprianou's representation was sufficiently negative that Ball suspected that a violent reception awaited any American troops entering the country over Makarios's objections—a conclusion the CIA reached independently. A pair of bomb attacks on the American embassy in Nicosia, accompanied by the wrecking of several American-owned cars, appeared to confirm the conclusion. Under these circumstances, Ball thought the United States should have nothing to do with a peacekeeping force. When the undersecretary explained his reasoning to Johnson, the president agreed.[12]

Inconveniently for reneging, the administration had already announced its support for the peace plan. But Ball believed he could use Makarios to get the administration off the hook. "I shall try to place on Makarios' shoulders the primary onus for our non-participation," the undersecretary cabled Johnson. Yet accomplishing this objective would not be easy. The archbishop was sly, and excessive subtley might be lost on the Turks, who still itched to solve the matter by force, and who considered American involvement in a peacekeeping venture necessary to prevent Greek double-dealing. "It will require the hazardous operation of walking on eggs with golf shoes," Ball told Johnson.[13]

III

Makarios turned out to be everything Ball had expected, which was a lot. The undersecretary's primary briefing on the archbishop came from Adlai Stevenson. During the Kennedy years the United Nations ambassador had once spent three days with Makarios in Nicosia. Stevenson departed with nothing but contempt for the man. The only way to deal with Makarios, Stevenson said, was by "giving the old bastard absolute hell." Stevenson added, "I have sat across the table from that pious looking replica of Jesus Christ, and if you saw him with his beard shaved and a push-cart, you would recall the old saying that there hasn't been an honest thief since Barabbas."

At Ball's first meeting with Makarios, on the porch of the archbishop's residence, Makarios greeted the undersecretary in the luxurious costume of the Orthodox prelate. But once the formal reception ended, the archbishop took Ball to his private study and performed what the undersecretary later described as "an astonishing striptease." Off came the headgear symbolizing his episcopacy. Off came the gold chain and medallion signifying his office as confessor. Off came the robes of the priest. Ball had expected to find a venerable ecclesiastic. "Now I found myself facing a tough, cynical man of fifty-one, far more suited to temporal command than spiritual inspiration." Makarios's actions during the next few days caused Ball to question part of this description. "He must be cheating about his age," Ball told Johnson. "No one could acquire so much guile in only fifty-one years."

Makarios attempted to convince Ball that the difficulties afflicting Cyprus were nothing out of the ordinary. "Mr. Secretary," he said, "the Greeks and Turks have lived together for two thousand years on this island and there have always been occasional incidents. We are quite used to this." Ball, dismayed at the violence occurring even as they spoke, responded, "Your Beatitude"—the honorific almost gagged him—"I've been trying for the last two days to make the simple point that this is not the Middle Ages but the latter part of the twentieth century. The world's not going to stand idly by and let you turn this beautiful little island into your private abattoir." Makarios countered, "You're a hard man, Mr. Secretary, a very hard man."

Later during Ball's visit, the archbishop addressed the undersecretary alone in his study. With disarming directness he said, "I like you, Mr. Secretary. You speak candidly, and I respect that. It's too bad we couldn't have met under happier circumstances. Then, I'm sure, we could have been friends." After a moment's pause, he continued, "We've talked about many things and we've been frank with one another. I think it right to say we've developed a considerable rapport. Yet there's one thing I haven't asked you, and I don't know whether I should or not. But I shall anyway. Do you think I should be killed by the Turks or the Greeks? Better by the Greeks, wouldn't you think?"

Ball, ever diplomatic, refused to commit himself. "Your Beatitude," he answered, "that's your problem."[14]

(Yet Ball thought Makarios's question not entirely inappropriate to the search for a solution to the Cyprus problem. A few years' exasperation later, Ball commented in confidence, "That son of a bitch will have to be killed before anything happens in Cyprus.")[15]

Matters of assassination aside, Ball found Makarios as opposed to a peacekeeping force from NATO countries as the undersecretary had anticipated. At the same time, the archbishop allowed a role for the American Mediterranean fleet, which he said he would "welcome" as a device to prevent Turkish intervention. Makarios indicated he would support a peacekeeping force from the United Nations, although he desired to limit its terms of reference to guaranteeing the independence and territorial integrity of Cyprus—in order, Ball acidly inferred, that he and the Greek Cypriots could "go on happily massacring Turkish Cypriots."[16]

Though Ball succeeded in attaining what had become the primary goal of his visit—to get the United States out of its commitment to a NATO peace force—a more pressing objective moved into view while he jousted with Makarios. "The situation of civil order has deteriorated markedly overnight," Ball cabled to Johnson on February 13. Pitched battles raged in the southern city of Limassol, where Greek Cypriot police units were firing heavy shells into the Turkish quarter. "It appears they have launched an all-out attack," Ball wrote. British troops had received orders to attempt to stop the fighting, and perhaps they would succeed, but it would be a close thing and might require more time than Turkey was in a mood to allow. Ball noted that his presence in Cyprus was exerting a restraining influence on Ankara. Unfortunately this influence would disappear as soon as word got out that there would be no NATO peacekeeping operation. Ball saw Turkish intervention lowering over Cyprus, with war against Greece looming in the background.

To forestall this outcome, the United States must act vigorously and at once. "I am convinced we are desperately short of time," Ball told Johnson. "I believe the bomb has already gone off at Sarajevo and the archduke is dead." Switching metaphors, the undersecretary asserted, "The governments and people of Turkey and Greece want peace, but they are like characters in a Greek tragedy. They cannot, by their own unaided efforts, avoid catastrophe. They can be pushed off a collision course only by some outside agency." The United States, acting in conjunction with Britain, must become that agency.

Because it appeared impossible to create a peace force that would win the consent of the two communities on Cyprus—the Turkish Cypriots placed no trust in a United Nations force for precisely the reason the Greek Cypriots supported it—Ball urged that the Johnson administration propose a force that did not require consent, namely one consisting of contingents from the three guarantor powers. He recommended the amalgamation of British, Greek, and Turkish soldiers into a joint command. All operations would be conducted together. Patrols would involve teams of three soldiers—one of each nationality—on jeeps. As soon as the tripartite force was installed, the five parties connected to the Cyprus dispute would be asked to agree on a mediator. Ball suggested former president and former NATO commander General Eisenhower. Once the mediator devised a solution, the United States should throw its full resources behind implementation.

Ball conceded the hazards of his plan, the most worrisome being the possibility that the Greek and Turkish elements of the peace force would start shooting at each other. The Greek Cypriots might resist the landing of the tripartite force, at least its Turkish elements. The plan, by short-circuiting the United Nations, would undermine the authority of the international body. Yet Ball thought the difficulties were manageable and the risks worth taking. The dangers of doing nothing were far greater.[17]

Ball reiterated his plan in person on arrival back in Washington. In a meeting with Johnson he emphasized the stakes of the situation, describing the present condition of affairs as "the most serious confrontation since the Cuban missile crisis." The Turks were not bluffing. While in the eastern Mediterranean, Ball had talked with Turkish prime minister Ismet Inonu, who said that another incident like Limassol would require Ankara to send in the troops. Ball had also visited Athens, where the Greeks vowed they would move if the Turks did. The chief of the Greek general staff informed him that if the current Greek government failed to act decisively, the Greek military would get a government that would. Ball told Johnson there was a fifty-fifty chance of a Greek-Turkish war. Such a war could easily flare out of control. A United Nations peace force, even if both Cypriot communities accepted it, would not form up quickly enough to do any good. Besides, Ball said, "A U.N. peacekeeping force can't shoot policemen, which is the heart of the matter."

Johnson agreed as to the gravity of the situation Ball described. The president remarked that a Greek-Turkish war was "inevitable" unless the United States sidetracked Ankara and Athens. He indicated his readiness to come down hard on the Turks if they appeared about to strike. "We might have to tell the Turks 'It's goodnight, nurse,'" Johnson said. The United States was in a position to force Ankara to settle peacefully. "If we told Inonu we'd cut off aid, he'd have to back down." The president wanted to study Ball's proposal more carefully, but in the meantime he directed the State Department to draft a message telling Inonu in "the toughest language possible" to hold still.[18]

IV

Before Johnson decided for or against Ball's tripartite proposal, the British vetoed it. The escalating violence and the war threats worried London, but not enough to buck up Whitehall and Downing Street for an effort to secure the necessary political support for a new initiative in Cyprus policy. As they had warned, the British requested that the United Nations assume peacekeeping responsibility on the island.

The United Nations organized a peace force, which the Johnson administration supported without enthusiasm or troops, although with money: approximately 40 percent of the cost of the force. Under the circumstances, a United Nations force appeared the least objectionable instrument for calming the Cyprus situation. Johnson explained his view to William Fulbright, with slightly less than complete candor. Replying to a letter from the Senate foreign relations chairman, Johnson wrote, "I fully agree with your basic conclusion that we must do everything we can to avoid humiliation of Turkey, and I also think it is important to do anything we can to keep the United Nations in the picture. That is what we have done so far, and we shall certainly keep on with it."[19]

The United Nations force arrived in Cyprus in March. The Turkish Cypriots remained distrustful of the United Nations, but there was little they could do to act on their distrust. In the event, the peace force had a momentary soothing effect. With Canadians and Swedes, and through them the world, watching, the Greek Cypriots scaled back their violence. Yet neither the peace force nor a Finnish mediator appointed by the United Nations accomplished anything to definitively defuse the situation, and when the newly-elected Greek prime minister, George Papandreou, called for Cypriot self-determination—a direct assault on the London-Zurich accords and an implicit attack on the Turkish Cypriot community—Ankara once more girded for war.

At the beginning of June, the American ambassador to Turkey, Raymond Hare, reported urgently that the Turkish military would soon invade and partition Cyprus. Johnson ordered Hare to contact Inonu at once, "calling him out of a cabinet meeting if necessary," to relay the president's "gravest concern" at the possibility of an invasion, and to urge restraint. In a cover cable, Dean Rusk added, "Use all arguments in your arsenal to pull them back."[20]

Within a few hours, Johnson augmented Hare's arsenal with a letter to Inonu that Ball characterized as "the diplomatic equivalent of an atomic bomb." After reading Johnson's letter, the undersecretary remarked, "That may stop Inonu from invading, but I don't know how we'll ever get him down off the ceiling." Rusk, who had shown Ball the letter, responded sweetly, "That'll be your problem."[21]

Johnson's letter left little to Inonu's imagination. By invading and partitioning Cyprus, the president said, Turkey would betray its obligations to the Atlantic alliance and to the United Nations. "There can be no question in your mind that a Turkish intervention in Cyprus would lead to a military engagement between Turkish and Greek forces." Such a conflict between two NATO allies was "unthinkable" and would comfort only the enemies of freedom. The Soviet Union, which already had taken advantage of the division in NATO ranks to offer support to Makarios, would be tempted toward direct involvement in any war that erupted. If the Kremlin succumbed to the temptation, the Turks had better look to their own devices. "I hope you will understand that your NATO allies have not had a chance to consider whether they have an obligation to protect Turkey against the Soviet Union if Turkey takes a step which results in Soviet intervention without the full consent and understanding of its NATO allies." Almost as an afterthought, Johnson added that Turkey could forget about further American aid. The president concluded, "You have your responsibilities as Chief of the Government of Turkey. I also have mine as President of the United States."[22]

Johnson's withering blast had the desired effect, stopping the Turkish operation dead. Inonu told Hare his government would consult with Washington before acting. At the same time, Inonu complained of the undue severity of Johnson's letter. Allies did not speak to allies thus, he said. He denied that Turkey intended to partition Cyprus. His government aimed only to protect Turkish Cypriots from the violence visited upon them by the Greeks. Inonu bitterly asked Hare to inquire of Johnson who would protect the Turkish community now that Turkey had been forced to stand down.[23]

While Johnson's ultimatum solved the immediate problem of keeping the Turks out of Cyprus, it created another problem in weakening Inonu's position with respect to hard-liners in Ankara. General Lemnitzer described the Turkish officer corps as "very strong for intervention" and indicated that although the military commanders accepted Inonu's decision to follow the American lead this time, they might not always do so. Hare remarked, regarding the atmosphere in Ankara: "Although Inonu's decision was one which to us would seem normal and rational, this is not the prevailing mood here, and as a consequence he has put himself in a very vulnerable position unless he can prove he is right and the majority wrong."[24]

Recognizing the need to help in the proof, the Johnson administration amplified its efforts to get Greece to discuss realistic terms of peace in Cyprus. Rusk cabled the American embassy in Athens to alert the Greeks to the narrowness of their escape. "It is important that the Greeks in Athens (and Nicosia) understand fully that the sword of Damocles is still hanging by a thread, now one fiber weaker," the secretary said. Johnson summoned the Greek ambassador in Washington to disabuse him of any notion that Ankara's embarrassment represented a victory for Athens. "Greece must avoid at all costs humiliating its ally Turkey," the president declared. "Even in the Cuban missile crisis, we always left the enemy a way out."

The president told the Greek diplomat that the American people were mystified at the Cyprus matter. They could not understand why two of America's allies were "growling at each other." Johnson suggested that the United States, which had come to Greece's aid during that country's civil war, deserved more gratitude than it was receiving. The hour had arrived for the Greek government to "show some statesmanship and get moving toward agreement." The president said the United States could not hold the reins on the Turks for much longer. "If I can't get you to talk, I can't keep the Turks from moving." To the ambassador's objection that talks were difficult under threat of Turkish invasion, Johnson rejoined, "Of course they're difficult. But it's more difficult to talk after an invasion. Get together and work something out." Reciting his favorite biblical passage, Johnson quoted Isaiah: "Come let us reason together." More specifically the president said, "What we want is for your prime minister to sit down with the Turkish prime minister and work out an agreement." To date, only the Turks had made concessions. "All I want is for the Greek prime minister to sit down and talk. This is not so difficult."[25]

In fact, it was so difficult—sufficiently difficult, at any rate, that Papandreou refused. The Greek prime minister visited Washington at the end of June. Johnson repeated his warnings of the dire consequences intransigence on the Cyprus issue might produce. He reiterated the need for compromise, making the obvious point that no settlement would fully satisfy everyone. He offered American good offices in achieving a settlement.[26]

The energy of the president's exposition had a strong, albeit negative, impression on the Greek delegation. Andreas Papandreou, the prime minister's son and a future prime minister, afterward described the White House session with disgust. "It had not been a discussion. It had been a monologue. . . . It became a brainwashing operation."[27]

Even so, the Greeks agreed to indirect negotiations with the Turks. Inonu, in Washington at about—but not exactly—the same time, likewise accepted a formula whereby Greek and Turkish representatives would meet separately in Geneva with an American intermediary.

To fill the role of intermediary, Johnson tapped the architect of the Truman doctrine, the grand old man of American relations with Greece and Turkey, Dean Acheson. The former secretary of state suffered fools no more gladly at seventy-one than he had a decade and a half before. If anything, he detected a sizable increase in their number. In this regard he was not the best choice for the job. Fools abounded on the Cyprus question, at least in Acheson's view. But Johnson hoped Acheson still possessed some pull with Ankara and Athens, and the president was willing to take a chance on Acheson's limited tolerance.

V

The Geneva talks began in July. The mere fact of the negotiations seemed a hopeful sign. "We put money in the bank with the passing of every week without serious fighting on the island," Ball told Johnson at the end of the month. Acheson, negotiating alternately with Greeks and Turks, devised a plan that would eliminate the Cyprus problem by eliminating Cyprus. Under this first Acheson plan, Cyprus as an independent country would disappear, divided between Greece and Turkey. Greece would get a piece of the island roughly proportional to the percentage of Greek Cypriots in the country's population. Turkey would get the rest. Because the partition would not cleanly separate Greek and Turkish communities, the plan included an elaborate scheme of minority guarantees. Of these the most noteworthy was a provision allowing Turkey to establish a military base in the Turkish zone, ostensibly to protect the approaches to Turkey proper but also to serve as a reminder of Ankara's continuing interest in the status of Turks in the Greek sector. American aid would help underwrite the arrangement.[28]

Sadly for the Johnson administration's hopes, the Greek government rejected the Acheson plan at once, calling it a denial of the rights of the Cypriot people—by which Athens meant the Greek Cypriot majority. The administration read utter cynicism into this response. "The Greeks know that time is on their side," McGeorge Bundy told Johnson. "They figure the worst they can come out with is a U.N. debate ending in a blessing for self-determination, which would result in enosis." The fact that Makarios also rejected the Acheson plan, as well he might have, contributed to Athens's opposition, since the plan would have to be implemented over the objections of the spokesman of the community the Greek government had politically committed to protect. (Makarios's opposition to anything Turkey might accept was the reason Washington did not include him in the discussions.)[29]

While Acheson was working on a revision of his plan, tension on the island again mounted to war-threatening levels. For some time, large numbers of Greeks, many with military backgrounds, had been entering Cyprus. The Greek government tried to explain that these were simply Greek Cypriots who had served in the Greek army and were either retiring or deserting to their homeland. Athens denied any infiltration of Greek army personnel. Yet a Greek general assigned to NATO conceded to the American ambassador in Paris that such individuals might constitute a training cadre for the Cypriot army. The Greek officer, while disclaiming any shipments of weapons from the Greek government to Cyprus, also admitted that extensive smuggling of arms was taking place.[30]

Heartened by this support, Greek Cypriot irregulars stepped up attacks on Turkish Cypriot villages. Turkey responded with air raids on Greek Cypriot communities. Makarios chose to appear helpless in the face of the fighting. "What can I do?" he asked Taylor Belcher, the American ambassador in Nicosia. Belcher knew Makarios better. The ambassador thought the archbishop was engaged in a calculated effort to force the United States to restrain Turkey again while the Greek Cypriots made Cyprus unlivable for the Turkish Cypriots.

Belcher appreciated the shrewdness of Makarios's strategy, since the ambassador thought the United States must do what Makarios wanted. Belcher understood the costs involved. "If the U.S. successfully stops the Turks on this occasion . . ." he said, "we will henceforth be held responsible for controlling the Turks and subject to this type of blackmail." Considering what was at risk, however, the American government had little choice. "I believe we must make every conceivable effort to obtain at least a temporary cessation of bombing in order not to create a holocaust on this island with all its ultimate consequences for our alliances not only in the area but elsewhere."[31]

From Ankara came indications that this time Turkey might not back down. The Turkish foreign minister made no attempt to disguise his government's role in the bombing of the Greek Cypriot towns. The government, he told Raymond Hare, hearing the "piteous appeals for help" from Turkish Cypriots, could not allow them to be destroyed by the Greeks. Turkey had cooperated with the United States, yet the killing continued. Every expedient had been exhausted. Nothing remained but the use of force. Turkey did not want war. It was acting only under "inescapable compulsion."[32]

When these reports reached the White House, Johnson responded with identical letters to Inonu, Papandreou, and Makarios, calling for "the greatest possible restraint at this critical moment." The president added, "To permit conflict to continue longer creates the imminent danger of widespread bloodshed, including the loss of lives of many innocent people, for which no statesman would wish to bear the responsibility."[33]

Aware that words were selling cheap in the eastern Mediterranean, Johnson prepared to complement his warning with measures more concrete. Since June, American intelligence operatives in the Cyprus area had closely monitored the activities of General George Grivas, the hero of the Cypriot revolt against British colonialism and the commander of the sometimes-terrorist organization EOKA. Grivas and Makarios had long been rivals, partly for personal reasons but principally because Grivas favored enosis while Makarios liked the idea of being president of an independent republic. During the summer of 1964 Washington received word that the Grivas-Makarios rivalry continued undiminished. Though Grivas was no more pro-American than Makarios, he appeared more predictable—in Ball's words, "easier to work with." Through Grivas's associate Socrates Iliades, the Johnson administration learned that Grivas was devising a scheme for overthrowing Makarios and attaching Cyprus to Greece, with an important proviso guaranteeing compensation to Turkish Cypriots who chose to leave Greece's new province and legal protection for those desiring to stay. Washington let Grivas know the United States did not look unfavorably on his plan.[34]

Perhaps Makarios got wind of American backing for Grivas. Perhaps that was the point. Perhaps Johnson's appeal for peace had an effect. Perhaps Athens became worried by the growing Soviet interest in Cyprus. From whatever combination of causes, Makarios chose to be persuaded by a United Nations appeal for a cease-fire. When Turkey joined the truce, the August crisis ended.

VI

Once more attention focused on Geneva. Soon Acheson unveiled a second Cyprus plan. This version differed from the American mediator's first proposal principally in providing for a fifty-year lease, rather than outright cession, of the northeastern portion of the island to be handed to Turkey. Ankara complained loudly, noting that while fifty years might seem a long time to Americans, in Turkey's neighborhood it was a blink of the eye. A lease would never do. Athens initially found the proposal attractive. For once, the Johnson administration thought the Greek government was adopting what Ball labeled a "serious negotiating position." Papandreou reportedly told skeptics, "Listen, we are being offered an apartment building and subletting only one penthouse to our neighbors the Turks." Ultimately, however, Papandreou refused to break with Makarios, who fervently condemned the proposal. The second Acheson plan died, leaving the negotiations where they had begun.[35]

Though the danger of war had receded somewhat, the Johnson administration recognized that without a genuine solution to the Cyprus problem, another crisis might occur at any time. American officials blamed Makarios for criminal intransigence, but they judged the Greeks almost equally culpable. Ball deemed Papandreou "hopelessly weak." Ball continued, "He knew how to oppose but had neither taste nor talent for positive action." Citing the need for and absence of Greek resolve to bring Makarios to a settlement, Ball told Johnson, "You can't hang a custard pie on a hook."[36]

Acheson, fuming over the opportunity for peace lost, summarized the American view. Writing to the American ambassador in Egypt—whose President Nasser had taken advantage of the Cyprus troubles for a few tweaks of NATO's nose—Acheson asserted:

> We came close to an understanding which might have cropped the Archbishop's whiskers and solved the idiotic problem of Cyprus to your Mr. Nasser's disappointment and chagrin. Our weakness was Papandreou's weakness, a garrulous, senile windbag without power of decision or resolution. He gave away our plans at critical moments to Makarios, who undermined him with the Greek press and political left. A little money, which we had, the Greek 7th Division in Cyprus, which the Greeks had, and some sense of purpose, which did not exist, might have permitted a different result. The Turks could not have been more willing to cooperate.[37]

Johnson felt the same way, only more so. The president vented his annoyance on the Greek ambassador in Washington. When the Greek envoy explained that even had Papandreou approved the American plans, the prime minister could not override the Greek parliament and constitution, Johnson burst out, "Fuck your parliament and your constitution. America is an elephant. Cyprus is a flea. Greece is a flea. If these two fellows continue itching the elephant, they may just get whacked by the elephant's tail, whacked good." Johnson warned, "If your prime minister gives me talk about democracy, parliament and constitution, he, his parliament and his constitution may not last very long." To ensure that the prime minister received the message (and indulging a standard Johnson ploy of feigned difficulty remembering troublesome individuals), the president concluded, "Don't forget to tell old Papa-what's-his-name what I told you. You hear?"[38]

Johnson possessed sufficient self-control not to employ such direct language unless it suited his purpose. In this case he desired to let the Greeks know that American patience was dissipating fast. At the same time, the president's eruption reflected genuine anger and frustration at what seemed a patent case of Greek politics taking precedence over Western and American security. This anger and frustration almost provoked Johnson to approve an operation that would have settled the Cyprus issue permanently—or at least as permanently as one could expect of problems involving Greece and Turkey.

"On Cyprus one act is ended and the next has not yet begun," Bundy wrote Johnson at the end of August. "Acheson's Geneva effort has broken down for the moment, but George Ball & Company have not yet settled on the next course." The next course turned out to be a sharp departure from the Geneva negotiations. As Ball and Acheson explained to a session of the Tuesday lunch at the beginning of September, further talks were a waste of energy. Only decisive action would solve anything. And action must take place swiftly. Soviet rumblings had frightened the Greeks and Turks into a mood as conducive to a solution as they would ever share, yet Athens's fright had convinced Makarios he could no longer count on Greek support, and consequently the archbishop was looking more earnestly to the Soviets. A solution to the Cyprus problem had to come "before the Makarios-Moscow axis is firmed up," as Robert Komer, agreeing with Ball and Acheson, put the matter. Komer added, "Makarios will try every trick he has, and the Soviets are now committed to make at least some trouble."

What Ball and Acheson proposed was lancing the Cyprus boil by immediate and radical surgery. The United States should give a nod to the Turks to invade Cyprus and seize the portion of the island the Acheson plans would have allotted them. If the Turks moved fast enough—and Ball and Acheson, who had great respect for the Turks' prowess at arms, thought they would—they could entrench themselves before the Greeks got into the action. Athens would have nothing to do but annex the rest of the island. While Makarios obviously would object, with his country gone his objections would not carry far.

Johnson was intrigued. What would happen to the Turkish Cypriots in the Greek sector?, he asked. Ball said that most would stay on the island. The really worried would move to the Turkish zone, yet with the Cyprus problem settled and both Greece and Turkey relatively satisfied with the outcome, the incentive for continued communal violence would diminish considerably. Would Athens really follow the scenario?, the president inquired. Ball believed so. Evidence indicated that Papandreou's government had a contingency plan for "instant enosis" in the event of a Turkish invasion, Ball explained. How about casualties during the operation?, Johnson queried. Acheson, who apparently had broached the subject with Turkey's leaders, said the military in Ankara could be relied on to carry out the operation with comparatively little violence. Acheson added that the invasion would entail minimal use of American weapons and such "internationally unpopular" tactics as air raids. Ball remarked, "With luck bloodshed would be limited."

Johnson wondered about the British reaction. Cyprus belonged to the British Commonwealth. What would London think about the disappearance of one of the Commonwealth's members? Acheson said he had talked with R. A. Butler, the foreign secretary, and Lord Mountbatten, the chief of the defense staff, on his way back from Geneva. Evidently he had not told them what he had in mind, and therefore could only speculate as to their reaction. Butler might be edgy, Acheson predicted, but Mountbatten would be "friendly." Acheson saw no reason to expect that either Turkey or Greece would interfere with British base rights.

Dean Rusk asked about the Russians. Would they respond aggressively to Makarios's calls for help? Ball was not sure Makarios would actually turn to the Soviet Union in a pinch. The undersecretary thought even Makarios would have problems bringing in the Soviets. He was an Orthodox archbishop, after all, and they were atheists. In any case, the Soviets would not interfere in a meaningful way. They would certainly make diplomatic noise, but things would happen too fast for them to meddle concretely, and they would not risk challenging a united NATO front. McGeorge Bundy wanted to know how the Turkish strike would begin. Acheson suggested that a normal rotation of Turkish troops from the mainland to the island, in accord with the London-Zurich agreements, would provide cover for commencing the invasion.

Johnson recapitulated the Ball-Acheson proposal as he pondered a decision. The two veteran diplomats believed that a forcible solution was inevitable. Sooner or later the Turks would move and partition the island. The essential issue, then, was whether partition should happen in an uncontrolled manner or according to a plan. The latter alternative allowed the opportunity to keep casualties down and to guide the outcome. Was this a fair summary of the proposal?, Johnson asked. Acheson said it was.[39]

The meeting ended with Johnson still pondering. The plan had obvious attractions but equally obvious risks. If it worked, it would solve the problem that had twice nearly produced an intra-NATO war and seemed certain to do so again. It would terminate the administration's need to walk a line that antagonized both Athens and Ankara. It would facilitate mending the rift that threatened to allow Soviet penetration into the eastern Mediterranean.

On the other hand, military operations were inherently unpredictable. Johnson liked to cite the surprise the Chinese sprang on MacArthur in Korea in 1950 as an example of how military planning often went awry. In this case the president had less confidence than Ball and Acheson that once large numbers of Turkish and Greek troops landed on Cyprus they would not start shooting at each other. Could millennia of hostility be kept so easily in check? How would the Turks know where to stop their partitioning? What if they got greedy? What would happen to Makarios? What if news leaked that the United States had conspired in the destruction of an independent country? The whole affair smacked of the seizure of the Sudetenland in 1938 and the division of Poland in 1939.

Finally, domestic politics weighed against approval of the Ball-Acheson proposal. Johnson had one war on his hands already in Vietnam. He also had an election coming up, now just weeks away. He was presenting himself as the peace candidate. The current period, he remarked, was "not a good season for another war."[40]

Johnson decided that on balance the risks and uncertainties were too great. He refused to give the word that would have set the Turkish invasion in motion. The Cyprus problem persisted, no nearer an end than before.

VII

For a while, Johnson's decision appeared a wise one. The impending catastrophe Ball and Acheson predicted failed to occur. Attacks on Turkish Cypriots diminished, perhaps because Makarios once more sensed he had pushed his luck to the limit. Pressure on Ankara to act forcefully decreased. During the next several months, and then the next few years, relative calm descended on Cyprus and its environs.

But the calm couldn't last forever while the underlying sources of tension remained. It didn't. In the autumn of 1967 the Johnson administration found itself back where it had been in 1964. The catalyst for crisis this time was a military coup in Athens that toppled the civilian government and installed a regime less diffident about the use of arms to settle the Cyprus dispute. So the situation in Athens seemed to the Turkish government, at any rate, which prepared to get in first licks.

The Johnson administration had not forgotten about Cyprus. An arms deal between Nicosia and Moscow that surfaced in 1965 kept American intelligence analysts on the alert. The State Department monitored mediation efforts by the United Nations. The White House heard from

Greek-Americans, who displayed continued interest in the Cyprus question. These voters got Johnson's ear, for example during the spring of 1966 when, at a moment of no extraordinary strain over Cyprus, the Greek Orthodox archbishop of Washington expressed a desire to speak to the president. The archbishop had been referred to the president by Everett Dirksen, whose goodwill Johnson always valued. Johnson had plenty of other affairs to occupy his time, but he decided that a few minutes with the archbishop and an accompanying group of Greek-Americans now might save considerable trouble later. Robert Komer captured the administration's opinion on the subject when he told the president, "I feel in my bones that receiving these excitable people may be better than letting a head of steam build up."[41]

The Turks were no less excitable than the Greeks, just less well represented in Washington. Yet when the State Department got word from Ankara in November 1967 that the Turks were once more mobilizing to invade Cyprus, the administration moved swiftly into action. George Ball had resigned by this time, so the White House had to find somebody else for war-stopping chores. Johnson turned to Cyprus Vance, one of the special envoys sent to the Dominican Republic in 1965 and a tenacious negotiator. Vance jetted off to the eastern Mediterranean.

Vance's efforts produced results only slowly. As he remarked afterward, the Turks were "*very* suspicious." But once he convinced them that restraint would not place them at a disadvantage with respect to the Greeks, and then when he convinced the Greeks of the same thing regarding the Turks, he believed he had the basis for an agreement not to fight. "Both countries found themselves in a situation where they had gone too far, and both were looking for a way to withdraw," he said. Vance provided the two sides a face-saving means of withdrawal. If hotheads in the two countries complained at being deprived of a war, the governments could blame the United States.[42]

Just as a solution appeared imminent, the CIA detected Soviet efforts to keep the crisis going. At the end of November, Richard Helms reported to Johnson: "The Russians are fishing in troubled waters by egging on the Turks." Vance, on the spot, noticed some unusual comings and goings between the Soviet embassy in Ankara and the Turkish foreign ministry.[43]

But however much Washington annoyed Turkey's leaders, the Russians—far closer and with a long history of nastiness toward Turkish peoples, both in Turkey and across central Asia—still frightened them. After a not very convincing feint in the direction of Moscow, the Turkish government made clear it preferred its superpower patron to be non-Russian.

At the beginning of December, Lucius Battle, assistant secretary of state for the Middle East, told Johnson that Vance was making solid progress toward ending the crisis. Greece and Turkey both had consented to reduce their forces on Cyprus. Makarios was calling for a total removal of foreign troops, but this was nothing new and would not prevent de-escalation between Athens and Ankara.[44]

Within days, tension in the eastern Mediterranean returned to its normal level. Though a definitive settlement remained as distant as ever, the Johnson administration was pleased at managing another narrow escape. "The problem of Cyprus will be with us for some time to come," Battle said. "But for the moment a Greek-Turkish war has been avoided."[45]

Four | # Suffer the General

Lyndon Johnson entered the White House during the twentieth anniversary week of the Teheran conference of 1943. At the Teheran meeting, Franklin Roosevelt, Winston Churchill, and Joseph Stalin had laid plans for the occupation of Germany and the temporary—so all pledged—division of the territory currently or lately controlled by Hitler's reich. Roosevelt went out of his way to be agreeable with Stalin, even at Churchill's expense. On his return to America the president reported that he and the Soviet leader had "got along fine."[1]

Roosevelt's successors did not get along so well with Stalin, nor Stalin's with Roosevelt's, and the temporary division became permanent. The division's center, Berlin, occasioned several crises during the next two decades. The most recent, at the time of Johnson's swearing in, had occurred in 1961. After bluster that consumed most of three years, Nikita Khrushchev had ordered East Berlin sealed off from the western half of the city and thereby from West Germany. Kennedy responded to the erection of the Berlin wall by reaffirming America's commitment to West Berlin's and West Germany's security, and by dispatching Lyndon Johnson to the front. Because vice presidents so often deliver condolences at foreign funerals, Bonn was less than reassured. Chancellor Konrad Adenauer, a man old enough to be Kennedy's grandfather, thought the Americans underappreciated the danger Germany faced. Kennedy mollified the West Germans in the summer of 1963 when he traveled to the wall himself and declared himself an honorary Berliner. Adenauer's fall from power a few months later, shortly before Kennedy's death, did not hurt chances for further improvement in United States-German relations. George Ball, as thoroughgoing a Europeanist as inhabited Washington, was glad to see Adenauer go. "The old man was mean," Ball declared afterward. "He really was mean."[2]

In 1963 the Atlantic alliance was not fifteen years old. The American commitment of troops to Europe was younger still. Though hope had largely vanished of a permanent and amicable answer to the German question and the larger issues of European security that hinged on Germany, Americans had not entirely accepted the idea that they must forever guarantee the security of Europe. Periodically Congress entertained resolutions to cut back the American presence in Europe, especially to withdraw some American troops and save the money their overseas deployment entailed. The Marshall Plan–induced revival of western Europe and of West Germany in particular contributed to the feeling in the United States that the Europeans should do more in their own defense.

Yet the German revival raised a countervailing fear among American strategic planners—as well as among strategists in Europe, on both sides of the Elbe. Germany had long possessed the technical capacity to produce atomic weapons. Until now Bonn had forsworn production, partly to allow continued concentration on the development of the German economy, partly to avoid frightening Germany's neighbors. The presence of American troops on the front lines in Germany, whose primary function in the event of attack from the east was to get killed and ensure determined American participation in the fighting, combined with Bonn's nuclear diffidence to form a largely unspoken pact between West Germany and the United States. The Americans would keep their uniformed hostages in Germany, and the Germans would not go nuclear.

The unspoken character of the pact complicated American relations with West Germany, and more generally with Europe, although not as much as a spoken pact of the same conditions would have—which was why it remained unspoken. No American president wished to admit he was placing American soldiers in the cannon's mouth partially from fear of Germany, an ally. In justifying the American presence in Europe, American leaders liked to cite the defense of liberty against communist aggression. In doing so they spoke the truth, but not the whole truth.

Occasionally during the late 1950s and early 1960s, certain Germans had indicated an inclination to cancel the arrangement with the United States. Their complaints reflected less a conviction that German self-respect demanded German military self-sufficiency than a desire to remind the nuclear members of the Atlantic alliance not to take Germany for granted. Responding to these complaints and wishing to preempt what they portended, American leaders devised a scheme for bringing the Germans into the nuclear club carefully and well chaperoned. The scheme centered on a nuclear multilateral force, or MLF, involving warships armed with nuclear weapons and manned by multinational crews.

The MLF was a politician's delight and a strategist's nightmare. All participants could claim nuclear status, because all would contribute to the development of the force and all would take part in decisions to use nuclear weapons. The latter consideration was what caused the cringing among the war gamers. Democracy had survived the committee approach to military conflict that the fight of the Grand Alliance against the fascists had involved, but the victory took years and exacted a tremendous cost. Modern nuclear war would not allow the time multilateral decision making required, and it would not tolerate the inefficiencies joint action inevitably entailed.

While the issues of both American troops in Europe and the MLF impinged most directly on American relations with West Germany, anything touching Germany necessarily touched Britain and France. Even more than the Americans, the British and French understood that the Atlantic alliance was about keeping Germany calm and peaceful—and divided—nearly as much as it was about containing the Soviets.

In dealing with the British, the Johnson administration experienced no inordinate difficulties. Such special relationship as had existed in the days of Roosevelt and Churchill, and to a lesser extent during the collaboration of Dean Acheson at the State Department and Ernest Bevin at the British Foreign Office, had eroded considerably under Eisenhower and Anthony Eden. Harold Macmillan repaired much of the damage done by the 1956 Suez crisis, working first with Eisenhower and then with Kennedy. Alec Douglas-Home did not remain in office long enough to have a significant impact, and his premiership coincided with the distracting transition from Kennedy to Johnson. Johnson found Harold Wilson a man after his own heart. On Wilson's visits to Washington the two politicians would dismiss their advisers and talk shop—Johnson relating his experiences as head of America's more-progressive major party, Labourite Wilson speaking from a similar position in British politics. For the British taken as a group, Johnson had high regard. David Bruce noticed it. "He's very admiring of the British people—it's absolutely genuine—of their best qualities, of their vigor under hardship, of the skill with which overall they conducted their world affairs and administered their global role for a couple hundred years or more," the ambassador to Britain remembered. "I would say he was strongly attuned to the British race, if you can call it that."[3]

For the French, Johnson had no such regard. There was reason for the lack. During Johnson's tenure, Charles de Gaulle threw himself directly in the way of American policy on European and international security. Embracing a worldview different from that animating American leaders, de Gaulle pursued policies that would have seemed challenging if practiced by a avowed enemy, but appeared intolerable in an ally. At times the general acted as though determined to end France's alliance with the United States. In the event, he stopped short of sundering the alliance, yet by withdrawing France from NATO—the military arm of the Atlantic alliance—de Gaulle demonstrated that Europe was entering a new era. Of necessity, the United States entered the same new era.

II

From the beginning, Johnson's relationship with de Gaulle promised to be trying. Since returning to power in 1958 the general had attempted by various means to increase France's influence in Europe and elsewhere, often at the expense of the United States. To Eisenhower and the British he proposed a kind of three-way condominium of the Western world, with Washington, Paris, and London agreeing to coordinate activities in mutual support. The State Department wanted nothing to do with the idea, which seemed a thinly disguised effort to gain American backing for France in that country's losing fight to keep control of Algeria. Eisenhower had doubts about de Gaulle's personal and national ambitions. On one occasion the American president remarked wryly, "De Gaulle merely wants to make France the first nation of the world with himself the first Frenchman." At another time, Eisenhower suggested that de Gaulle considered himself "to be, by some miraculous biological and transmigrative process, the offspring of Clemenceau and Jeanne d'Arc."[4]

When Eisenhower rejected de Gaulle's tripartite proposal, the French leader ordered American nuclear forces out of France and withdrew the French fleet from NATO. Though neither action materially weakened Western security, both signaled Paris's dissatisfaction with the status quo on the continent and in its environs, and hinted at more drastic measures in the future.

De Gaulle sent additional signals in short order. Following the successful test of a French atomic bomb in the Sahara in 1960, he committed France to the development of an independent nuclear deterrent, the *force de frappe*. De Gaulle plausibly contended that the United States would never risk its own existence in defense of France's sovereignty. If the contention had not been true to begin with, de Gaulle's cantankerousness quickly made it so. He attacked the United States indirectly by hitting at America's closest ally, Britain. Judging the Anglo-Saxons to be all of a kind, and deeming Britain an American stalking horse, he waged a diplomatic offensive against London, with the central thrust being a resounding *non* against Britain's application for membership in the Common Market. Meanwhile he engineered an entente with West Germany, capping his campaign in January 1963 with a Franco-German treaty of cooperation, mutual support, and strategic and tactical coordination.

"I can hardly overestimate the shock produced in Washington by this action or the speculation that followed, particularly in the intelligence community," George Ball remembered of the news of the Paris-Bonn axis. "There were wild rumors of a plan to pave the way for France, with Bonn's assistance, to negotiate with Moscow for a whole new European arrangement. We compared and supplemented our intelligence reports with bits and pieces gathered by the British. We looked at all possibilities of a Paris-Bonn deal with Moscow, leading toward a Soviet withdrawal from East Germany to be followed by some form of confederation between the two parts of that severed country. That would, of course, mean the end of NATO and the neutralization of Germany."[5]

The shock had diminished, but the suspicions of de Gaulle's designs persisted when Johnson took control of American foreign policy. Charles Bohlen, the American ambassador in Paris and Johnson's principal guide to handling de Gaulle, prepared for the new president a detailed analysis of French policy. Bohlen was a consummate diplomat, having served in numerous difficult positions, including Moscow twice, since his initial posting to interwar Prague. Bohlen also cultivated a lifelong interest in the French language and culture. His relations with de Gaulle were professionally exemplary, but the ambassador had little sympathy for either the French president or his purposes. "The character of de Gaulle is completely formed by his education, experience and his own characteristics, which are highly egocentric and with touches indeed of megalomania," Bohlen wrote. "Insofar as I can ascertain from conversation and reading, he has never been induced to change any of his basic views by conversations with others nor as a result of concessions or favors done him by other countries." Consequently American efforts to improve relations with de Gaulle, undertaken for the purpose of improving relations rather than with some more concrete objective in mind, would be ill advised and probably counterproductive.

De Gaulle irritated purposefully, Bohlen asserted. "Relations with the U.S., unsatisfactory as they are to us, are in my opinion the way that de Gaulle wants them." De Gaulle realized that France no longer enjoyed the freedom of international maneuver it once had. Precisely for this reason he jealously guarded its remaining independence. "He undoubtedly feels that too close a relationship between a relatively small country (which he bitterly recognizes to be the case in regard to France) and the U.S. could in his view lead only to an actual derogation of the weaker country's sovereignty." Much of de Gaulle's refractoriness resulted from his wish to resist this tendency. "Many of his acts, particularly in some of his public statements—where action is not possible for France—have been motivated purely by a desire to strike out a difficult line for French policy." Bohlen thought it significant that French relations had not improved with any country since de Gaulle had come to power.

Bohlen did not exclude West Germany from this last statement. The recent Franco-German accord, the ambassador believed, had much more about it of symbolism than of substance. Even the Germans were often put off by de Gaulle's arbitrariness. Bohlen dismissed lurid guesses that the accord presaged a shift in the European balance that would materially damage the Atlantic alliance. "I believe personally that there is no possibility that de Gaulle would try to double-cross the Alliance by a deal with the Soviet Union. He undoubtedly looks forward to a time many, many years hence when the Soviet Union will cease to be a cause and merely become a Russian country. At that time obviously he would expect France to welcome Russia back into the community of European nations. This, however, is problematical and so far into the future as to be unnecessary even to consider." Bohlen was convinced de Gaulle recognized the value of the Atlantic alliance to French security, although the general made no secret of his disagreement with certain aspects of the alliance's military organization. Nor, for all his sniping at the United States, did he wish the Americans to withdraw from Europe. He was not even putting pressure on the thirty-five thousand American troops scattered about 180 installations across France—at least not yet.

Bohlen reminded Johnson that de Gaulle was not France, whatever the general might think or hope. As de Gaulle aged, he withdrew increasingly into himself. "He quite literally has no close friends or even associates," the ambassador asserted. De Gaulle would leave no heir committed to his policies. French officials taken together and officials in the French foreign ministry in particular were "favorably disposed" toward the United States. "They are disquieted and some are alarmed by de Gaulle's attitude and frequently go out of their way personally to show that they do not share his attitude." Among the French population at large, there existed a not insignificant group of dedicated Gaullists who echoed the worst aspects of de Gaulle's anti-American sentiments, yet, as the recent and continuing outpouring of sympathy over Kennedy's assassination demonstrated, America remained well-considered in France. "In all the times I have been in France I have never seen more genuine pro-American sentiment among the people of France as a whole."

The United States ought to keep this underlying goodwill in mind in dealing with de Gaulle. Relations with the general should be irreproachably correct and marked by utmost courtesy. Washington must exercise great care in avoiding critical statements de Gaulle could turn to his own purposes. But the United States must give away nothing. "He will accept any concessions or courtesies as a natural right and as a recognition of his 'greatness.'" Washington could do little to improve relations with France until de Gaulle changed his mind or departed the scene. The latter course appeared likely to precede the former. Since neither seemed imminent, the American government must reconcile itself to some protracted prickliness.[6]

The prickliness increased when de Gaulle announced that France would recognize the communist government of China. Recognition itself was not especially provocative. Britain had recognized Beijing more than a decade before. But the timing seemed—to Washington at least—calculated to antagonize. At a moment when the United States was trying to defend South Vietnam against aggression abetted by the Chinese, de Gaulle's decision could not help raising doubts regarding the solidarity of the West in the face of the communist challenge.

Bohlen deemed de Gaulle's China move completely consonant with the general's overall diplomatic strategy. "Recognition is primarily an act demonstrating French independence of American control in foreign affairs," the ambassador asserted. Bohlen thought it noteworthy that de Gaulle had refused even to discuss the matter of recognition with any of France's allies. This disdain for discussion demonstrated de Gaulle's insistence on total freedom of action. "The fundamental and basic element in de Gaulle's foreign policy is his strongly held and unchangeable conviction that the nation (the state and not the people) represents the permanent unit in international affairs. Its authority and sovereignty must under no conditions be watered down or weakened in any way." Fearing that France had lost some of its sovereignty by adhering too closely to the United States, de Gaulle insisted on breaking loose.

Bohlen explained that unlike American leaders, who saw regular political coordination as a natural and necessary concomitant of the military provisions of the Atlantic alliance, de Gaulle felt no ongoing responsibility to his partners in the pact. "His concept of alliance is the old-fashioned 1914 type, i.e., the alliance operates only in times of crisis." This conception explained France's solidarity with the United States and the other alliance members during the periods of tension with the Soviets over Berlin and Cuba. It also explained de Gaulle's unwillingness to coordinate when tension eased, his antagonism to the military institutions of the Atlantic alliance, and his aversion to anything smacking of European or Atlantic integration. In the general's mind, the alliance was a dead letter absent a direct threat to French security. "De Gaulle has in effect withdrawn France from the Alliance in a political or diplomatic sense and, to a large extent, in military matters."

Bohlen reiterated his opinion that the United States must harbor no illusions about bringing de Gaulle around to America's perspective. "No concession or bribe of any kind will affect de Gaulle's attitude or policies. He would regard any such gesture on our part as confirmation of the correctness of his views and his just due without seeing any necessity to change his position at all." Relations between the United States and France were strained because the general liked them strained. As far as possible, President Johnson should adopt a low profile, staying out of de Gaulle's line of fire. Though perhaps difficult to maintain, a posture of patient restraint would eventually pay off. "It should always be borne in mind that de Gaulle cannot have very many more years in power, and the present indications are that a very large portion of the objectionable features of current French policy would disappear with his departure from power."[7]

Bohlen's interpretation of de Gaulle's actions gained reinforcement from C. Douglas Dillon. The treasury secretary remarked that the general's actions were "largely based on his messianic belief in the glory and importance of France and thus are not subject to reasoned argument." Dillon concurred with Bohlen that the French president could not be appeased. "Attempts to propitiate de Gaulle are unlikely to succeed and would probably only serve to increase the level of his demands." Dillon stressed the importance of distinguishing between de Gaulle and France—though the general often did not. "We should operate on the assumption that de Gaulle's leadership of France is temporary and that he will be succeeded by a government more responsive to public opinion, hence more favorable to NATO, United Europe and the United States." Meanwhile, however, the United States must give de Gaulle no cause for complaint about his treatment at American hands. "We should lean over backward to be polite and friendly to France, to de Gaulle personally and to all French government officials."[8]

The singular–de Gaulle theory Bohlen and Dillon proposed did not pass unchallenged in Washington. The assistant secretary of state for Europe, William Tyler, accepted much of what Bohlen said but thought the ambassador overemphasized the uniqueness of the general's position among the French. Tyler pointed out that the decision for a French national nuclear force antedated de Gaulle's return to power, and he suggested that the majority of the French people liked the idea of independence in nuclear affairs. "Even those who don't feel that a national nuclear force amounts to much militarily are inclined to support it on the grounds that it confers on France the possibility of playing a greater role on the international scene." Tyler believed they were right. Similarly he contended that though many French citizens might judge de Gaulle's style needlessly abrasive, most sympathized with his desire to confirm France's diplomatic independence. Indeed a similar sentiment existed across much of Europe. The sentiment existed despite—and because of—an inexorable trend toward European integration. Tyler described the reaction of an American official, following de Gaulle's veto of British entry into the Common Market. This official lifted his hands to the heavens and cried, "Tell me, ye gods, how is it possible for one lonely, elderly ruler of a small country to frustrate the desires and aspirations of 250 million other Europeans?" Leaving aside that France was by no means small, Tyler commented, "The answer is, of course, that the situation is not quite as simple as that implied in the question, and that de Gaulle's ascendancy rests to some extent on his ability to express sentiments which his fellow countrymen and many Europeans recognize and with which they associate themselves."[9]

Averell Harriman challenged Bohlen's assertion that Washington would waste its time trying to moderate de Gaulle's behavior. "I have known de Gaulle since 1941," Harriman said, "and have seen him under many conditions. I do not agree that he is unaffected by outside and personal approaches." Harriman related an incident in December 1944 when de Gaulle had indicated his intention to recognize the Soviet-backed Lublin government of Poland. "I went to de Gaulle directly and told him quite bluntly that he had a choice of the good will of Moscow or Washington. He changed his position." Harriman conceded that de Gaulle enjoyed greater latitude for independent action in 1964 than he had in 1944, but still Harriman thought the general susceptible to American pressure. "It is my guess that General de Gaulle is amazed that he has gotten away with his disregard for American interests so far without strong reaction to him from us." Harriman did not advocate threats. The United States lacked the ability to impose sanctions against France that would not do more damage than good. "But he should understand that he faces American displeasure, including that of the President." Such understanding should be conveyed at once. "We are only encouraging future difficulties if we allow this to continue."[10]

A French view of de Gaulle arrived at the White House courtesy of journalist Adelbert de Segonzac, whom McGeorge Bundy described to Johnson as the "most knowledgeable French foreign correspondent here in Washington." De Segonzac had recently returned from Paris, where his contacts, some of whom occupied positions high in the French government, had apprised him of de Gaulle's strategy and plans. "His method, they say, is similar to throwing a stone in the water and then to wait and see how the ripples grow, and act accordingly," de Segonzac explained. "His aim essentially is to push France in the forefront and to demonstrate, by the initiatives he takes, that she is one of the leading powers of the world." De Gaulle's decision to recognize China reflected not so much an intent to strike at the United States as a belief that the West must eventually accept Beijing and a desire to place France in a position to broker the rapprochement. Yet de Segonzac granted that de Gaulle thought in terms of a "cold war of ideas" between the United States and France, with the British lining up with the Americans. One of de Gaulle's closest advisers, whom de Segonzac declined to identify, was thoroughly convinced that the Anglo-Saxons were out to dominate Europe and that France must resist this threat as stoutly as it would resist a Russian invasion. De Segonzac said de Gaulle had by this time laid out

all of his basic policies. "There will be no more surprises." Because
relations between the United States and France could hardly get worse,
de Gaulle expected them to get better. But on certain questions, notably
NATO, improvement would be only marginal. The general had told a
visitor not long before that the treaty organization was "dead." It would
not revive.[11]

Johnson, faced with these conflicting views regarding de Gaulle and
French policy, chose to follow Bohlen's advice, with certain exceptions.
As the president told George Ball, he refused to get into a "pissing
match" with the general. Publicly the president was all smiles and un-
derstanding. He described his approach in an interview with a group of
French reporters. "I keep mum," Johnson said. "If you hear something
nasty about de Gaulle, it has not come from me or anybody in the White
House. I told everybody in the government to be polite to President de
Gaulle. Just tip your hat and say, 'Thank you, General.'"[12]

To former French prime minister Paul Reynaud, who visited
Washington early in 1964, Johnson explained that he simply ducked de
Gaulle's "beanballs." Johnson said he understood that de Gaulle was a
"proud, egotistical man who closely identifies himself with France." But
whatever de Gaulle's self-perceptions, exchanges of ugly words would
help no one. Patience eventually would take care of present difficulties.
A historic friendship connected the American people and the people of
France. It would continue to do so. "In any time of crisis with the chips
down, we will be able to rely upon our French friends to be at our side."[13]

In a conversation with Dwight Eisenhower some time later, Johnson
recounted a meeting he had had with de Gaulle. "He was, as always,
immensely courteous," Johnson said. "But at the moment there does
not appear to be a great deal of substance that we can constructively say
to one another." Johnson did not hide his annoyance at de Gaulle from
Eisenhower, who commiserated. "Nevertheless," Johnson added, "I
have judged it important for our country and for the Alliance that we
minimize public discussion of our differences."[14]

III

Initially Johnson took a somewhat more forceful approach in private.
Responding to a proposal by de Gaulle that the problems of Vietnam be
settled by neutralizing the country, as had been attempted with Laos,
Johnson directed Bohlen to have a "frank discussion" with the general.
Neutralization, Johnson said, would work to the advantage of North Vietnam and to the detriment of South Vietnam. For a variety of reasons,
among them the ability of a totalitarian state to ruthlessly marshal resources for aggression, Saigon required American assistance to defend
itself. Johnson left to Bohlen's discretion the manner of communicating
the message to de Gaulle. "But you should make it clear that we expect

France, as an ally, to adopt an attitude of cooperation rather than obstruction in this critical area of United States interest."[15]

Johnson judged Vietnam a test of the political cohesion of the Western alliance, and de Gaulle failed the test contemptuously. As Bohlen
had pointed out, the general held a different view of the nature of alliances. He thought France owed the United States nothing regarding
Vietnam. France must look out for the interests of France. De Gaulle
was thoroughly unmoved by the prospect of a communist victory in Vietnam, since he was unconvinced that a communist victory in Vietnam
would lead to communist victories elsewhere in the region and he was
unperturbed by the prospect of an American loss of face in the transaction. He saw no reason to follow American leaders into the mire from
which France had extricated itself a decade before.

In the opinion of certain American officials, de Gaulle had more in
mind than avoiding entanglement in a losing cause. The CIA suggested
that the general was trying to regain some of the influence the French
had lost in Indochina in the wake of the Dienbienphu debacle. This
desire, the intelligence agency indicated, accounted for de Gaulle's neutralization proposal. To be sure, the general was happy to exploit any
chance to distance France from the United States. "The current difficulties of U.S. policy in Southeast Asia provide him with another opportunity for asserting French independence and countering U.S. influence." Further, de Gaulle might feel some lingering resentment at
America's refusal in 1945 to allow French forces immediately back into
Indochina. But on the whole his neutralization plan rested on the conviction that neutralization, by terminating the preponderance the
United States enjoyed in South Vietnam, and that China and the Soviet
Union enjoyed in North Vietnam, would allow at least a partial restoration of French influence. De Gaulle realized that France could not
match the two superpowers and China in material resources applied to
Southeast Asia. Therefore France must move in a fashion to capitalize
on what the French deemed their special competence—the result of
historical and cultural ties, personal contacts, and the like—in the affairs
of the area. Neutralization, by shifting the contest from the battlefield
to the political arena, would play to French strength.

De Gaulle's plans for Southeast Asia figured in his decision to normalize relations with China, the CIA said. The French president believed
that any negotiated settlement in Vietnam would require Chinese participation. By opening to Beijing, de Gaulle hoped to increase his
chances of mediating a settlement. In addition better Sino-French relations would yield increased commercial opportunities to French businesses. For all his grand diplomacy de Gaulle did not neglect the needs
of French merchants and industrialists.[16]

French actions regarding Indochina doubtless made sense from the
French perspective, but Johnson did not like them a bit. The president

sent Bohlen a cable stating, "The difficult and dangerous situation in South Vietnam is being made still more difficult and dangerous by impressions and rumors of French policy which have larger than lifesize effect because of a French social, economic and cultural presence there." Johnson had just received disquieting reports from American officials in Vietnam. "Rumors of intrigue by French agents are now commonplace in Saigon," Johnson said. "There are some who ascribe the recent barbaric attacks against Americans to French inspiration and direction." Johnson conceded room for differences between the United States and France on certain aspects of policy toward Asia. "But neither General de Gaulle's purpose nor ours is served by the kind of confusion and mistrust which have been spread in recent weeks." Unless the French changed their approach toward Vietnam, "the influence of France will work toward disintegration of the only forces which can prevent a Communist take-over there."[17]

To get this message across to de Gaulle, Johnson dispatched George Ball to Paris in June 1964. Johnson directed the undersecretary to impress four points on the general. First, American efforts in Vietnam had the purpose of preserving peace and the independence of a country under outside attack. The United States had no interest in dominating Southeast Asia. Second, notwithstanding America's desire for peace, the currently increasing threats from North Vietnam and China might force the United States to undertake more extensive military action. Third, both France and the United States would suffer if confusion between Washington and Paris contributed to a need for American escalation. Fourth, the United States government persisted in the belief that in a crisis it could rely on France as it had during the Cuban crisis of 1962. This last point was really a question. Johnson commented, "If by any chance I am wrong on this point, it is a matter of great importance that we should know it now."

Johnson authorized Ball to fill the French president in on such details of American thinking about Vietnam as would bolster the case for cooperation. But Ball should not spill the whole story. "I hope that you will not weary General de Gaulle with an account of our virtues which he must have heard before," Johnson said in his letter of instruction, "and I also hope that you will not be so specific about our contingency planning as to tempt him to any indiscretions with those whom he now seeks to cultivate."[18]

Ball had no luck with de Gaulle. After the undersecretary gave a lengthy exposition of the American case regarding Vietnam, the general very politely expressed his skepticism. He said he respected America's motives. The United States had taken on itself responsibilities France had borne in the past. In each case the aim was honorable. Unfortunately, he continued, the means the United States had chosen did not suit the circumstances. "I do not believe that you can win in this situation even though you have more aircraft, cannons and arms of various kinds." Washington approached the trouble in Vietnam in military terms, but the contest there was primarily political and psychological. De Gaulle reiterated an opinion he had expressed to President Kennedy, that Vietnam was a "rotten" country in which to fight. Answering Johnson's query regarding French policy in the event Vietnam led to a larger conflict, the general said France would not get involved in any war in Asia, as an ally of the United States or otherwise.

De Gaulle urged the Johnson administration to reconsider its opposition to a negotiated solution to Vietnam's troubles. To Ball's objection that negotiations under present circumstances would risk undermining the government of South Vietnam, the general answered, "All policy involves risks. If it is a policy that does not involve risks, there is no choice of policy." De Gaulle went on to suggest that the undersecretary overestimated the dangers. What the French government had in mind was a broad diplomatic initiative that would include such noncommunist Asian countries as India and Japan, in addition to the United States, the Soviet Union, China, and France. This broad participation would afford the Vietnamese people—if not necessarily the government of General Khanh—a "sense of support and assurance for the future."

De Gaulle sympathized with President Johnson regarding the predicament confronting the United States. France had experienced similar problems before. De Gaulle said he understood the frustration such problems produced. He simply wished the Americans would refrain from blaming France for America's difficulties. France was not seeking to hinder the United States in whatever actions the American government felt required to take. The Americans should recognize this fact and cease their complaints.

Ball assured de Gaulle that whatever complaints he had heard did not represent the views of the American government. The undersecretary prevaricated, "You, Mr. President, are held in the highest esteem by everyone in the administration, beginning with the president himself."

On his return to Washington, Ball reported his conversation to Johnson. "Both Ambassador Bohlen and I had the impression that General de Gaulle is merely waiting for events to come his way," Ball wrote in a memo. "He is confident that they will." De Gaulle expected the Americans in the not distant future to see the wisdom of a negotiated solution, at which time his maneuvering would pay off. "He quite likely assumes that we will then ask the French to take soundings with the Chinese and the North Vietnamese." As Ball interpreted French thinking, negotiations would result in the departure of the Americans, leaving France again foremost among the Western powers in the region.[19]

IV

Events came closer to proving de Gaulle right regarding Vietnam than to vindicating the Americans. De Gaulle's timing was a bit off, for he gave insufficient credit to Johnson's determination to avert a communist victory. As de Gaulle continued in his belief that a negotiated settlement offered the only answer to the Vietnam question, he continued to generate suspicion among American officials. French criticism of Johnson's decision to escalate the war in 1965 provoked Bohlen to assert, "The one cardinal interest that the French have in the Vietnamese matter is that of enhancing the prestige of General de Gaulle." Bohlen added that most French diplomatic moves regarding Asia in the previous few years could be explained "by the desire of de Gaulle to promote France and of course himself into a big international fixer." In 1968 Washington would grudgingly accept de Gaulle's offer of Paris as a location for the negotiations he had long advocated. But before then the Americans had some other matters to work out with the general.[20]

The Johnson administration had kept close watch on de Gaulle's restiveness at the status quo in Europe for some time. De Gaulle's discontent had shown in a 1964 decision to withdraw French naval officers from NATO's planning division and in repeated calls for restructuring the Atlantic alliance. Though neither Johnson nor other high administration officials would say so for the public record, they believed de Gaulle intended to move France away from the United States to a position more nearly neutral between the superpowers. Then, they reasoned the general reasoned, the Americans and the rest of the world once more would take France seriously. A State Department paper of 1964 identified de Gaulle's principal international goals: "(1) reasserting France's national independence in foreign policy and defense and thereby terminating what he considers its excessive dependence on the United States; and (2) recovering France's traditional role as a great power, that is, as a full participant in world and Western alliance decision-making, from which he feels that France has been excluded since 1940."[21]

De Gaulle's efforts to forge close French-German ties appeared part of the general's master plan. So, seemingly contradictorily, did his determined opposition to the multilateral force, which Germany presumably desired. Regarding the MLF, one could never do more than presume, since no country would commit to the force without knowing its composition and terms of reference, while composition and terms of reference required commitments. Consequently the MLF lingered in the realm of political ideas, approaching strategic reality only occasionally. "We see the MLF as a means, not an end," McGeorge Bundy wrote Johnson in December 1964. The end was the prevention of German nuclear unilateralism; the means was nuclear multilateralism. "We believe that some responsible participation is necessary for Germany in the long run, and the mixed-manned surface fleet is the best system we have found for this purpose. But we are not rigid or doctrinaire."[22]

George Ball explained the rationale behind the MLF in slightly greater detail. The American aim, Ball said, was "to tie Germany irrevocably to the Western world by giving the Germans the feeling that they are respected, first-class members of the Atlantic Alliance and are not being excluded from responsible participation in their own defense." A secondary objective was to spike in advance the guns of nations that might follow Germany into the nuclear age—as Ball phrased it: "to set a pattern for the management of atomic weapons by collective action rather than by the proliferation of individual national deterrent systems."

Ball made these points just prior to a meeting between Johnson and Harold Wilson. Britain had demonstrated decided ambivalence toward the MLF, and the undersecretary worried that a sudden conversion of Wilson would raise doubts among the other members of the alliance. "We must avoid any impression that the Anglo-Saxons are doing a deal that the continental Europeans will be pressed to accept. The Germans, Italians and Dutch are all watching with great interest, much anxiety and some suspicion the outcome of the Wilson visit." At the same time, Ball warned, President Johnson should not let British reservations scuttle the MLF, at least not without ensuring that London got blamed for the sabotage. "If, because of British stickiness, the present effort should break down, we want to make clear—not merely in the eyes of our own people but of the Europeans—that the blame falls squarely on the British and not on the United States."[23]

Following a pattern typical of MLF planning, Wilson came and left without settling anything. McGeorge Bundy interpreted the nonsettlement as evidence that the Johnson administration ought to find other causes to sponsor. "I am reaching the conclusion that the U.S. should now arrange to let the MLF sink out of sight," the national security adviser said. "I reach this conclusion because it seems increasingly clear that the costs of success would be prohibitive." American officials would have to override the reluctance of a skeptical Britain, of an unenthusiastic ruling party in Germany, and of an Italian government under severe—even by Italy's high standards of severity—political stress. Congress had shown scant support for the MLF. Moreover, the MLF might derail the administration's efforts toward nuclear nonproliferation. Last but not least was the French factor. Bundy predicted that a debate within NATO over the MLF would "provide justification for further Gaullist outrages against the organization." The total came to too much, Bundy asserted. "The MLF is not worth it."[24]

Bundy and others in Washington might judge de Gaulle's actions outrageous, but officials of the Johnson administration never doubted the shrewdness of the general's methods. American analysts argued that if de Gaulle managed—with help—to torpedo the MLF, he would force Bonn to turn to Paris for nuclear help. "De Gaulle's grand design for Europe involves the creation of a loose confederation of states capable of providing its own nuclear defense and determining its own destiny," the CIA asserted. "Essential to such a confederation would be the closest of Franco-German collaboration in all fields." Creating this European confederation did not necessarily involve breaking French ties with the United States and the Atlantic alliance. "But it does involve an end to direct U.S. political and military controls in European defense matters and the end of NATO in its present form."[25]

De Gaulle judged the MLF an instrument intended by the Americans to forestall development of an independent European nuclear deterrent, a CIA report declared. This judgment accounted for his opposition. Until lately the general had relied on indirect methods to oppose the MLF. Footdragging on the part of potential members and confusion as to what form the MLF would take allowed him this luxury. Recently, however, indications of motion toward the MLF had caused him to mount what the CIA report called "a frontal attack" on the force. French foreign minister Maurice Couve de Murville and French prime minister Georges Pompidou had denounced the MLF as inimical to the interests of France and the security of Europe. De Gaulle himself decried the MLF as destructive of European autonomy.

Should the MLF come into being over France's objections, the CIA predicted, the general would feel required to take further action. He might pull France out of the Common Market, on the ground that unwillingness to maximize European independence in defense matters rendered all European institutions meaningless. More probably he would strike out at NATO. At the least he would scale back French participation in military exercises and strategic planning. He might withdraw France from the military apparatus altogether. At the extreme he would abrogate France's portion of the Atlantic alliance. The CIA did not think he would go to the extreme. Despite progress on the *force de frappe*, de Gaulle appreciated the protection the American nuclear umbrella afforded France. Yet because he deemed a Soviet assault on the West unlikely, he had few compunctions about measures calculated "to hasten the disintegration of NATO in its present form—his ultimate goal."[26]

As part of a campaign to frustrate de Gaulle, the Johnson administration sought to cultivate the more cooperative members of the Atlantic alliance. The cultivation required care, since the United States must not be seen as anti-French. Nor should American officials give way to exasperation and encourage other members to speak out against France. This would only goad the general and could conceivably push him to leave the alliance. "Such a course might be emotionally satisfying but is unlikely to profit the Alliance in the end," a briefing paper for a NATO ministerial meeting explained. "France is too strategically located, too much the geographical center of NATO, too important in the Western political constellation, for a severance from the Alliance to serve its objectives. Unless and until such a break were absolutely inescapable, we should not provoke it." The United States should steer a careful course between capitulation and confrontation. American officials should point out the benefits NATO members derived from membership, not only militarily but in the political and economic fields. Though de Gaulle would ignore such arguments, leaders of other NATO countries would not. Above all, the United States must make clear that de Gaulle could not wreck NATO. American officials should "calmly but firmly insist that NATO operations must continue despite any obstruction from a single source," and they should work "to build up precedents for action by those willing to cooperate for Alliance purposes while objectors stand aside."[27]

This last train of thought sounded reasonable, but it collided with the earlier observation regarding the geographic and strategic centrality of France to NATO. France could not simply stand aside from the organization. If de Gaulle withdrew France, NATO would have a gaping hole where its heart once lay. A French pullout would almost certainly entail the ejection of foreign troops from French soil, raising serious logistical problems. De Gaulle might restrict NATO overflight privileges, which would produce additional troubles. Further, French noncooperation would generate anti-French feeling in the remaining NATO countries, making the task of holding the Atlantic alliance together more difficult still.

V

The possibility of an anti-French backlash—which might become an anti-European backlash—figured crucially in Johnson's thinking about NATO. As a senator, Johnson had taken part in the great debate of 1951 that sent American troops to Europe. During the next decade he helped the American presence on the continent acquire an apparent permanence. But Johnson understood that appearances might deceive, and he recognized the persistence in America of the feeling that the Europeans ought to look after themselves. If he had started to forget this feeling, recurrent efforts in Congress to bring some or all of the boys home would have reminded him.

The bring-the-boys-home sentiment gained strength during the first half of the 1960s. The strengthening followed largely from Europe's economic revival, which provided the Europeans with the economic means to defend themselves. Now nearly as rich as the Americans, they could pick up their defense burden—if they would. Moreover, after the Marshall Plan, Americans not unnaturally felt the Europeans owed the

United States something. Americans did not ask for money, at least not directly. What they expected was greater cooperation, especially in the area of defense against the common communist enemy.

From the standpoint of domestic politics, Johnson must have been tempted to take on the supremely uncooperative de Gaulle. The general had few defenders among the American people. American francophilia had mostly died with Thomas Jefferson, and though the Lafayette Escadrille of the First World War had hinted at a resurrection, the cynicism of Clemenceau, the appeasing tactics of Daladier, and the collapse of French resistance before the German blitzkrieg renailed the coffin lid. De Gaulle momentarily won American favor by refusing to collaborate with the Nazis and eventually rallying the French against the invaders. But his monumental ego and his undisguised antipathy to most things American boiled the friendliness away. While William Fulbright disagreed with Johnson on most important aspects of international affairs, the Arkansas senator concurred with the president regarding what Fulbright called de Gaulle's "very disruptive influence upon the progress toward a much closer association of the Atlantic countries." Fulbright went on to say that de Gaulle was "tending to return to the nationalism which contributed certainly to two world wars."[28]

But Johnson refused to use de Gaulle as a foil. The president judged an American commitment to Europe necessary to world peace and American security, and he feared arousing popular feelings against the American commitment. Should France decide to leave NATO, American support for the countries remaining in the organization would matter more than ever. Americans, if riled against one set of Europeans, might extrapolate their annoyance to the rest. A reaction like that after the First World War, though not really probable, could not be ruled out. In any event, a major fight to maintain the American commitment to Europe was the last thing Johnson needed. He had plenty of other fights on his hands. So Johnson ducked de Gaulle's beanballs and bit his tongue.

In the spring of 1965 de Gaulle threw one high and tight, engaging in what the debonair Bohlen, who chose his images from a different pastime, characterized as a "rather indecent flirtation with the Soviet Union." Bohlen had just spoken with informed but nongovernment French sources who reported that the general felt dissatisfied at the results of his previous two years' diplomacy. De Gaulle's high hopes for a Franco-German entente had collapsed. That issue was "completely dead." De Gaulle's rapprochement with China had likewise produced little. Paris had expected to broker an end to the war in Indochina, but no end impended. "The French have placed this question on ice and are not planning, for the moment at least, any further initiatives."

Now de Gaulle was looking to Moscow. The Soviets evidently were looking back. On good authority, Bohlen asserted that the Kremlin had proposed a renewal of the wartime alliance with France. De Gaulle had turned the Russians down, but not before appearing to give the proposal some thought. The Russians recently were taking a different tack, aiming for French recognition of East Germany. Such recognition would irretreviably ruin de Gaulle's chances of reviving his German policy. But because he had written Bonn off, this did not pose a problem. Bohlen thought that from "reasons of prestige or exaggerated vanity" de Gaulle just might go with the GDR.[29]

Bohlen also detected a speedup in de Gaulle's timetable for action against NATO. In June 1965 Bohlen flatly predicted a French withdrawal. On the basis of conversations with well-placed informants, the ambassador said the French government planned to propose revisions to the organization's charter, probably in the spring of 1966. The revisions would be intentionally unacceptable to the other members. De Gaulle would then seize on the rejection as a pretext for hitting at the Atlantic treaty. "In the event these proposals are not accepted he intends to denounce the treaty itself and to replace it with a series of bilateral defense agreements with at least the U.S., the U.K. and Germany." Eviction of American and other foreign troops from France would follow. "It is clear that by 1969 the latest he intends that there be no installations or forces on French soil not under French command and subject to French law." Bohlen thought de Gaulle would wait until after the French elections scheduled for year's end before making his move. But the demise of NATO as currently constituted was approaching rapidly. "We must engage in some hardheaded planning for alternative arrangements."[30]

A short while later de Gaulle himself seemed to confirm Bohlen's prediction. The general told George Ball, visiting Paris once again, that the Atlantic alliance must adapt itself to changing circumstances. De Gaulle pointed out that at the time of the Atlantic treaty's signing the Soviet Union had no atomic bomb. Even for several years afterward the Russians lacked the capacity to threaten the United States with significant destruction. Now things were different, and the difference necessitated rethinking in Europe. Would the United States really risk its own annihilation to protect interests intrinsically important only to Europeans? De Gaulle did not wish to cast aspersions on the American character, but such a sacrifice was more than anyone could reasonably ask. Consequently the European members of NATO required an independent nuclear deterrent, which France was attempting to create.

A second consideration prompted France to reevaluate the nature of the Atlantic alliance. In 1949 the countries of western Europe had remained weak and disoriented from the war. They required American assistance and American leadership. For this assistance and leadership they would ever be grateful. But now Europe had regained its strength and self-confidence. It was only natural that the Europeans should wish to reassert their autonomy and take larger responsibility for their security. The alliance must become less a coalition of one superpower and several protégés and more a body of equals. In its present form NATO subordinated the other members to the United States. France could no longer accept such subordination. Hence it must leave NATO. Likewise NATO must leave France. French troops did not occupy the United States. Neither could American troops, or troops from other NATO countries, occupy France.[31]

VI

After this warning, if de Gaulle's announcement of an end to French participation in NATO arrived as a thunderbolt, the bolt did not blast out of a blue sky. Perhaps de Gaulle had wished the other members of the alliance to provide him a *casus exitus* by approving the MLF. They failed to oblige. Through 1965 the multilateral force dithered along, befriended still by the foreign ministries in the likely candidate nations but increasingly opposed by the finance and defense ministries. The pattern did not exempt Washington, where the State Department's George Ball and Walt Rostow remained among the last apostles of the MLF. "I am sorry to say that this has become an obsession with Walt, as also with George Ball," Bundy told Johnson at the beginning of 1966. "These enthusiasts for a collective force have been a zealous lobby within the government for five years, and it is always quite a job to keep a proper eye on them. Dean Rusk does not do it, so the job has fallen to me."

Bundy continued convinced that the costs of creating an MLF outweighed the benefits. Harold Wilson in Britain would require substantial prodding. The Germans, for whose benefit the project was conceived, were hardly demanding its accomplishment. The French were as set against as ever. The Soviets, who of course lacked a vote within the Atlantic alliance but who nonetheless could make their annoyance felt, had not diminished their violent opposition. Congress had other projects to spend money on—as Johnson intended it should. Bundy conceded that the United States might create the MLF by main force. But such creation would require "a kind of pressure which would be extremely unpopular both in Europe and on the Hill."[32]

While Johnson's political instincts inclined him to agree with Bundy, his historical memories prevented him from agreeing too quickly. "The Germans have gone off the reservation twice in our lifetime," the president told Gerard Smith of the State Department, "and we've got to be sure that this doesn't happen again, that they don't go berserk." The MLF promised to help keep the Germans on the reservation. Yet, finally, Johnson had to admit that Bundy was right and that the good the MLF might do came at too high a price. By the beginning of 1966 the momentum of his domestic reforms was decreasing, smacking against the escalation of the war in Vietnam. Under different circumstances—if his advisers had unanimously supported the MLF, for example—he might have pushed it. As matters stood, he decided not to, and the idea died.[33]

De Gaulle did not need the excuse of the MLF to break ranks with NATO, as he readily demonstrated. In line with Bohlen's forecast, the general waited until after the French election of December 1965 to move. The stunning strong showing of challenger François Mitterand, who forced the presidential incumbent into a runoff and gave him a scare there, caused some observers to suggest that de Gaulle might change his mind about disrupting the European status quo. This only showed how little they knew de Gaulle, Bohlen said. "General de Gaulle has never considered that voices other than his, no matter how numerous, represent the will of France." The offensive would proceed, the ambassador predicted.[34]

Sure enough, at the end of February 1966 de Gaulle held a press conference to read the Americans and the other NATO members a notice of their eviction from France. He put the notice in writing a few days later, telling Johnson that France would "reassume on her territory the full exercise of her sovereignty, which is at present impaired by the permanent presence of allied military elements." De Gaulle added that France would withdraw its troops from the organization. At the same time, he reaffirmed France's commitment to the Atlantic alliance, saying that when the treaty came up for renewal in 10969 France would sign on once more.[35]

Though de Gaulle's decision against NATO came as no surprise to American officials, until he announced it there remained the possibility he wouldn't. Now the French withdrawal pushed all other NATO issues aside. Johnson responded immediately to de Gaulle's letter with one of his own. Johnson did not criticize the general's action, and he did not try to get de Gaulle to reconsider. He thanked de Gaulle for his courtesy in notifying Washington of the French decision and the reasons therefor. Yet neither did Johnson understate the significance of the French move. "I would be less than frank," he wrote, "if I did not inform you that your action raises grave questions regarding the whole relationship between the responsibilities and benefits of the Alliance." The president, however, gave no indication of American countermoves beyond saying he would consult with the other members of the alliance.[36]

Administration officials differed over how to handle de Gaulle's demarche. No one questioned the general's right to order American and other NATO forces out of France. As Johnson told Robert McNamara, "When a man asks you to leave his house, you don't argue; you get your hat and go." But some in Washington thought the United States ought to take its time leaving. Dean Rusk argued for an interpretation of the Atlantic treaty granting the United States two years to complete its withdrawal from France. Lest de Gaulle contend that such an interpretation evinced an American desire to challenge France's sovereignty, the secretary of state advocated bringing a few troops home at once. General Maxwell Taylor, formerly Joint Chiefs chairman, advised against any quick pullout, on the reasoning that such haste would call into question America's resolve to remain in and defend Europe. McNamara, on the other hand, cited the expense involved in keeping American troops overseas and pointed out that since the troops would be coming home sooner or later, the sooner the cheaper.[37]

Beyond recommending that the United States take its time departing France, Rusk wanted the president to declare forthrightly the American belief that de Gaulle was endangering the peace of Europe by his imprudent action. The secretary drafted a speech for Johnson in which the president would rebut de Gaulle by saying, "There are those who argue that conditions have changed. We are told that we can relax our guard, dismantle our system of integrated defense and avail ourselves of the luxury of working out our destinies each alone. I am deeply convinced that this is the counsel of error. To tread such a retrograde path would surely diminish our present security and lead us to division among ourselves."[38]

Johnson refused to speak so strongly. He allowed George Ball to indulge a little annoyance in an interview with *Le Monde,* in which the undersecretary said that France's course could represent "a step backward toward a disastrous past." But for himself the president hewed to the high road. He rejected Rusk's draft in favor of an innocuous assertion of hope for the future: "We look forward to the day when unity of action in the Western family is fully reestablished and our common interests and aspirations are again expressed through institutions which command universal support among us." A short while later the president added, "We regret very much that General de Gaulle has felt it necessary to express himself as he has. We have accepted what he has said more in sorrow than in anger."[39]

The State Department did not discourage easily, though. For many months American diplomats had heard European complaints that Vietnam was distracting the United States from the central—that is, European—theater of the Cold War. The State Department thought de Gaulle's attack on NATO afforded America a chance to reaffirm its commitment to Europe. At the same time Washington could steal a march on de Gaulle by coming out strongly for European integration. Henry Owen of the department's Policy Planning Council wrote a memo that Walt Rostow, now national security adviser, placed in Johnson's night reading. "The NATO crisis offers an opportunity to dramatize the U.S. commitment to European unity," Owen declared. "This concept has more political sex appeal in most Common Market countries (and increasingly in the U.K.) than Atlantic partnership." Taking an idea that had originated in the Italian foreign office, Owen suggested offering the

Europeans access to the latest American developments in science and engineering on condition the European countries establish a joint framework for putting the knowledge to good use—sort of a high-technology Marshall Plan. The Europeans had been agitating to diminish the technology gap separating them from the Americans. Meanwhile, many Europeans—among the non-Gaullists, that is—had been seeking a method of bringing Britain nearer to the continent. "A positive European response would therefore be assured," Owen predicted.[40]

For the moment, Johnson declined new initiatives. The administration, he believed, must concentrate on closing NATO's ranks behind the French departure. "Our task is to rebuild NATO outside of France as promptly, economically, and effectively as possible," the president told Rusk and McNamara. Further, any novel policy now would appear aimed against France. As before, Johnson refused to get into a public fight with de Gaulle. "I wish the articulation of our position with respect to NATO to be in constructive terms. I see no benefit to ourselves or to our allies in debating the position of the French government."[41]

Johnson's preference for soft speech caused him to discourage other leaders of the Atlantic alliance who wished to hold a fourteen-way summit to deal with the French departure. With Rusk, the president feared that such an unwieldy session could easily get out of hand. Besides, even if Johnson confined himself to the blandest of remarks, he would appear part of a group that at least implicitly was ganging up on de Gaulle. Instead Johnson opted for a gathering of foreign ministers.[42]

Preparations for the session, which took place in May 1966, elicited further reflection within the administration regarding how to handle the French pullout. The always assertive and frequently combative Rostow declared, "De Gaulle is trying to gut us," and advocated a strong response. In a memo written with aide Francis Bator, Rostow granted that the administration could not do anything within reason to force de Gaulle to back down. But it could and should make unmistakable where responsibility lay for such damage as the alliance suffered from the French withdrawal. Western European security had certain minimum requirements, such as rights of overflight across France. The United States should press de Gaulle to guarantee these requirements. If relations broke down because he refused, too bad for him. "It will be clear to the world that de Gaulle alone is responsible for the breakdown—that the monkey is on his back."[43]

As matters transpired, no one got saddled with the monkey. De Gaulle had demonstrated his point about French sovereignty, and now he chose to cooperate with the other members of the alliance on matters not infringing that sovereignty. He allowed alliance planes to overfly France, although he stressed that this was a privilege and not a right, and in other areas made himself almost agreeable. In June 1966 the general visited Moscow, where he discovered that the Soviet Union and France had less in common than American officials worried they might. As he told Bohlen on his return, the Kremlin at present had no desire to tangle militarily with the West, but it might develop the desire later. France recognized this fact and, without hindering French independence, wished to keep the Atlantic alliance strong.[44]

VII

But some damage was already done, notably in the United States. De Gaulle's ejection of American troops produced the kind of anti-European backlash Johnson had feared. For several months, a small group of retrenchers led by Democratic senator Stuart Symington of Missouri had been agitating for a curtailment of America's transatlantic presence. Now the agitation became a groundswell. In August 196 forty-four senators, including Democratic leader Mansfield, embraced Symington's idea, cosponsoring a resolution calling on the president to reduce substantially the number of American troops in Europe.

Though difficulties with de Gaulle and France precipitated this latest manifestation of incipient neo-isolationism, the troop issue also touched American relations with Britain and West Germany. Since the late 1950s the stationing of American troops in Germany had created balance-of-payments problems for the United States. Monies spent by the American government on supplies for troops and dependents and by those troops and dependents on various necessities and luxuries of life caused a sizable net outflow of dollars to Germany. The growth of German exports to the United States accompanying the rebuilding of the German economy ballooned the deficit further. During the Kennedy years, the American and West German governments papered the matter over by a series of ad hoc arrangements whereby Bonn pledged to offset American expenditures in Germany with purchases of American military equipment. The papering reflected Kennedy's sputtering interest in economic affairs. George Ball remembered that the president's concern over the bottom line was closely related to his distance from his businessman father. "Every weekend he went up to Hyannis Port," Ball said, "he came back absolutely obsessed with the balance of payments." Consequently Ball, a lawyer who cared little about economic matters, tried to get Kennedy to decide issues regarding the balance of payments during the winter when he took vacations elsewhere.[45]

The American-German offset agreements worked reasonably well for a time, but within several months after Johnson assumed the presidency they came under strain from budget shortfalls in Bonn. While the German trade surplus did not show up, at least not directly, on the ledgers of Chancellor Ludwig Erhard's government, the budget deficit did. Because the offset purchases composed a hefty portion of the latter, they pinched politically. Besides, many Germans wondered why their country, with its tradition of manufacturing excellence, should be purchasing weapons vital to national defense from abroad.

The British, who also had both a large number of troops in Germany and balance-of-trade problems, had negotiated similar offset agreements with the Germans. Though the British public, viewing the field from closer range than the American public, had an easier time distinguishing France from Germany, a certain anticontinentalism comparable to American anti-Europeanism crept into British politics, and when de Gaulle chucked NATO out of France more than a few in Britain argued for a retreat across the English Channel. Britain suffered additionally from chronic pressure on the pound, the result of sagging productivity and burgeoning public services. Wilson's Labour government could no more dismantle the welfare state that formed his party's raison d'être than Johnson would subvert the Great Society, but unlike the American president, Wilson lacked the financial depth to pay for guns and butter simultaneously. Consequently in July 1966 the prime minister announced that unless Bonn consented to offset completely the cost of maintaining British troops in Germany, he would bring many of them home.

The timing of the British announcement did Johnson no favors. In the wake of de Gaulle's eviction declaration, France and Germany had begun bilateral talks regarding the future of French troops in Germany. If the French withdrew from Germany, and now if the British followed, the United States would be left holding the German bag. Johnson well remembered the arguments of depression-era isolationists who said that if the Europeans did not care for their freedom enough to defend it themselves, they deserved whatever they got. American Asia-firsters, especially Republicans, had consistently argued that America should pay less attention to Europe and more to Asia. By taking the United States so deeply into Vietnam, Johnson had played into their hands. Now, as the Mansfield resolution demonstrated, Democrats were joining the chorus of anti-Europeanism. A forced flight from the continent—forced by an unlikely coalition of French Gaullists and American neoisolationists—appeared a real possibility.

At the end of August, Johnson wrote Wilson and Erhard proposing trilateral negotiations to seek a solution to the offset difficulties and related problems. "I have become increasingly concerned during the past few weeks about the dangers of an unravelling in NATO which could easily get out of hand," the president told Wilson. Between the anti-European rumblings in America, the strain on sterling, and Erhard's budget crunch—"all against the background of the General's antics"—there existed "danger of serious damage to the security arrangements we have worked so hard to construct during the last twenty years."[46]

Johnson spoke none too soon, for within days Mansfield issued a strong statement calling for a fundamental reassessment of the Ameri-

can position across the Atlantic. "Western Europe has long since reha-
bilitated itself after the devastation of World War II," the majority leader
declared. "It is now a thriving and dynamic region of greatly expanded
economic and political and potential military capacity. That factor alone,
in my judgment, would justify a revision of the fifteen-year-old level of
deployment whereby the greatest share of Western Europe's defense is
borne by the United States as though the former were still war-weakened,
exhausted, and incapable of an equitable defense effort of its own."
France's decision to leave NATO demonstrated the degree to which tensions between East and West had diminished. Under the circumstances,
the American presence on the continent outstripped the need. The burden that presence put on the American economy added to the demerits
of the case. "It is wholly unwarranted to sustain an unnecessary dollar
and dollar exchange drain." No less an authority than Dwight Eisenhower had asserted that reductions were in order. Who would gainsay
the man who rescued Europe and first commanded NATO forces?, Mansfield asked.[47]

Wilson agreed at once to Johnson's proposal for trilateral talks, but
Erhard wished to delay a decision until after a visit to Washington scheduled for September. Though this was but weeks away, American officials
feared that the problem could not keep that long. George Ball described
Erhard's procrastination as "totally unrealistic." The undersecretary
worried that Wilson's position might crumble meanwhile. "The pressure
for action in Britain due to the instability of sterling cannot be dealt with
in a leisurely fashion," Ball complained. Robert McNamara, projecting
an annual $500 million shortfall in American accounts with Germany,
pointed out that while British taxpayers were contributing 6.8 percent
of their country's gross national product to defense, and Americans 8.8,
the Germans were taxing themselves for defense only to the amount of
5 percent. Moreover, the rapid growth of the German economy made
Germany better able than the United States or Britain to expand defense
spending. The president should remind Erhard of these facts, McNamara suggested. They certainly had not been lost on proponents of the
Mansfield resolution.[48]

Much as Johnson might have wished to apply his famous treatment
to get Erhard to bend on the offset issue, the president understood the
limits of persuasion. Erhard had gone far in German politics preaching
friendliness toward the United States, yet this friendliness had raised
criticism for leaning too much in Washington's direction. Any more
leaning might topple his government. The chancellor had to balance
carefully. As Francis Bator said, "For Erhard it may be essential to political survival to prove that he remains a special friend of Lyndon Johnson,
but at the same time that he can successfully stand up to the Americans
on bread-and-butter issues." Whether Erhard could accomplish the feat,
Bator would not predict. While the administration should do nothing

to undermine the chancellor, it must prepare for all eventualities. "For us it is important—even more than Erhard's survival—that we not appear the culprit if he falls. Otherwise charges of American selfishness and unreliability—and the whole question of German relations with the U.S. and the West—will become major issues in the struggle for succession." The president should not let Erhard depart Washington without leaving a commitment to trilateral negotiations. Beyond this—toward a commitment to a particular level of offsets, for example—Johnson would be pushing the administration's luck.[49]

Erhard handled himself superbly during his September visit. At his White House session with Johnson, the chancellor seized the initiative from the president by immediately delineating the troubles he confronted in Germany. Those who demanded distancing from the United States constituted a minority, he said, but a minority that generated "a lot of noise." Some of the noisemakers held influential positions in the German parliament and could not be ignored. All Germans knew how much they owed the United States for its uncompromising defense of liberty in the past, as during the several Berlin crises, and at present, as in Vietnam. Within the limits of financial and political capacity, the government of the Federal Republic had worked hard to redeem its obligations to America. Erhard reminded the president how Germany had made early payments of certain sums due the United States in order to ease budgetary difficulties in Washington.

Johnson responded that the problem his own administration faced was most grave. He personally had always tried to be a friend to Germany. He intended to continue being a friend. But he could not overlook the feelings of many Americans who doubted the need for a large American presence in Europe. Those many certainly were not overlooking the economic burden the American presence had become for the United States. He and the chancellor must work together to solve this problem lest it produce "regrettable results" for both. At the same time, Johnson said he did not want to be too inflexible lest he "win the argument and lose the sale."

Erhard consented to begin trilateral talks and added a bonus: approval of a communiqué strongly supporting the American stand in Vietnam. Happy to get this, Johnson declined to press for a further commitment regarding offsets.[50]

Having signed up Erhard, the president now had to persuade Wilson to hold the fort until the talks got going. In a letter to the British prime minister, Johnson characterized his meeting with Erhard as "strenuous but useful," adding that while the chancellor agreed to nothing beyond negotiations, this represented a signal step in the right direction. Johnson expressed his appreciation for the weight Wilson was carrying in resisting troop cuts in Germany. Yet if he could just hang on for a little while more, the worst would cease. For the British government to give in to demands for retrenchment would create "serious trouble" for the alliance and badly damage the Western position in Europe. On the other hand, success in the talks would open the way to a revivification of NATO. Looking beyond the current difficulties and particularly beyond Charles de Gaulle, Johnson wrote, "It can give us a start in reconsolidating the alliance for the longer pull—and on a more sustainable basis—until the day when the French will be ready to rejoin the fold."[51]

To handle the American end of the trilateral negotiations, Johnson appointed John McCloy. The choice was an inspired one on grounds diplomatic, personal, and political. The former postwar high commissioner commanded a respect in Germany no other American could match. His devotion to European security was unassailable, but he also possessed a reputation for countenancing no nonsense from the Europeans. His banking background had prepared him well for the intricacies of what at the least complicated would be an intricate business. No one on Capitol Hill would be able to accuse the administration of letting the Germans slip something past Jack McCloy.

Perhaps most important, McCloy's appointment bolstered those in the administration who insisted on drawing a line against Mansfield and the other withdrawers. Eugene Rostow thought this aspect of the appointment decisive. Noting that the Pentagon was offering the president little help in withstanding congressional calls for a pullback, since McNamara hoped to make up at least some of what he was spending in Vietnam with savings from Europe, the undersecretary of state for political affairs said afterward that the McCloy appointment was designed to prevent economics from dictating military strategy. "It was part of a policy of strengthening the hands of the people within the government who wanted to stay and not have any change unless the Russians changed," Rostow asserted. On this issue at least, Rostow accounted Johnson a master of bureaucratic politics. "It was a very deep and well-conceived thing to try to fortify his own position within the government, because McNamara was pressing very hard for a massive cut."[52]

As a bonus, McCloy, a critic of de Gaulle known for speaking his mind, provided just the right balance of distance and stature to register American concern at French policy without directly implicating the president in the manner of registration. A few weeks after de Gaulle's announcement of France's departure from NATO, McCloy gave a background, not-for-attribution briefing to a group of American correspondents in Bonn. McCloy lashed de Gaulle for inconsistency, indeed "contradiction," in questioning America's readiness to defend Europe while taking steps that made such defense more difficult. If the general genuinely wanted a stronger American commitment, he should not be throwing American troops out of France. NATO existed largely because the United States wished to make clear in advance how it would respond to hostilities initiated by the Soviets. No one had ever adduced credible evidence indicating an American retreat from the commitment NATO represented. Despite de Gaulle's action the organization would continue to exist. The United States government was "not going to participate in the destruction of NATO or acquiesce in it."[53]

The trilateral talks opened in October 1966. Reporting on the initial round, McCloy informed Johnson, "All are keenly aware that the outcome of the talks is extremely important for the cohesion and future of the alliance." So far the British and Germans seemed friendly but businesslike. McCloy hoped for a settlement, yet he recognized the pitfalls on the road and cautioned that progress would come slowly.[54]

Progress came more slowly than McCloy had guessed. Hardly had the caterers cleared the glasses from the first round of talks than Erhard's rickety coalition in Bonn collapsed. Erhard's demise seemed certain to set the talks back several weeks and to establish a sinister precedent for German governments that got too cozy with Washington. It also dismayed Wilson, who, if he did not read his own future in Erhard's, at any rate complained more loudly than before that he could not hold on forever while his government drowned in red ink.

To buck up the prime minister, Johnson proposed to offset British expenses in Germany with American funds. In November the president offered to buy $35 million worth of British military equipment in exchange for Wilson's continued support in the trilateral talks. Johnson let Wilson know that the purchase wouldn't be easy. "I may get some heat in Congress on this and cannot move definitively until I have talked it over with some of my people on the Hill," Johnson wrote. But speaking as one politician to another, the president assured Wilson that the money would be found.[55]

When Wilson accepted Johnson's offer, the president's attention returned to Bonn, where Erhard's would-be successors were grabbing after the prize of power. The winner of the chancellor's cup turned out to be Kurt Kiesinger, who headed a coalition of Social and Christian Democrats. Kiesinger prudently put some distance between himself and the Americans. At the same time, he began reaching out toward de Gaulle. Neither move much pleased Johnson, although the president kept his annoyance private. In a meeting with German reporters, he told the correspondents he considered Kiesinger "a charming man."[56]

While waiting for Kiesinger to settle on a stance on the trilateral negotiations, Johnson and the administration worked to consolidate their own position. Rusk and the State Department, more concerned with the diplomatic and strategic repercussions of an agreement than with the cost, argued for only minor cuts in American troop strength in Germany. The State Department proposed achieving these cuts through a dual-basing arrangement whereby the personnel of one ground division and three air wings would rotate between the United States and Germany. Projected savings would come to $100 million annually. The Defense

Department and the Treasury, more worried about costs than about diplomacy, advocated deeper cuts. McNamara forwarded a plan for rotating and dual-basing two divisions and six air wings, with savings of $200 million per year. John McCloy opposed any cuts, though when pressed he hinted he might swallow the rotation of one division.[57]

The contending sides in this dispute lobbied for the president's support. McCloy argued from national interest. In a February 1967 letter to Johnson, McCloy declared, "Any unilateral withdrawal would only stimulate the further loosening of U.S. ties to Europe, weaken the whole concept of Atlantic security and also further shake German confidence." McNamara countered by pointing to the advantages of saving an extra $100 million per year. Besides deflecting the Mansfield resolution, these advantages included easing Vietnam-induced budget troubles and freeing money for domestic programs. Rusk rebutted by noting that a single repetition of a crisis like the Berlin confrontation of 1961 would quickly erase any savings. Besides, security should always come before financial considerations. Francis Bator, who as White House point man on the issue tried to mediate between the cutters and the stand-fasters—while his sympathies lay with the latter—summarized the alternatives in a memo to Johnson. "Whatever we do during this period of difficulty in our relations with Germany and Europe," Bator wrote, "a lot of people will blame the trouble on the U.S., and a lot of Republicans will try to pin it all on you." Should the United States slice its forces in Europe, critics would lump the president with de Gaulle as an abandoner of NATO and an endangerer of European security. "NATO as such doesn't carry many votes, but the charge that the president helped scuttle it would lend some spurious respectability to the general foreign policy attack we will face in 1968." Knowing his boss, Bator added, "The key to all this is your judgment of the domestic politics."[58]

Johnson procrastinated. Convinced of the necessity of a strong American position in Europe, the president wanted to keep as many troops as possible in Germany. Yet he also understood the necessity of placating Congress. To appear uncompromising on the issue would provoke the cutters and might lead to a new era of American isolationism. "The congressional position is three to one for substantial cuts," Johnson told McCloy in March. "The only one who will slug it out to hold out is Dirksen—he has German grandparents as I have. The rest of them will run just like turkeys." McCloy thought the situation might not be as grim as the president indicated. He reminded Johnson of a recent breakfast meeting at which congressional leaders had stated only mild objections to keeping the troops in Europe. Johnson disagreed. "I have dealt with those babies for thirty years," the president said. "The breakfast went the way it did not because that's the way the Congress really feels but because it's the way I managed it. First of all they were my guests, eating my breakfast. Then I laid out a very hard line, more arbitrary than

I like, which made it difficult for them to disagree with the president of the United States. And I first called on McCormack who I knew would take a hard, stand-up-straight line."

The attitude of West Germany made Johnson no more optimistic than did the mood of Congress. Recently Kiesinger had complained publicly that Washington was not consulting Bonn often enough. Johnson judged the complaint unfounded. "If I had a dollar for every time I consulted the Germans I'd be a millionaire," he said. As for Germany's willingness to compromise on the offsets, Johnson looked at the German character and saw little cause for hope. Stretching the city limits of Fredericksburg to make his case, he declared to McCloy, "I know my Germans. You know I lived in Fredericksburg, grew up in Fredericksburg. They are great people. But by God they are stingy as hell."

McCloy asked if Johnson would let him tell Kiesinger the president would talk to him personally. Johnson replied, "Not until he gets off this lecturing me in public speeches about consulting. That is no way to make me feel good about him. In any case I think he is shopping around, seeing where he can make deals with de Gaulle and the others. Maybe when he gets done with that and figures that in the end he's got to come and deal with the United States. Then, maybe. But only after the lectures stop."

The president returned to the hazards awaiting the alliance in America. "Between now and 1969 we face dark and dangerous ground," Johnson said. "Romney is running all over the country with his shirttails out. Nixon is taking his overseas trips. Bobby Kennedy is all over the place. I've got the liberals beating me on one side, the Southerners on the other. I have to try to maintain a position where we can hold a position that the real majority wants. But it's hard."

Johnson said he would stand tough if he had help. "I'll try to hold this alliance together longer than anybody else will, longer than the British will and longer than the Germans. But they have got to put something in the family pot. I can't do it alone." He directed McCloy to pass this message along to London and Bonn. "Tell them that every time they put in a dollar I am willing to put in several dollars. But they have got to put in some money. If they don't the pressures will really build." For Kiesinger the president had a particular message. "Say to him that you think that if he'll exercise some courage and leadership and determination to preserve the alliance, I'll try to do a little bit better than he and Wilson on everything they do."

McCloy once more suggested that the president was too pessimistic about Congress. Johnson replied, "You are wrong. You are an old and good banker. But it's one thing what a banker will tell you at a cocktail party. It's another when you go up to the window. I know damn well I don't have the votes. They're not there."[59]

By this time the Germans were ready to start talking again. Their readiness came just in time, for Johnson's $35 million payment to London purchased Wilson's cooperation only until June, now mere weeks away. The president sent McCloy back to the bargaining table with instructions that troop levels in Europe must be determined "on the basis only of security considerations, broadly construed." The last two words were crucial, for they allowed a role for politics—which in the end would decide the issue, as Johnson fully recognized. McCloy's letter of instruction went on to assert that "just as security is a common problem which can only be solved by cooperation, so are the monetary difficulties created by security efforts a common problem which can only be solved by cooperation." As a reminder for McCloy to pass to the Germans, the president added, "In the absence of a financial solution, and especially in light of the large German payments surplus, Congressional and public pressure would be intense, and the Germans should recognize that the situation might get out of hand."[60]

To give McCloy maneuvering room, Johnson allowed that West Germany might make up a portion of the payment deficit by purchasing American bonds rather than American military hardware. This was an important concession, since the United States would have to buy the bonds back eventually, with interest.

The bond device eventually proved the key to a settlement, but McCloy had some heavy work to do with both the Germans and the British to close the deal. "Although from time to time the trading instincts of your Fredericksburg Germans cropped out in the FRG representatives," McCloy cabled Johnson, "I am not certain that the subtler but still acquisitive instincts of the British are any less formidable."[61]

At the end of April the three sides cut a deal. The United States announced that it would place thirty-five thousand troops and ninety-six aircraft, or somewhat less than one division and three wings, on rotation between bases in America and Germany—although a State Department release asserted that the troops and planes remained "fully committed to NATO." The British would withdraw an essentially token number of troops. Germany pledged to purchase $500 million of medium-term American notes and would refrain from activities that put pressure on the dollar, in particular from converting dollars to gold. Washington and Bonn agreed to split the $40 million payments gap between Britain and Germany.[62]

IX

The conclusion of the trilateral talks released without explosion most of the pressure that had built up in America for a major reconsideration of policy toward Europe. It also established, or at least confirmed, a spirit of cooperation that carried over into other aspects of U.S.-European relations. Two months after the trilateral pact, the Johnson administration and its European partners (as well as Japan, that honorary Western power) concluded the so-called Kennedy round of trade talks. These talks, survivors of the Kennedy years, involved billions of dollars of exports and imports among the United States and the other countries. The talks went slowly at first, vexed by the usual objections and special pleading that typically attend commercial negotiations—and which, on the American side, Johnson took care to counter by constant monitoring of congressional opinion. The talks speeded up as a deadline of June 30, 1967, imposed by Congress on Johnson's special negotiating authority, approached. During the last weeks, the American team and its counterparts from the other countries pounded out a bargain that reduced tariffs on a wide array of items. The final details were arranged on the last day, and the required signatures were exchanged only hours before the deadline.[63]

The successful completion of the Kennedy round disposed all involved to cooperate again a few months later when a minor crisis developed in the structure of international finance. The offset deal of the previous spring had temporarily alleviated the strain on Britain's economy, but the same forces that had been besetting Britain continued to operate. In November 1967 the British government tried to ease the situation by devaluing the pound. The devaluation triggered a wave of speculation, especially among gold buyers who believed that the shock to the system would force the American government to back away from its commitment to sell gold at $35 per ounce. Johnson immediately reassured the markets, pledging to hold the line on the gold price, and asked Congress for a tax increase and spending cuts to diminish the American deficit. "If we don't act soon," he melodramatically told a group of legislative leaders during the third week of November, "we will wreck the Republic." He added, "With sterling devalued, the world depends even more on the dollar."[64]

Meanwhile Johnson directed administration officials to get together with representatives of the other major gold states to design a common-front strategy against the speculators. The French proved no more helpful here than elsewhere at the time, and de Gaulle effectively abetted the speculators by announcing France's withdrawal from the international gold pool and directing his finance minister to purchase large quantities of gold. But the other major economic powers stuck together, and by the end of November put in place a program to defend the $35 price against upward pressure.

This afforded only a respite, however, and when bad news regarding the American balance of payments triggered another run on gold, Johnson was forced to more vigorous action. He explained his thinking in identical letters to Chancellor Kiesinger of West Germany and Prime Minister Aldo Moro of Italy. "We have the means at hand to overcome the dangers caused by the disorder in the gold market," Johnson declared in March 1968. "The speculators are banking on an increase in the official price of gold. They are wrong." The president went on to call for close cooperation among the allies. "These financial disorders, if not promptly and firmly overcome, can profoundly damage the political relations between Europe and America and set in motion forces like those which disintegrated the Western world between 1929 and 1933." On his own, Johnson decreed tighter regulation of American overseas investment and the trimming of certain foreign aid programs. In conjunction with the leaders of the other gold pool countries, he ordered the severing of the connection between the private gold market and the gold reserves of the United States and the other countries. Henceforth the speculators could do what they would with the price of gold on the private market, but most of the world's major economic powers would protect their gold supplies separately.[65]

The gold settlement mitigated the economic confusion just in time for Washington to witness an uproar of a different kind in Europe. Not since 1848 had the continent undergone such outbreaks of popular protest as it experienced in the spring of 1968. Violent demonstrations jolted universities across Germany, indicating that Germans—young Germans at least, who knew nothing else—sought more from life than peace and prosperity. British students lent their voices to the uproar, prompting Harold Wilson to try to buy their silence and perhaps their votes by pushing through the parliament a measuring lowering the franchise threshold to eighteen years.

Louder noises emanated from Paris. What began with student complaints against crowded apartments and stultifying academic regulations escalated to street warfare. By the middle of May, students controlled much of the Left Bank and eyed the Elysée. Workers added their grievances to the cause, often over the orders of their frightened union leaders. At month's end, de Gaulle's regime verged on disintegration. With a characteristic sense of drama, the general disappeared for a day— partly, as it was learned later, to gauge the support he still commanded among army leaders. Reassured, he reappeared and dissolved the national assembly, calling for snap elections. This brilliant stroke disarmed the radicals, who lacked the organization to convert their protest into electoral victory. The "events of May" concluded in balloting in June that returned a large Gaullist majority. But de Gaulle's triumph turned out to be temporary, and within a year the general was forced into retirement—only months after a similar fate claimed Johnson.

Even more momentous events took place in Czechoslovakia. Complaints against the status quo in Prague centered on the oppressive weight of Soviet-installed communism. Under Alexander Dubcek, the Czechs carefully sought to work their way out from beneath Moscow's burden. For several months their efforts succeeded, and by the late spring of 1968 the Czech capital had regained much of the liberal atmosphere that had marked it between the wars. Yet despite Dubcek's repeated disclaimers that internal reform had nothing to do with foreign relations, and despite his promises of unflagging adherence to the Warsaw Pact, the "Prague spring" failed to warm the cold hearts in the Kremlin. In August the Soviet Union and four allies—Romania refused to take part—poured hundreds of thousands of troops across the border in the largest military operation in Europe since 1945.

A short while later Leonid Brezhnev provided the ideological justification for the invasion. Under what pundits quickly tagged the "Brezhnev doctrine," the Soviet leader declared the road to communism a one-way street. Whenever socialism came under threat in one country, he said, other socialist countries had the right and obligation to defend it.

More persuasively than anything Johnson could have said, Brezhnev's statement and actions argued the continued need for NATO. No one expected Soviet armor to continue rolling from Czechoslovakia into western Europe, but the danger of additional military action by the Soviets could not be dismissed. Johnson told a group of congressional leaders, "It concerns us that they might want to clean up Yugoslavia, Romania and Czechoslovakia at the same time." If that occurred, there was no telling where the violence might spread to. "The next target might be Austria," Johnson said.[66]

Johnson never entertained thoughts of materially opposing the Soviet suppression of Czechoslovakia. As Earle Wheeler of the Joint Chiefs advised the president on the day the invasion began: "There is no military action we can take. We do not have the forces to do it." Rusk added that the Russians might respond to even diplomatic initiatives with a move against Berlin. "Khrushchev called Berlin the testicles of the west," Rusk reminded Johnson, "And when he wanted to create pressure he squeezed there." Hubert Humphrey summarized Johnson's lack of options. "All you can do is snort and talk," the vice president said.[67]

In certain respects, the Czech blowup complicated Johnson's European policy. It particularly upset relations with the Soviet Union—the offstage presence in American relations with France and Germany and Britain, as well as in various other areas of American policy. During Johnson's time in office, the war in Vietnam, in conjunction with the persistent Democratic fear of seeming soft on communism, had prevented much motion toward defusing the confrontation between the superpowers. Yet Johnson, who realized that crises with the Kremlin would distract Congress from his Great Society even more than the Vietnam War already was doing, did manage to achieve modest improvements in dealings with the Soviets. He oversaw negotiation of agreements restraining the production of fissionable materials, forbidding the deployment of nuclear weapons in outer space, initiating commercial air travel between the United States and the Soviet Union, ironing out disputes between American and Soviet fishing interests, and facilitating cultural exchanges. He also succeeded in diminishing the stridency of the rhetoric that often had marked dealings between the United States and the Soviet Union. In this last endeavor he received important help from the noisy Khrushchev, who got himself deposed in 1964.

Most significantly, through steady and quiet diplomacy Johnson brought to the signing stage a treaty committing the United States and the Soviet Union (and Britain) to nuclear nonproliferation. Building on the partial test-ban treaty of 1963, the nonproliferation treaty embodied the desire of the three original members of the nuclear club to maintain their group's exclusivity. Nonmembers who signed the pact consented not to petition for entrance to the club, in exchange for a pledge by the nuclear signatories to work toward disarmament. For the Americans, the NPT would keep nuclear weapons out of the hands of reckless Third World leaders. For the Soviets, the NPT would indefinitely preclude German acquisition of the big bombs. In July 1968, after years of painstaking negotiations, representatives from the United States, the Soviet Union, Britain, and more than fifty other countries met at the White House to sign the treaty. Johnson called the NPT "the most important international agreement since the beginning of the nuclear age"—a statement that wasn't more exaggerated than many the president uttered. To celebrate the achievement, and to start the push beyond nonproliferation to the stipulated curbs on superpower weaponry, Johnson scheduled a trip to Moscow for the autumn. He also prepared to submit the NPT for Senate ratification.[68]

The Soviet invasion of Czechoslovakia upset the president's plans. Clark Clifford commented to Johnson, "Obviously the Soviet leaders were prepared to jeopardize their relationship with us, with the rest of the world outside the Eastern European Bloc and with the members of that Bloc in order to shape the Czechoslovakian internal situation more to their own liking." The president himself wondered aloud, "Can we talk now after this?" Dean Rusk, hoping to keep the administration's diplomacy on track, said the administration could. The secretary acknowledged that the administration could hardly proceed as though nothing had happened. "On the other hand," Rusk said, "Soviet action against Czechoslovakia has not eliminated many major world strategic problems involving the U.S.S.R. and the U.S." Life went on. But Johnson had a better grasp than Rusk of the political realities of the situation. He had hoped to end his tenure as president with quick ratification of the nonproliferation treaty. He even considered calling a special Senate session for the purpose. Soundings on Capitol Hill, however, convinced him that sufficient favorable votes would be difficult to find. He decided to leave the treaty to a more propitious moment rather than risk its defeat.[69]

Johnson thought somewhat more seriously about trying to reschedule the summit. He felt he might be able to make important progress toward strategic arms control, and he did not want to let the opportunity slip away. But by the time the thunder out of Czechoslovakia died down, Richard Nixon was president-elect, and the Kremlin decided to wait for the new man. Moscow withdrew its invitation to Johnson.[70]

If the Soviet suppression of reform in Czechoslovakia upset certain aspects of Johnson's European policy, it simplified others. It immediately silenced calls for further reductions in the American presence in Europe. Moscow's willingness to use force to assert itself in Europe convinced Congress and nearly all Americans that Washington needed to be ready to do likewise.

In addition, the Soviet invasion confirmed the wisdom of Johnson's approach to France and de Gaulle. A few months earlier the general had seemed to get the better of the Franco-American relationship when Johnson announced a desire to negotiate an end to the Vietnam War and accepted de Gaulle's offer of Paris as a venue. (Johnson had tried mightily to find a different site but had had to settle for the French capital when the North Vietnamese would not agree to any other location the Americans proposed.) Yet when nearly half a million Warsaw Pact troops entered Czechoslovakia—an especially symbolic place, in Johnson's and America's Munich-dominated view—the president's warnings that the prerequisites for peace had not changed much since the 1930s carried greater credibility than de Gaulle's description of a Europe "from the Atlantic to the Urals." At a time when America's Southeast Asian flank was crumbling, Johnson could congratulate himself that in Europe the center held. It was not as strong as it had been, but it held.

| # When the Twain Met— Head-on

THE ADJUSTMENTS the Johnson administration was forced to make in policy toward Europe reflected significant changes in the structure of power in the Atlantic region since the early Cold War. France's withdrawal from NATO indicated an accurate perception in Paris that the Soviet Union, notwithstanding its crushing of reform in Czechoslovakia, posed no serious active military threat to the West. Whether it ever had posed such a threat was—and is—an open question. But regardless of the underlying reality, the altered perception fostered a loosening of the Atlantic alliance almost unthinkable a decade before. The trilateral negotiations with Bonn and London indicated West Germany's coming of age as an economic great power. Though West Germany still relied on the United States for military protection, this reliance was now a matter of political choice rather than economic necessity. Should the Germans decide to follow the example of de Gaulle and broaden their diplomatic horizons to the east, Washington would have little alternative but to make the best of another unsettling development.

Yet important though the changes in Europe were, they represented modifications to a status quo that on the whole remained intact. The changes took place within the bipolar framework constructed at Yalta in 1945, and for all the restiveness in both the capitalist and socialist camps—most obviously in France in the West and Czechoslovakia in the

East—the fault line between the camps continued to run along the Elbe, where it had for a generation.

In South Asia, by contrast, the fault line had shifted stunningly since the Cold War geopoliticians had first mapped it. One branch still fractured Kashmir and what had been Bengal, separating sternly authoritarian and Islamic Pakistan from riotously democratic and secular India. But where the Himalayan uplift once had functioned principally to divide Pakistan from the communist territories to north and east, with Indians and Chinese embracing amid professions of nonviolent mutual respect, recently the Himalayas had served as a battleground between India and China, with the Pakistanis moving close to Beijing.

Though it was too soon to know, developments in the Indian subcontinent during the 1960s presaged the revival of global balance-of-power politics that followed Richard Nixon's opening to China in the early 1970s—an opening facilitated, not coincidentally, by Pakistan's persistent desire for reinsurance against India. In Europe the rumblings of change in the tectonics of the Cold War yet emanated from deep within the mantle, and though the opposing plates ground against each other, they remained separately intact. In South Asia, by contrast, the plates had each been ripped apart. The pieces were now careening about, colliding and reamalgamating in ways few could have predicted ten years before.

The biggest smash of the Johnson years occurred in September 1965, when India and Pakistan again went to war over Kashmir. The war provided the focus for American diplomacy toward the subcontinent during the 1960s, and it tested Washington's ability to deal with the changing nature of world politics. The aftermath of the war, moreover, included a grave threat of famine in India. As usual, New Delhi looked to the United States for assistance in averting mass starvation. In a rare case of overruling the unanimous counsel of his top advisers, Johnson refused for several months to release American grain. Only after letting congressional support for Indian aid build to such a degree that it nearly forced his hand did the president allow the grain ships to load and leave. The consequences of Johnson's gamble included both a major nudge to India's ultimately successful efforts to get off the international dole and severe straining of the U.S.-Indian relationship.

II

Although it took American leaders more than a decade to recognize the full significance of the falling-out between China and the Soviet Union during the late 1950s and early 1960s, the effects of the split showed up in the politics of the Indian subcontinent almost at once. For several years after 1954, when Zhou Enlai had traveled to New Delhi to announce China's friendship for India, the Chinese premier—probably the most adroit diplomat of his day—carefully concealed his dislike for India's founding prime minister, Jawaharlal Nehru. "I have met Chiang Kai-shek, I have met American generals," Zhou said later, "but I have never met a more arrogant man than Mr. Nehru." Yet as the rift between Moscow and Beijing opened, so did a divide between Beijing and New Delhi. India's cordial relations with the Soviet Union contributed to China's hostility toward India; at the same time, that hostility resulted equally from a Chinese desire to establish its claim to leadership in Asia against India's competing claim. Border troubles consequent to China's 1959 suppression of nationalist turbulence in Tibet might have been handled smoothly, yet conspicuously weren't. In part to signal its disdain for India, China improved relations with Pakistan, significantly negotiating an accord rectifying frontiers in the part of Kashmir that India claimed but Pakistan controlled.[1]

Tensions broke into war in the autumn of 1962. Chinese troops occupied large portions of Indian territory, threatening at one stage of the contest to carry the fight down onto the Assam plain. India appealed desperately for help to the United States, figuring Washington was a better bet than Moscow, then in the process of blinking in the Cuban missile crisis.

Many in the United States couldn't resist gloating at the troubles of Nehru, who for years had told the Americans they worried too much about the Chinese. Even Kennedy, on the whole an ardent Nehru supporter during his days in the Senate, where he sponsored an important India aid bill, and equally friendly since becoming president, could not resist a gentle jibe. "You have displayed an impressive degree of forbearance and patience in dealing with the Chinese," Kennedy wrote to the prime minister. "You have put into practice what all great religious teachers have urged and so few of their followers have been able to do. Alas, this teaching seems to be effective only when it is shared by both sides in a dispute."[2]

The slight smugness passed quickly as Kennedy's advisers calculated how to turn India's setbacks to America's account. "We may have a golden opportunity for a major gain in our relations with India," Robert Komer remarked. Since the early 1950s, India had received a large amount of American economic aid. This afforded Washington some leverage over Nehru's foreign policy, but not much. The Indian prime minister made it a point of national pride to reject conditions on aid, and the specter of starving millions deterred the Truman and Eisenhower administrations from insisting. Military aid, largely absent until now, was another matter. Humanitarian distractions did not enter the picture, and with India's back to the wall the United States might expect a bit more cooperation than Nehru had delivered to date. "The sheer magnitude of India's reverses on the Chicom border may at long last awaken Delhi to the weakness of its position," Komer said.[3]

Nehru, in fact, was wide awake by this time, as evidenced by his decision to fire his closest friend in government, V. K. Krishna Menon, a longtime apostle of Indo-Chinese cooperation and Indo-American enmity. Menon's ouster eased the way for Kennedy to approve a sizable program of weapons aid. The guns, trucks, and radios had no effect on the fighting in the Himalayas, which ended before the equipment began arriving, but it did promise to create a new form of Indian dependence on the United States. While wheat is fungible, bullets aren't. American arms require American ammunition and spare parts.

The Kennedy shift toward India, not surprisingly, annoyed Pakistan. In 1954, when Eisenhower and Dulles had agreed to send American weapons to Pakistan in exchange for pledges of undying opposition to Allahless communism and, more concretely, for the use of Pakistani airfields and listening posts, Karachi had held a near monopoly on America's official affections in the subcontinent. The second-term, looking-to-history Eisenhower rethought this tilt and tried to improve relations between the world's most powerful democracy and the world's most populous. Pakistan bridled as America's eye wandered. The Pakistani mood didn't improve when Khrushchev responded to the 1960 shooting down of Francis Gary Powers's U-2 plane, launched from Peshawar, by threatening to rain Russian rockets across the Hindu Kush. Most of the rockets existed only in Khrushchev's bluster, but the Pakistanis, not privy to the photographs the U-2s took, did not know that.

When Kennedy assumed office a pair of alliances still legally bound the United States to Pakistan. CENTO, the successor to the ill-fated Baghdad Pact, nominally contained the southward thrust of the Soviet Union. SEATO purported to do the same with regard to China. But both organizations, hatched in a manic moment of American alliance building, and comprising countries—Pakistan being a prime example—with other aims than fighting communism, were growing moribund. The two weightiest non-American SEATO signatories, Britain and France, showed no inclination to join the American effort to protect South Vietnam, with de Gaulle's Paris positively obstructing American aims in the region SEATO was designed to defend. CENTO survived primarily because the United States did not wish to concede that the bloody coup that liquidated the Hashemite monarchy in Iraq in 1958, thereby knocking Baghdad out of the Baghdad Pact, could knock over a whole treaty system as well. But the remaining regional members—Turkey, Iran, and Pakistan—had little in common besides a taste for American aid, and the alliance as a whole possessed hardly more life than SEATO. While American officials were not willing to declare the condition of either pact terminal, they hoped less for recovery than to slow the decline.

Pakistan certainly wondered what good either alliance did Pakistan. Though diplomatic discretion had always required Karachi to declare that the arms America provided were for use against the communists,

neither they nor honestly observant Americans—and still less the Indians—had any doubt whom the Pakistanis judged their foremost enemies. Nehru's government generally avoided provocative statements
about Pakistan's right to exist, yet Indian irredentists irritated the issue
sufficiently to keep the Pakistanis nervous about the possibility of being
reabsorbed by their enormous neighbor. The dispute over Kashmir ag-
gravated the nervousness. As it had since the 1947 partition of the subcontinent, the Pakistan government called for a plebiscite in the con-
tested province, counting on the Muslim majority to deliver the prize.
India based its claim on the desire of the last Hindu maharajah to attach
Kashmir to India, and on its judgment of the general illegitimacy of
Pakistan. New Delhi found various excuses for opposing a plebiscite. The
United States refused to choose sides, thereby galling Karachi, which
argued that allies ought to back allies and that a country presuming to
preach democracy ought to support self-determination. (That domestic
self-determination did not obtain within Pakistan did not lessen the ve-
hemence of the Pakistan government's argument.)

Karachi had long rejected the argument made by India that American
military aid to Pakistan constituted a greater threat to India than American economic aid to India did to Pakistan. The Pakistanis correctly contended that American dollars to purchase food freed Indian rupees to
buy guns. After Kennedy initiated military aid to India, the Pakistanis
didn't have to bother equating bread and bullets, since India now received both from Washington. (So did Pakistan, but far less bread and,
before long, fewer bullets too.)

Despite the provocation involved in arming Pakistan's principal rival,
officials of the Kennedy administration thought they could keep Pakistan on America's side even as they courted India. Robert Komer, who
headed the White House effort on South Asia in both the Kennedy and
Johnson administrations, was excited at the prospect of what he called
"a long-needed readjustment in our policy" in favor of India. Komer
told Bundy, "And we can achieve this without losing our Pak assets and
alliance if we play our cards right."[4]

The hand was a difficult one. American officials explained to Karachi
that American weapons for India would not be used against Pakistan,
only against China. The Pakistanis were no more reassured than the
Indians had been when Washington said the analogous thing to New
Delhi. They had reason to be far less reassured. They lacked India's
numbers, India's industry, and India's internal lines of communication.
A study for the American Joint Chiefs of Staff, completed the month
after Johnson entered office, delineated Pakistan's problems and concluded, "Finding itself in such a disadvantageous strategic posture, Pakistan is deeply and genuinely afraid of Indian aggression." Referring
to the regime of general and president Mohammad Ayub Khan, the
chiefs added, "Our U.S. pledge to help the government of Pakistan to

resist such aggression is politely noted but Ayub and those around him openly express their fear that U.S. aid in a crisis would be delayed by the possible ambiguities of the situation, by the delays or hesitation inherent in the decision-making process of democracy, or by the distances to be traversed by our reinforcements."[5]

As placation to the Pakistanis, Washington continued weapons shipments to Karachi even while it armed India. Johnson recognized the political logic of the situation as that of the lobbyist who contributes to both candidates in a close electoral contest. No matter who wins, the lobbyist has a foot in the door. The strategy possessed a certain geopolitical logic as well. American arms to India diminished Delhi's dependence on Soviet supplies, while aid to Pakistan kept Ayub from sidling closer to China. Unfortunately the United States did not offer enough to either party to persuade it to cut ties altogether to its communist patron. As the Soviet-Chinese split widened, Moscow and Beijing carried their competition by proxy to the subcontinent, exacerbating the tension preexistent between India and Pakistan. In the process, the American foot in each door came to resemble a foot on either side of a widening crevasse.

III

After their liberal treatment at Kennedy's hands, the Indians worried that his death foreshadowed a conservative shift back toward Pakistan. Chester Bowles, the American ambassador in New Delhi, explained the Indian reaction: "In India President Kennedy was looked upon as a special friend who was author of the Indian resolution in Congress, who had frequently spoken in behalf of Indian aid, and who had singled India out as a great experiment in democracy." Bowles continued, "With the loss of this friend, Indians now believe we favor Pakistan in its disagreement with India and discount India's importance."[6]

Bowles concurred with the Indian view on this issue, as on most others. The ambassador, on his second tour in New Delhi, having also headed the embassy under Truman, exemplified to the point of caricature the liberal American infatuation with India. He and his family embraced India enthusiastically. Mrs. Bowles adopted Indian dress, and the Bowles children attended Indian schools. The embassy entertained by hiring Indian artists and musicians. Receptions and dinners featured Indian food—to the dismay of Indian guests hoping for a break from curry and rice. The White House often groaned at the predictability of Bowles's adherence to the New Delhi line. Komer got so fed up with Bowles that he suggested that Johnson appoint an "ambassador from us to India, not from India to us."[7]

Bowles and India both were wrong in perceiving a preference for Pakistan in the new president. Johnson, probably to a greater degree than any other tenant of the Oval Office during the first forty years of subcontinental independence, treated India and Pakistan evenhand-edly. Truman had detested Nehru—who reciprocated the feeling—and set in motion the chain of events that led to the Eisenhower alliances with Pakistan. Kennedy tilted toward India. Nixon tilted back toward Pakistan, to the extent of engaging in nuclear-gunboat diplomacy against India during the 1971 Bangladesh war. Following the Ford hiatus, Carter first warmed to India, largely to chide Pakistan for human rights abuses, but after the fall of the Iranian shah and after the Soviet invasion of Afghanistan, Carter swallowed his scruples and cultivated the regime of General Mohammad Zia ul-Haq. Reagan, with no observable human rights scruples to swallow, amplified the trend, sending billions of dollars directly to Islamabad (Karachi's successor as Pakistan's capital) and more via rakeoffs from aid to the Afghan mujahideen.

Yet evenhandedness is rarely in the eye of beholders, at least those with a stake in preference. Johnson's balance was lost on both parties. At the same time that the Indians braced for a post-Kennedy wave of Texas reaction, the president remonstrated with Pakistan's foreign min-ister, Zulfikar Ali Bhutto, for making it nearly impossible for the United States to help his country. In a meeting with Bhutto just after Kennedy's funeral, Johnson pointed out that Pakistan's ties to China did not win votes in Congress, the wellspring of American aid. "The strongest men in Congress in favor of Pakistan are also the strongest against the Chi-nese communists," the president said. The foreign minister must realize that if it came to a choice between helping Pakistan and opposing China, the latter course would win.[8]

In a message to Ayub a week later, Johnson reiterated his concern at actions by Pakistan that worked in China's favor. "Regardless of Pakis-tan's motivations, which I can understand but frankly cannot agree with, these actions undermine our efforts to uphold our common security interests in the face of an aggressive nation which has clearly and most explicitly announced its unswerving hostility to the Free World." Most members of Congress would not understand Pakistan's motivations, let alone agree with them. Having asked the legislature to support a war in Vietnam to contain Chinese communism, the president could hardly request major aid for a country that increasingly acted like a Chinese ally.[9]

Johnson's admonitions combined with ongoing shipments of arms to India to convince Ayub that the Kennedy preference for India contin-ued. Among some in the administration it did. Komer, for all his disdain of Bowles, deemed excessive concern regarding Pakistan's sensibilities wrongheaded and counterproductive. After all, he told Johnson, India, not Pakistan, was "the major prize for which we, the Soviets and Chicoms are competing in Asia." Komer argued that American obligations toward

Pakistan, while real, must not subvert larger aims. "We can and should protect Pakistan against India, but we cannot permit our ties to stand in the way of a rational Indian policy."

Komer made this case shortly after Nehru suffered a major and debilitating stroke. Though reports soon had the prime minister mending, he was an old man and probably could not survive much longer. The next months might be crucial, Komer said. "With India heading into a succession crisis, we have to watch our step. If India falls apart we are the losers. If India goes Communist, it will be a disaster comparable only to the loss of China. Even if India reverts to pro-Soviet neutralism, our policy in Asia will be compromised. These risks are not just Bowlesian hyperbole; and if they prove real, Pakistan loses as well."[10]

In fact, talk of India going communist *was* hyperbolic, as Bowles himself would have admitted. For all Nehru's tolerance of communism in the Soviet Union and—at least formerly—in China, the prime minister had consistently kept a sharp eye on Indian communists. Rude and occasionally troublesome, the Indian communists nonetheless represented no threat to either the entrenched position of Nehru's ruling Congress Party or the stability of Indian democracy. While a successor would lack Nehru's unchallengeable prestige, comparisons with China were farfetched and would remain so.

Komer may simply have been trying to get Johnson's attention for his more reasonable scenario: an Indian reversion to pro-Soviet neutralism. The mutual antagonism of Moscow and New Delhi for Beijing rendered collaboration between the Soviet Union and India nearly unavoidable. How much this collaboration would mean now that Mao credibly challenged the Kremlin's claim to revolutionary leadership remained to be seen. It certainly mattered less than in the days when India formed the potential third corner of a Soviet-Chinese-Indian triangle that would dominate Eurasia, at least in population. But it could hardly serve American purposes.

Consequently Komer worked to keep India before Johnson's gaze. In April 1964 Nehru's daughter Indira Gandhi visited the United States. Johnson could be gracious on cue, yet he begrudged the time diplomacy required him to spend chatting with what sometimes seemed every cousin of half the world's heads of state. Naturally, the less significant the state, the more significant a visit with the American president was for consumption back home. White House aides resisted requests from the State Department for such meetings, partly because they recognized better than the diplomats the other demands on the president's time, partly to keep the diplomats in their place, and partly because when overbooked the president took out his annoyance on those nearest at hand.

But Johnson's aides thought he ought to make time to see Indira Gandhi. "We've been valiantly keeping Indians off your schedule for months now," McGeorge Bundy and Komer informed Johnson. "They're beginning to get edgy." Further, Gandhi was a special case. She was not simply the prime minister's daughter. "Though out of government now, taking care of her father, she's a political wheel in her own right and could well be the next foreign minister or even prime minister."[11]

Johnson agreed to meet Gandhi, but he did not much like the idea. The one sin Johnson found hardest to forgive in foreign leaders was criticism of the United States on American territory. Like most southerners—including Texans, in this case—Johnson believed in personal courtesy. Just as guests to a person's house should not insult their host under the host's roof, so guests to the United States should not insult the American government on American soil. Johnson grew greatly upset when British prime minister Alec Douglas-Home, attending a White House dinner, mentioned Britain's decision to finance the sale of 450 buses to Castro's regime, in contravention of the American-organized embargo of Cuba. Johnson found Douglas-Home's decision offensive enough, but the fact that the prime minister raised the issue at the White House itself struck Johnson as the depth of poor taste. The incident lived in the administration's institutional memory as shorthand—"selling buses on the White House steps"—for a surefire way to get Johnson mad.[12]

A statement by Indira Gandhi to the *New York Times* that "the West is on Pakistan's side no matter what" fell into the bus-sale category. Not only did it violate the president's code of courtesy, it flew in the face of continuing shipments of American weapons and economic aid to India, which both currently and over the years totaled more than American assistance to Pakistan. Johnson politely took issue with Gandhi's assertion when he received her at the executive mansion. The United States did not play favorites between Pakistan and India, he said. His administration sought and hoped for good relations with both countries. If it was any consolation, he added, the Pakistanis were far more upset with the United States than the Indians were.[13]

Nehru's death a short while later distracted Gandhi and India from their dissatisfaction with America. Gandhi's opponents ganged up to keep her out of power, and the premiership passed to Lal Bahadur Shastri, a compromise candidate whose principal qualification was his lack of enemies. By temperament a conciliator and by circumstance a place-filler, Shastri seemed a solid bet not to undertake any startling foreign policy initiatives. From the American perspective this offered an advantage and a disadvantage. On the positive side, Shastri probably would not unduly exacerbate problems with Pakistan. On the negative, he likely would not make much progress toward a settlement of the Kashmir dispute.

IV

Rarely has South Asia per se attracted much American attention. Other regions have generally claimed precedence on the ladder of America's priorities. Considering what close American attention has accomplished over the years in places like Central America and the Caribbean, this relative uninterest may have been a blessing to the inhabitants of the region. Plenty of Mexicans, Guatemalans, Nicaraguans, and Cubans would gladly have traded their countries' locations for those of India or Pakistan. From the time of India's and Pakistan's independence, the subcontinent has usually entered American calculations in the context of relations with the Soviet Union and China.

The exceptions to this rule, and not complete ones at that, were the occasions when India and Pakistan went to war. In 1947, in 1965, and in 1971 American leaders were forced to take notice of events in South Asia. Their notice resulted partly from a humanitarian desire to stop the killing, but more from fear that conflicts would spread and destabilize much of Asia. From such chaos, Washington believed, no good and great bad would result.

The September war of 1965 could not have caught the Johnson administration with less attention to spare for South Asia. While Vietnam caused Johnson the greatest concern, he had several other headaches as well. American troops still patrolled Santo Domingo, de Gaulle threatened to bolt NATO, and Sukarno's Indonesia seemed heading rapidly down the same Chinese road Ayub's Pakistan was apparently treading. And the Great Society, the apple of Johnson's eye, was only just seeing the light of day. Yet if the renewed fighting in Kashmir caught Washington otherwise occupied, it did not catch Washington by surprise. Remarkably, considering all the other demands on his time, Johnson kept carefully attuned to the situation in South Asia. Recognizing the untenability of the American position of supporting opposite sides in a regional cold war that might get hot fast, the president sought to limit America's liability, or at least get more for America's money.

Shortly after his 1965 inauguration, as he prepared to escalate the war on poverty in the United States and the war on communism in Vietnam, Johnson had decided to shake up India and Pakistan. Shastri and Ayub were scheduled to visit the United States, presumably with the goal of receiving more American aid. Shastri had particularly in mind a shipment of American F-5 jets to counter Pakistan's American F-86s and F-104s. The White House staff pushed hard for approval of the F-5s. Harold Saunders acknowledged that approval would raise Ayub's ire, but he judged the repercussions manageable. "I think we ought to do what makes most political and military sense in India and then adjust to Pakistani pressures as necessary when we see how Ayub responds," Saunders said. The advantages in terms of tying the Indian military to the United States, meeting India's security requirements against China, and displacing the Soviets, who had outfitted New Delhi with MiGs and probably would be happy to send more, seemed to overbalance the disadvantages in terms of cost and Pakistani irritation.[14]

Johnson disagreed. The president not only refused to approve the F-5 transfer but abruptly canceled—in diplomatic parlance, "postponed"—both the Shastri and Ayub visits. A recent trip by Ayub to Beijing precipitated the president's disinvitation of the Pakistani leader, which he felt compelled to balance by doing the same to Shastri. Yet the decision reflected more than momentary pique. Aid for India and Pakistan had come under fire in Congress, and Johnson desired to head off a cutoff by showing that he was reviewing the situation himself. Additionally, he wanted to shock the beneficiaries of American assistance into realizing that the United States expected cooperation at the receiving end. Dean Rusk, relaying Johnson's decision to cancel the visits, told the American embassies in New Delhi and Karachi to proffer the appropriate apologies. At the same time, the secretary said, the Indian and Pakistan governments should be apprised of the deep American dissatisfaction with their inability to make progress toward a settlement of the Kashmir quarrel.[15]

Johnson's move had the desired effect of waking up the governments of Pakistan and India. Ayub's disinvitation made the Pakistan leader's life difficult, in that it provided brickbats to Pakistan's anti-American left, led by Foreign Minister Bhutto, to hurl at him. Shastri's disinvitation had an even greater impact on the Indian prime minister. Indian immoderates had decried Delhi's growing dependence on the United States for half a decade. The attacks had glanced harmlessly off Nehru, but to the precariously perched Shastri they threatened mortal damage. Bowles, on receiving Rusk's cable, predicted that the prime minister would feel "profound shock and resentment" at the summary treatment. "Believe me," the ambassador said, "there is no earthly way that I or anyone else can plausibly explain this action in India at this time and under these conditions." He urged Rusk to ask the president to reconsider.[16]

Rusk would have complied had he thought it would do any good. The secretary had enough experience of diplomacy to understand the ramifications of such a slap to a country the United States was trying to cultivate. But he also knew Johnson well enough to realize he would be wasting his breath. The president had made up his mind how he wanted to handle South Asian policy, and that mind wouldn't be easily changed.

Not for the last time, Johnson surprised his experts with his ability to get away with outrageous actions. Shastri survived the blow, as, more predictably, did Ayub. "All things considered, deferral of the Ayub/Shastri visits went off quite well," Komer conceded in the middle of

April. Though it was too early for a definitive assessment, the Indian and Pakistan governments seemed to have understood the president's message. "Both will hopefully reflect on the moral that Uncle Sam should not just be regarded as a cornucopia of goodies, regardless of what they do or say."[17]

Yet if yanking back the welcome mat got the attention of the Indians and Pakistanis, it had no measurable effect in disposing them to settle their differences. Evidence, in fact, pointed in just the opposite direction. In May, Indian and Pakistani forces skirmished in the Rann of Kutch, a salt marsh east of the Indus delta where the India-Pakistan frontier had never been carefully surveyed, for the sound reason that neither country considered the region worth the expense. Pakistan's American tanks pushed several miles into indisputably Indian territory before a British-sponsored cease-fire took hold.

The Johnson administration adopted an agnostically pacifist attitude toward the affair—the same attitude the United States had taken toward India-Pakistan quarrels from the start. "In all candor we find it difficult to judge the merits of the Pakistani and Indian positions on the Rann of Kutch," the president wrote Ayub shortly after the fighting started. "But there can be no question as to the terrible consequences of a war between your two countries." Johnson urged Ayub and Shastri to move as quickly as possible to settle the dispute before it escalated.[18]

Johnson threatened no sanctions, such as a cutoff of American aid, at this point, and the fighting ended before he felt he had to. But the Kutch affair, despite its relatively harmless outcome, soon proved merely to foretell further conflict. The Indians, humiliated in 1962 in battle with the Chinese and now humbled by the Pakistanis, ached for revenge. The Pakistanis, having long derided the war-making ability of the Indians and now more convinced than ever that the Indian army was a pushover, desired to have it out once and for all with their overbearing infidel neighbors. The Pakistanis may also have believed it made sense to fight before India got any more American weapons.

The shooting in the marshes of Kutch had hardly stopped when sniping began in the hills of Kashmir. Early in the summer of 1965 Indian troops crossed the truce line and seized three Pakistani outposts. The Pakistanis responded by infiltrating "volunteers" into the territory controlled by India—the tactic that had touched off the first Kashmir war in 1947. By the end of August, few observers expected a long wait until the area erupted.

During this period of growing tension Johnson decided to suspend new aid to both countries. The State Department and the NSC staff advised against a suspension, believing it would diminish American influence at a time when Washington needed all the influence it could get. But Johnson held firm. "I'm not for allocating or approving $1 now unless I have already signed and agreed," Johnson scribbled across the bottom of an aid authorization request from McGeorge Bundy. "If I have, show me when and where."[19]

The president ordered a comprehensive review of American assistance to South Asia. The review served two purposes. First, it facilitated a cost-benefit analysis of American policy toward the region. Congress always appreciated evidence that foreign aid was accomplishing worthwhile tasks. Second, it afforded a pretext for procrastination. With conditions daily growing more disturbed in the subcontinent, discretion dictated delay. Such, at any rate, was Johnson's judgment.

At the beginning of July, Robert McNamara sent Johnson a summary of the military aid program for South Asia. In the current fiscal year, the United States was providing India roughly $100 million in weapons, spare parts, and ammunition. Shipments to Pakistan amounted to slightly less than half this. Despite the two-to-one ratio in favor of India, a cutoff of American military assistance, should the president deem this desirable, would more seriously affect Pakistan. India's arsenal still contained large quantities of weaponry from other countries, including the Soviet Union, that continued to supply New Delhi. Pakistan had depended primarily on the United States for a decade, and recent outsourcing to China had done little to change the situation there. "U.S. logistics is the lifeblood of the Pakistan armed forces," McNamara told Johnson. Because the Pakistan government would find it impossible to locate alternative suppliers of ammunition and spare parts on short notice, cancellation of American shipments would have a "very serious impact" on Pakistan's ability to wage war. The defense secretary thought Ayub's small squadron of F-104s would be grounded within a few weeks after the onset of hostilities. A larger group of F-86s would last perhaps a few months, depending on how quickly Pakistan's mechanics resorted to cannibalization.[20]

Economic aid provided less fast-acting leverage than military assistance, but its long-term potential was probably greater. Komer reported that as of the middle of 1965 the United States had provided more than $3 billion to Pakistan exclusive of arms. The figure for India was more than $5 billion. In contrast to the situation with military aid, India was more susceptible to pressure on economic aid than Pakistan. While Pakistan had come to rely on American weapons, India had developed a continuing need for American wheat. Regarding both countries, however, a suspension of economic aid would have little effect in averting or ending a war. Fighting would start and stop before the pressure from a loss of American food and dollars took hold.[21]

Komer still believed the administration should approve new aid to both countries, but he recognized that Johnson would not accept a business-as-usual policy and he shaped his counsel accordingly. With respect to India he suggested measures designed to ensure that American economic aid contribute to rendering itself unnecessary. In particular the

Indian government should free up the private sector of the economy and place greater emphasis on boosting agricultural productivity. Komer's recommendations regarding Pakistan were less economic than diplomatic and political. "Our immediate problem is how to keep Pakistan from leaning so far toward Peiping that Congress cuts off U.S. aid," Komer told the president. The administration must somehow "convince Ayub that he can't have his cake and eat it too." Doing so would not be easy, but Pakistan's desire for American help would incline Ayub to listen to American complaints about Pakistan's foreign policy.[22]

Johnson continued to block new aid for India and Pakistan on the last day of August, when the CIA reported that the Pakistan military had had enough of Indian incursions across the truce line in Kashmir and would strike back in force within hours. The prediction proved out the next morning, as Ayub's armor rolled into action against Indian positions.[23]

The White House responded deliberately at first. Komer, relaying to the president the CIA prediction of the Pakistani attack, conceded that the move would trigger "a critical Pak/Indian crisis." But the crisis would be "still one big step short of a Pak/Indian war." Komer added, "There's a case for sitting back a while longer and letting both Paks and Indians face up to the awesome risks involved. These might make both more malleable vis-a-vis us." Komer suggested that Dean Rusk call in Ayub's ambassador and tell him that the president knew what the Pakistanis were up to and would hold them to account. But beyond this, Komer thought the United States should confine its official interest in the affair to supporting a United Nations or perhaps a British peace initiative.[24]

Unfortunately for this approach, the Pakistanis made a low American profile impossible by using American-made and American-financed tanks in their incursion across the Kashmir line and by providing air cover with American jets. New Delhi, which brought some of its partially American-equipped units to the front but did not immediately commit them to battle, naturally cried foul. The Indian ambassador, B. K. Nehru, reminded Rusk of Washington's oft-reiterated pledges that American military aid to Pakistan would not be turned against India. The United States must take measures to make Pakistan stop at once.

Chester Bowles heard the same demands in New Delhi. Bowles had opposed American military aid to Pakistan during the early 1950s, which was one reason Eisenhower and John Foster Dulles sacked him. Bowles now urged Johnson to announce an immediate cutoff of military supplies to Pakistan. The president might demonstrate nonfavoritism by warning India against misusing American aid and threatening a similar suspension. Because Bowles judged Pakistan the aggressor, and was pro-Indian anyway, he was pleased to know that even if the president embargoed arms to both side the Pakistanis would hurt more and first.[25]

Johnson rejected Bowles's entreaty. The president directed the State Department to concentrate its efforts on the United Nations, where the Security Council was calling for a cease-fire. Rusk, informing Bowles of the rejection, explained the president's reasoning. "Given the existing strains on our relations with both parties," Rusk said, "we do not believe such further action as threats to suspend military aid along the lines you suggest are likely to halt fighting at this time." Not to leave Bowles totally without recourse, Rusk indicated that the ambassador might inform the Indian government that the administration was prepared to rethink its aid policy "when responses to the UN approaches emerge."[26]

Part of Johnson's equanimity resulted from an expectation that the fighting would not go much beyond the Kutch scrap of the previous spring. On September 2 the CIA sent the White House a cable containing information from a Pakistani source to the effect that Pakistan's objectives were limited to teaching India a lesson not to tamper with the truce line in Kashmir. Once India was chastened, Pakistan would halt.[27]

India did not take to the teaching easily. As Pakistan's forces threatened to cut Delhi's supply line to Kashmir, India retaliated by crossing the international boundary into Pakistan proper. With Indian tanks approaching Lahore, the official in charge of the American desk in the Indian foreign ministry conveyed his government's resolve to the American embassy. "Either we fall or they fall," he said. "It's come to that."[28]

Actually it hadn't. Nor would it. But the Johnson administration feared it might. The most alarming scenario, from the American—and Pakistani—perspective, involved an Indian decision to widen the fighting from the Kashmir-Punjab area to East Pakistan. Such a move would threaten the integrity and very existence of the Pakistan state. It might also force China to come to the aid of Pakistan, either by invitation or by inability to resist the urge to hit the Indians when India was otherwise engaged. Where that would leave the United States, still formally tied to Pakistan, no one in Washington could tell. None wished to find out.

Various noises from China added to the American alarm. Beijing's foreign minister Chen Yi traveled to Karachi to pledge his country's firm support of Pakistan. The American National Security Agency, the top-secret bureau charged with listening to all the world's radios, reported an unusual number of encrypted, high-priority messages being transmitted from Chinese stations close to the Indian border. The CIA noted indications that Chinese troops in Tibet and Xinjiang had gone on alert.[29]

The prospect of Chinese intervention forced the Johnson administration to take a larger role in trying to stop the fighting. The president wrote directly to Ayub and Shastri, advising them most energetically to heed the United Nations call for a cease-fire. When this produced no favorable response, he ordered Rusk to announce that all deliveries of American weapons to both sides, whether supplied under agreement with the American government or purchased commercially, were being halted immediately.[30]

Johnson appreciated the likely consequences of his decision. The Indians resented the implication that they shared equally in responsibility for a war Pakistan had started. (Pakistan, of course, had a different view of the war's origins.) The Indians' resentment as well as their instinct for survival would push them to greater reliance on the Soviet Union, which would be delighted to further bolster an enemy of China. And sure enough, not a week had passed after the American cutoff before the CIA relayed the news that Delhi had requested expedited delivery of Soviet tank ammunition and three squadrons of MiG-21s.[31]

The impact of the cutoff on Pakistan was expected to be even greater. American estimates gave the Pakistanis no more than three weeks until their tank engines needed rebuilding and their planes blew out their landing-gear tires. Absent American resupply, their army and air force would have to stop fighting. In that time they might be persuaded to seek peace, but they might also take desperate steps to break out of their predicament, such as requesting Chinese intervention. They would definitely be driven closer to Beijing, and they might formally repudiate their alliances with the United States.[32]

One reaction Johnson did *not* expect, but which occurred, was an appeal by Pakistan to fellow CENTO members Iran and Turkey for military aid. The Iranians and Turks shared a common devotion to Islam with Pakistan, of course. More to the point, they shared the Pentagon's military specifications. If Washington would not rearm Pakistan, perhaps Teheran and Ankara, which possessed large stores of similar and sometimes identical equipment, would.

As soon as the Johnson administration got wind of Ayub's maneuver it quietly but very clearly told Iran and Turkey to stay out of the matter. Should the two countries undertake to transfer any materials supplied under agreement with the American government, they could forget about getting more.[33]

In slamming CENTO's side door in Ayub's face, the Johnson administration recognized that it could be knocking the whole house down. The CIA guessed that Ayub might respond by pulling Pakistan out of CENTO. The agency described a withdrawal by the Pakistan government as "one way of venting its frustrations stemming from the war with India." A memo prepared for the president added, "Such a move would just about finish the weak CENTO organization." The member countries likely would look to bilateral rather than regional arrangements for defense and mutual support, and they would probably avoid anti-Soviet military pacts. American activities in the area would certainly suffer.[34]

Beyond denying military equipment to Pakistan and India, the Johnson administration warned both sides about the dangers inherent in continued hostilities. George Ball lectured the minister of Pakistan's embassy in Washington on "the momentous problem of choice" Pakistan faced. Pakistan must resist the temptation to bring China into its conflict with India. Indeed, Pakistan must actively discourage Chinese intervention. If the Chinese did attack India, even without Pakistan's invitation, most Americans would consider the attack the logical result of the growing closeness between Pakistan and China during the last few years. It went without saying that the American Congress would never countenance aid, military or otherwise, to a country that fought side by side with the Chinese.[35]

While telling Pakistan to keep China out of the war, American officials told India to reflect on what would happen if China came in. India's assault on Lahore had put the Pakistanis on the defensive, but Chinese intervention would reverse the situation at once. Rusk directed Bowles to tell Shastri that in order to prevent this the Indian government must seek a cessation of the fighting. "Continuation of the conflict is likely to plunge India more deeply into the cross currents of the cold war and internal Communist bloc conflicts," Rusk wrote. "The Chinese Communists will be certain winners. It is difficult to see how either India or Pakistan could benefit regardless of the outcome." India must heed the appeal of the United Nations for a cease-fire and withdrawal behind frontiers. To do otherwise risked "sheer disaster."[36]

Disaster drew nearer on September 11 when Beijing demanded that India demolish certain fortifications the Indians had erected on the Sikkim-Tibet border. The demand had no strategic connection to the war between India and Pakistan. The Chinese simply wanted to cause trouble. When India ignored the demand, Beijing stiffened it by setting a deadline. New Delhi had until September 19 to remove the offending posts.

The Indians hinted that they would appreciate an American warning to China to back off, but Johnson declined to respond. The president had not the slightest desire to involve the United States in India's problems with China. Besides, smart money guessed that the Chinese were bluffing: that unless conditions changed drastically Beijing would content itself with psychological warfare. China appeared more intent on keeping India off balance than on delivering a knockout blow.

The smart money guessed right. When India continued to ignore the Chinese demand, Beijing simply announced that the installations had been demolished (the region was off limits to would-be verifiers) and declared victory.

Meanwhile reports filtered in to Washington that Pakistan had decided it had made its point and considered further fighting undesirable. The Indians agreed in principle, although like the Pakistanis they wanted their opponents to seek peace publicly first. An Alphonse-Gaston routine occupied a few days. The killing continued, but at a reduced pace. Eventually, on September 23, with Indian and Pakistan forces each still occupying opposition territory, a cease-fire took effect.[37]

V

The Johnson administration's handling of the India-Pakistan war fell short of success, if success is defined as maintaining or improving American prospects in a given part of the world, in this case South Asia. Indians were outraged at the equation, implicit in the embargo of American weaponry to both sides, of their actions with those of Pakistan. More damningly, many Indians charged the United States with responsibility for the deaths of Indians killed by Pakistani-wielded American weapons, arguing that Washington had willfully neglected to restrain Pakistan from abusing American aid. Bowles remarked later that "even the most pro-American Indians went on an emotional binge" against the United States.[38]

In Pakistan the fallout was more poisonous still. Ayub and his associates lost no opportunity to pin blame for Pakistan's frustrations on Washington, which they denounced for failure to support an ally. The government-controlled press castigated the Americans as "false friends" and agents of Hindu imperialism. Crowds assaulted American nationals and ransacked buildings housing operations of the United States government.[39]

Yet if the Johnson administration incurred the wrath of both parties to the war, it was difficult to see how the administration might have handled the situation better. An earlier cutoff of weapons would have appeased no one. Longer delay would have made matters worse. A one-sided embargo against Pakistan would have gratified India while totally alienating Pakistan, and it would have done nothing to refute criticism that the real crime against India was sending arms to an irresponsible and aggressive regime for a decade before the war. A one-sided embargo against India was unthinkable in light of Pakistan's responsibility for starting the serious shooting, and it would not have accomplished anything to halt the conflict.

The outcome for the United States was not unrelievedly bleak. One observer describing the war commented that the conflict "came at the right time for the U.S. in the way that some calamities are welcomed by debtors and bigamists." The pro-India policy Kennedy had grafted onto Eisenhower's pro-Pakistan approach had impossibly overcommitted America in South Asia. The 1965 war simply revealed what had been implicit in the American position in South Asia for some time. And in revealing the contradiction of trying to win the favor of both sides in an abiding and murderous struggle, the war made unmistakable the need for a change. In commenting on the situation facing the Johnson administration at the end of the war, Bowles stretched the time frame to suit his preference for India, but his essential point was well taken: "For ten years we have been travelling down a dead-end street in South Asia. Now we have an opportunity—perhaps our last opportunity—to make a fresh start." Yet a fresh start wouldn't be easy, for in certain respects the situation in South Asia was more complex than ever. As William Bundy only half-facetiously remarked to George Ball, "It comes down to the old saying, 'The road to Taipei lies through Rawalpindi.'"[40]

Johnson's decision the previous spring to cancel the Ayub and Shastri visits and to tighten up on American aid indicated that the president had already started on a new approach. In some ways the war merely expedited what Johnson probably would have done anyway. He desired to revamp American aid policy toward India and Pakistan. He was determined that the Indians and Pakistanis understand that aid was not an entitlement but a conditional gift. The most convincing method to convey this message was to deny aid for a while. The war provided an excuse for the denial, at least with regard to military aid.

Johnson went beyond halting military aid, as part of his effort to put the fear of Congress into America's South Asian beneficiaries. About the time he refused to see Ayub and Shastri the president placed the principal American food assistance program, labeled PL-480 after the 1954 congressional act that created it, on a short-term basis. Like many other government programs, PL-480 had a convoluted rationale and a complicated constituency. In part it was designed to feed hungry foreigners. In equal measure it was intended to keep American farmers prosperous and happy, to provide business for American railroads and truck lines and shipowners, and to provide jobs for teamsters, longshoremen, and members of the American merchant marine. The operation of the program was simple enough. The government purchased wheat, corn, milk powder, and other items from farmers and commodity brokers, then sent the goods to needy countries overseas. During the decade of its existence PL-480 had generated powerful support for foreign aid among a sizable array of interest groups, more than a few of which had no idea where the aid was going and didn't much care.

Working through the PL-480 program had particular attraction for Johnson. Though the president liked and trusted Dean Rusk, he never lost his suspicions of most of the State Department, which he judged full of elitist professionals and Kennedy holdovers of dubious loyalty to his administration. PL-480 had the signal advantage of being administered by the Agriculture Department—bureaucratic territory Johnson felt much more comfortable in. Agriculture Secretary Orville Freeman, while a Kennedy choice, was reliable and politically safe. Freeman didn't frequent Georgetown cocktail parties, and when he did show up he didn't leak compromising information.

Further, the terms of the PL-480 program required the president to make periodic judgments regarding the progress of recipients in lessening their dependence on donated food. Countries like India submitted blueprints for raising agricultural productivity. If the blueprints failed to pass muster or if the countries' actions failed to match the blueprints, the program's enabling legislation mandated a suspension. Because the president was the one judging success and failure, he possessed enormous power over this key area of foreign policy. Whatever Johnson knew about the commodities PL-480 dealt in—quite a lot, in fact—the one commodity he understood best was power.

During the summer of 1965 Johnson began refusing to approve extensions of current PL-480 agreements for more than two months at a time. With good reason, the president felt that India was not living up to its side of the bargain. In 1954 when the PL-480 program had begun, India imported 1 million tons of grain. Ten years later imports had risen to 5.5 million tons. Growth in population accounted for some of the increase in imports, but more important was a conscious policy pursued by the Nehru government of emphasizing industrial development at the expense of agricultural development. To provide a ready labor force for Indian factories, the Indian government took measures to keep food prices low. Low prices discouraged domestic production, as did government direction of resources away from the manufacture of chemical fertilizer, pesticides, and farm equipment. Acceptance of foreign food aid also depressed prices and production.

The figures on American aid were discouraging—if not to the American farmers and others PL-480 subsidized, at least to the taxpayers who funded the program. Rather than render India independent of American aid, PL-480 was doing just the opposite. A 1960 agreement had pledged the United States to provide India sixteen million tons of grain over four years. In the spring of 1965 India requested a new agreement for another fourteen million tons, spread over just two years.

At this point Johnson said no. He said no again when the Agency for International Development submitted a scaled-back proposal, one specifying six million tons for one year. In denying this second request, the president inaugurated what came to be called his "short-tether" policy. Rather than committing the United States to four years or two years or even one, he authorized the shipment of just two months' worth of grain—one million tons. Later in the summer of 1965 and through the fall he cut back to one month at a time.[41]

A veteran of the foreign-aid wars of the 1950s, Johnson understood the skepticism with which many members of Congress viewed assistance to countries like India—countries that, while refusing to ally with the United States or even support American diplomatic initiatives, appeared to expect something for nothing. As the expectations grew larger, so did the skepticism. In 1965 Johnson's domestic reform programs were claiming an unprecedented portion of the American budget. Vietnam was

eating up much of the rest. Congress was far from eager to pour money into India, and Johnson preferred to avoid a fight on the issue.

In addition the president wanted to shake India into recognizing that it would have to make faster progress toward self-sufficiency if it hoped to receive more American help. Such progress would lessen opposition to aid in Congress, beyond providing obvious benefits to the Indian people. Shastri's government had shown promising signs of realizing that Nehru's centrally directed emphasis on industry would have to be changed. Johnson wanted to ensure that reform efforts not flag.

The September war knocked plans for reform backward several paces. National security naturally came before rectification of the economy. Moreover, with Indian passions against the United States inflamed by the fighting and the events surrounding it, Shastri could not easily implement measures that had the appearance of being dictated by Washington, however intrinsically beneficial he thought them to be.

Some American officials believed that the growing anti-American sentiment in India required a loosening of the short tether on PL-480 aid. McGeorge Bundy noted that the fighting with Pakistan had touched off severe communal violence between Hindus and Muslims in India. Should food run low as a result of the war and of the stringent American policy, bread riots and general panic might ensue. The United States almost certainly would receive blame for deaths thus suffered at Indian hands, to go along with the blame it was already receiving for deaths attributable to American bullets fired by Pakistanis.[42]

Robert Komer argued for increasing the amount doled out to India from a one-month's supply to two months'. Aside from easing India's problems, such an increase would save the administration considerable work. A sixty-day approval schedule involved only about half the red tape a thirty-day schedule did. "The Indians know we're jockeying them," Komer told the president. "A two month extension still leaves Shastri in no doubt that he's on a short tether."[43]

Johnson stuck to one month. He did not want the food to stop, but he wanted to make the Indians think it might. They did. "Why are you starving us?" asked B. K. Nehru of Komer just before the war ended. The Indian ambassador exaggerated. India was far from starving. Yet the question why was apt, and the answer was the same as before. Johnson wished Shastri to recognize the importance the American government placed on the Indians' efforts to get themselves off the American welfare rolls.[44]

In the wake of the renewal of fighting over Kashmir, the president also thought motion toward a resolution of that draining dispute was in order. The arms race between India and Pakistan was costing the United States lots of money. Moreover, what India itself spent on weapons it could not spend on fertilizer, agricultural training programs, and the like.

Officially the administration did not link food aid to a settlement on Kashmir. Johnson understood that food was a political weapon that could as easily detonate in the hand of the user as against the intended target. It should be employed, if at all, only where it promised a reasonable chance of success. In the aftermath of the September war the chances grew increasingly unreasonable. The White House was forced to conclude that a problem that had lasted a generation and caused two wars—not counting the Hindu-Muslim strife that reached back centuries—would not soon yield to threats from the United States. Besides, an agreement could be no stronger than the governments that concluded it. Under current circumstances Shastri's government not could make the compromises necessary to a settlement without guaranteeing its own demise. Bundy and Komer summarized the administration's view at the beginning of October 1965. "We should not kid ourselves about any early Kashmir settlement," they said. "Any American fidgeting over Kashmir will only make us trouble with India and arouse false hopes in Pakistan."[45]

Yet Johnson could not resist a little push—knowing that Congress would find aid for India much easier to accept if the prospects for regional stability improved. The president made precisely this point in a meeting with India's agricultural minister Chidambara Subramaniam. Johnson said he was having great difficulty with Congress. The legislature was very tight with money for foreign aid. India must recognize this fact. As the American president, he did not presume to dictate Indian policies, but as a friend of India he thought the Indian government should know what Congress would look for in considering aid. First would be improvement in agricultural productivity. Second would be assurance that other countries would contribute as well. In this area India might start working on potential donors. Third would be progress toward a Kashmir settlement and a comprehensive peace with Pakistan.[46]

VI

Before long the administration realized it would do well to get two out of three. Though the cease-fire of September 23 theoretically stopped the shooting, in practice violations occurred so frequently that many observers wondered why either side had bothered accepting the truce. Thoughts of a Kashmir settlement evaporated, and optimists hoped merely for a nonresumption of regular war between India and Pakistan.

In this unpromising atmosphere, Washington sought to reconstruct relations with the two combatants. For once, for the moment at least, America's two-timing worked to its advantage. India and Pakistan each wanted to get back into the habit of receiving American aid, and each worried that the other would gain an edge in doing so. As a consequence Ayub and Shastri each hinted broadly at a desire to visit the United States in the autumn. In mid-October, Komer reported with satisfaction, "Both our fish are nibbling furiously."[47]

The administration landed Ayub first. Or perhaps he landed the administration. American officials had long respected the Pakistan leadership, always for its fighting spirit if not always for its judgment. Eisenhower had tilted to Pakistan in 1954 not least because John Foster Dulles and Vice President Nixon returned from visits to Karachi thoroughly taken with the martial virtues of Pakistan's government and people. The pacifism of Nehru and the Indians paled by comparison. Johnson toured Pakistan on his 1961 circuit of Asia and came away similarly respectful. Johnson described Ayub to Kennedy as "a singularly most impressive head of state." Johnson continued, "He is seasoned as a leader; confident and straightforward. He is frank about his belief, offensive as it is to us, that the forms of representative government would only open his country to Communist take-over at this time." Johnson remarked that Pakistan knew its value to the United States—or for that matter to the Soviet Union or China. "Ayub is wisely aware of Pakistan's strategic position," the vice president said.[48]

Ayub's strategic wisdom underlay his attempts to balance China against the United States. The balancing continued past the 1965 war with India. American officials hoped the cutoff of American arms and new American economic aid would draw Ayub and his associates back toward the United States. Yet the administration recognized that such pressure might have the opposite effect. Komer commented, "It's by no means outside the realm of possibility that an irredentist, inflamed Pakistan could slip away from its Western friends and join the Chicoms and Indonesians in further squeeze plays on India and the West." Komer added, "It's even possible that Ayub could be replaced by a less sensible leader. It's amply clear by now (from Peiping's attempt to meddle in the recent fracas) that Mao sees Pakistan as a very useful lever against India."[49]

Ayub kept the administration guessing which direction he would move during the months after the September war. Bundy reported to Johnson at the beginning of October: "Ayub's attitude toward us seems to have hardened; we're not quite sure why." Indications a week later of Ayub's interest in a trip to Washington reassured American officials. Indeed, Komer began sounding overconfident. Ayub, Komer said, had "no place else to go." The administration might even "let him sweat awhile." At October's end Komer told Johnson, "There are many signs that Ayub is coming our way."[50]

Ayub came, at least as far as Washington. Cordiality and cooperation marked his public appearances with the president, who imaginatively described the United States and Pakistan as being quite similar. Each country, Johnson explained, had commenced its national existence as an experiment few thought would succeed, and each had proved the skeptics wrong. Slightly less incongruously the president likened Ayub to himself personally. "President Ayub is a rancher as I am," Johnson told a gathering at one reception. "His home district is a country very much like Johnson City, Blanco County, where I live."[51]

Less warmth characterized the private conversations. The joint communiqué of the discussions classed them as "frank," which meant Johnson and Ayub agreed on nothing of substance. They differed over Johnson's refusal to take Pakistan's part on Kashmir and against India generally, and over Ayub's refusal to distance Pakistan from Beijing. Johnson complained that when the United States had needed Pakistan, Ayub "was in Peking or Moscow." Komer read back the positions of the two sides in a note to the president: "Ayub has used all his charm to convince us that if only we get Kashmir arbitration and cut back Indian arms all would be rosy. You in turn have told him that we admire him but that we can't get in bed with China." Johnson did agree to send a team of medical and scientific advisers to Pakistan, but he refused to resume military aid and he was sparing with economic assistance. Ayub talked about his desire for good relations with the United States, but he made no commitment to compromise on Kashmir or to cool his relations with China.[52]

This outcome, though disappointing, was not entirely unforeseen. While Ayub had submitted himself to the voters earlier in the year and been reelected, the dissatisfaction that would drive him from office a few years later was growing. His foreign minister Bhutto alternately declaimed and conspired for a policy leading Pakistan closer to China and farther from the United States. Johnson's treatment of Pakistan during and after the September war strengthened Bhutto and made concessions from Ayub exceedingly difficult. American officials understood the pressures facing Ayub, although they hoped he could overcome them and guessed he might. Their hope was not unreasonable, but their guess was wrong.

After Ayub left Washington the Johnson administration anticipated better luck with Shastri. If the Indian prime minister confronted political constraints no less confining than those vexing Ayub, Shastri's greater need for American help, especially in food, appeared to make him more willing to cooperate. "Shastri has to tread the difficult line between looking politically independent and knowing he is politically dependent," Komer wrote Johnson." The always hopeful Chester Bowles spied a silent majority Shastri might tap in turning toward Washington. "There is every evidence that thoughtful Indians in and out of government are deeply conscious that many criticisms of the U.S. during the past months have been grossly unfair," the ambassador asserted, "and there is evidence on every side of a desire to bring the situation back into balance."[53]

Before Shastri traveled to America, however, he had to journey to the

Soviet Union, to the Uzbek capital of Tashkent. There in January 1966 he and Ayub negotiated a formal end to the 1965 war. The Tashkent accord produced a withdrawal of forces behind international frontiers and behind the Kashmir truce line, but it accomplished nothing toward a settlement of the underlying Kashmir quarrel and hardly more toward amity between the two quarrelers.

For all that the Tashkent conference took place under the Kremlin's auspices and implicitly acknowledged the leading position of the Soviet Union among outside powers in South Asian affairs, the Johnson administration acted the good sport about it. The prospect of a Russian role in brokering an end to the war had hardly overjoyed American officials, but better the Soviets than no one, and should the negotiations fail, as evidence indicated they well might, better the failure be Moscow's than Washington's. George Ball captured the administration's mood when he remarked, "If the Russians can settle it, all the good. Let them try their hand. Let them break their lance."[54]

When Tashkent succeeded, the administration predictably credited India and Pakistan rather than the Soviet Union with the success. Johnson sent a congratulatory cable to Ayub. "I have greatly admired what you and Prime Minister Shastri did at Tashkent in the cause of peace," the president said.[55]

Johnson would have sent Shastri a similar message but the Indian prime minister suddenly died of a heart attack hours after signing the Tashkent accord. India mourned the loss of an honorable man who had done his best in the impossible task of following the founder Nehru. Then India turned to the founder's daughter, Indira Gandhi, who despite holding only a junior position as Shastri's minister of information had earned a reputation, in one cheeky commentator's words, as "the only man in a Cabinet of old women."[56]

Gradually Gandhi would emerge as one of the most formidable world leaders of her day, defying the great powers as well as the Indian constitution. But during the Johnson years she consolidated her authority carefully, giving the impression—to Washington at least—of being relatively complaisant. To be sure, Johnson had her at a disadvantage. She was new in office while he was thoroughly settled in. She headed a poor country while he led the richest country on earth. She needed foreign aid and he had it to give, or withhold.

India's need increased significantly just as Gandhi assumed the premiership. Ever since partition of the subcontinent in 1947, which had handed much of the subcontinent's most productive land to Pakistan, India had faced the annual possibility of famine. In good years the country's harvest, supplemented by imports from America and elsewhere, kept the wolf at bay. In bad years—which didn't have to be very bad in terms of weather to be very bad in human terms—either imports rose dramatically or millions went hungry. The Nehru government's policy of concentrating on industrial development exacerbated the problem.

The year of Gandhi's accession threatened to be a bad one. The 1965 monsoon had failed in much of India's prime grain country, resulting in a dismal harvest. The American Agriculture Department, monitoring the situation, calculated a decrease of twelve million tons in primary grain crops from the previous year's eighty-seven million tons. India's population had grown rapidly as production was falling, stretching supplies even thinner. In December 1965 India's granaries contained a supply equal to less than one month's ordinary demand.[57]

The looming famine gave the United States extraordinary influence over the actions of the Indian government. At the beginning of 1966, even as the Soviet Union hosted Shastri and Ayub at Tashkent, Orville Freeman reported to Johnson that India would have to import ten to fifteen million tons of grain during the coming year to avoid mass starvation. The agriculture secretary added that the United States had the supplies available to relieve the bulk of the deficit. No other country did.[58]

As India's supplies ran down, Johnson's PL-480 short tether tightened almost to the choking point. The president recognized the fact and intended to turn it to use. In February 1966 he summoned Indian ambassador Nehru, who had been asking repeatedly when India might expect a decision on enough aid so that the Indian government could plan rationally for the future. The one-month approach was playing havoc on Indian efforts to distribute food fairly and efficiently. Johnson, as before, took refuge behind the recalcitrance of Congress. Exaggerating for effect, the president said he faced an "incipient revolt" in the legislature over foreign assistance. He told Nehru he could not go to Capitol Hill with an aid package before talking to Prime Minister Gandhi, scheduled to arrive in Washington in March. When he did present a measure it would stand a far greater chance of passing if American aid were folded into a multilateral package including contributions from other countries. Nehru responded that the United States was the only country that possessed the surplus food India needed. The ambassador added that because it was surplus it would cost the American government next to nothing. Johnson corrected him quickly, reminding that the American people had to pay for "every nickel's worth of wheat" India received. All these nickels were what was upsetting Congress. Johnson cited William Fulbright as an example of a congressional leader who made trouble for the administration. Fulbright, Johnson said, was berating him as president for being "ostentatious and dictatorial" in not sufficiently consulting the legislature. Should the administration not prepare Congress carefully before asking for more money for India, the administration risked a reaction that would leave India worse off than currently.

Nehru listened patiently, then stated that regardless of the political situation in America, the people of India needed five million tons of grain to get them through June. Otherwise speculation and hoarding, already a problem, would grow unmanageable. Johnson, interpreting Nehru's statement as a bargaining position, which to some extent it proved to be, said he could not provide such a large amount without congressional approval. At the moment Congress had its hands full with other business. Johnson said he might be able to scrape together some extra grain on an interim basis. He directed Nehru to meet with Komer and work out a sensible program. But he could not go beyond the "utter minimum."[59]

Two days later Johnson signed off on an additional one-month's supply, but he held the line there. Johnson's advisers thought he was pushing the Indians too far. Komer reiterated that while multilateral aid sounded good in theory, in practice only the United States had grain in sufficient quantity to meet India's needs. Toward the five million tons required—Komer accepted Nehru's figure—it would be "terribly difficult" to get more than two million from other sources. Canada had some surplus, but this was inaccessible since most Canadian ports were frozen and would remain so for months.[60]

During the next few weeks Komer, who at this time was the acting national security adviser, worked up a plan by which the administration would agree to continue food aid at the six-million-ton level of the previous year and provide an additional two and a half million tons on a matching basis with other countries. If other countries lacked grain, which most did, they could pay cash and the United States would convert it into food. The result would land eleven million tons in India during 1966, toward the low end of the ten-to-fifteen-million-ton range Orville Freeman had specified as necessary to avert famine.[61]

Johnson, to the distress of the White House staff and the State Department, rejected the proposal. To their greater distress he announced he would continue to deny anything more than stopgap requests until he could place the entire issue before Congress. When that would be, he did not say.[62]

At the end of March, Gandhi arrived for the state visit Shastri's death had delayed. Johnson exhibited his customary courtesy. He expressed a keen interest in the prime minister's sons, including the future prime minister Rajiv. After one meeting he personally walked Mrs. Gandhi to her quarters in Blair House. He violated protocol in the interests of friendliness by attending a reception at the Indian embassy and staying unexpectedly for dinner.

In discussions with Gandhi the president reiterated the points he had made to B. K. Nehru. He explained, again overemphasizing, his difficulties with Congress. Again he underlined the need for India to show good faith by efforts to improve agricultural productivity. Such efforts, he said, would go far toward convincing the legislature of the soundness of aid as an investment in India's future. Congress would insist on a multilateral approach to aid. Consequently India should work closely with the World Bank.

Gandhi replied that she had political problems too. At the end of 1965 India's agricultural minister Subramaniam had worked out an agreement with Orville Freeman specifying the changes India would make in shifting from an industrially oriented economic policy to one stressing agriculture. In exchange for the agreement Johnson had released a larger than usual supply of grain. Gandhi appreciated the grain and said she believed the Subramaniam-Freeman plan was prudent. But many in the Indian parliament felt that concentrating on agriculture would consign India to perpetual underdevelopment, and the fact that the plan bore the USDA seal of approval made it that much harder to sell. Already many on the Indian left had branded her government a tool of the United States and the World Bank. She could move ahead only with caution and at political cost.

Even so, the costs of not moving ahead—and not receiving American aid, though she did not put the issue to Johnson in such terms—were larger. "Asia is in an explosive state," Gandhi said. "Now that independence has been gained, people have come to expect more than the past has offered. New horizons have been opened up which are still beyond their reach. They are impatient for change."[63]

Johnson was impatient too, and he thought the changes he had in mind were not incompatible with, and indeed were necessary to, the needs of the millions of India's hungry. A few weeks earlier the president had declared a global "war on hunger" but left the details of the campaign imprecise. Now he filled them in, especially as they related to India. In a message to Congress at the culmination of Gandhi's visit, Johnson congratulated India on what it had accomplished thus far, and encouraged it to make further progress. He explained that in 1965 the United States had supplied India six million tons of wheat, an amount equal to two-fifths of domestic American consumption. In 1966, he announced, the American government would furnish six and a half million tons, along with large amounts of corn, cooking oil, and milk powder. The United States would provide an additional three and a half million tons of grain subject to matching contributions from other countries. Moreover, the United States would send teams of agricultural technicians to India to assist Indians in raising crop yields.

America had no desire to meddle in the internal affairs of India, the president said, but he emphasized that the American aid must be put to most effective use. "The record of the last fifteen years is sufficient proof that we ask only for results." Johnson reemphasized results when he spoke about continued American support. "We are naturally concerned with results—with insuring that our aid be used in the context of strong and energetic policies calculated to produce the most rapid possible

economic development." With cooperation from the Indian government and the Indian people, as well as from other countries, the United
States would help eradicate hunger from the planet. "Can it be said that
man, who can travel into space and explore the stars, cannot feed his
own?"[64]

Between his conversations with Gandhi and his statement to Congress, Johnson calculated that he had gotten his message across. The
president assessed the situation in a letter to British prime minister Wilson. He described his talks with Gandhi and said, "We made no hard
and fast decisions, and it will probably take several weeks to know exactly
where we go from here. However, I am confident that she understands
the nature of the economic changes we would like her government to
work out." Should Gandhi convince the World Bank and the principal
member governments—of which Wilson's was one, this being the reason
for Johnson's letter—that India was progressing steadily with efforts to
make India self-sufficient in food, the United States would support a
major aid program. As a hedge and a reminder of the critical audience
India had to satisfy, Johnson appended to his promise the qualifier:
"subject of course to Congressional appropriations."[65]

VII

For a time administration officials thought they might have gone too far
in assuaging Indian worries. At the beginning of April, Dean Rusk wrote
Bowles in New Delhi of his concern that "euphoria" not set in after
Gandhi's visit. "It is crucial that this not happen," Rusk explained, "because while we do not intend to abandon the Indians, we mean what we
say about self-help. We want to bring the Indians back to earth gently
but quickly."[66]

Rusk needn't have worried. If there was any crashing to earth, it occurred in Washington. While the Johnson administration continued
publicly to disclaim a connection between American aid to India and
India's position on questions of international relations, administration
officials tended to think that discretion, if not necessarily gratitude,
would cause the Indian government to at least mute its differences with
the United States on sensitive issues. Gandhi evinced no gratitude and
hardly more discretion, or so it appeared to Johnson, when on a trip to
Moscow she joined the Kremlin in denouncing the American "imperialists" in Southeast Asia.

Of course, Gandhi had her reasons, among them a desire to maintain
India's standing in the nonaligned world and to fend off opposition
from Indian anti-Americans, including the comeback-bent Krishna
Menon. B. K. Nehru told Walt Rostow that Indian criticism of the United
States contained "strong elements of gross irrationality," but in India as

in the United States even the irrational had votes, and the government could not ignore their feelings.[67]

Moreover, the steps Gandhi was taking to satisfy the World Bank and potential aid donors were causing considerable pain in India. At the end of May, the prime minister had announced a devaluation of the rupee and a relaxation of import restrictions. The latter reform injured Indian manufacturers, an important government constituency whose members did not hesitate to declare their dissatisfaction. Devaluation hit nearly everyone, but it did particular harm to many millions living close to the poverty line.

Whatever Gandhi's reasons for attacking American policy in Indochina, Johnson was not pleased. Yet as in his dealings with de Gaulle, the president demurred from public criticism. Rostow explained the strategy when he said presidential criticism of Gandhi would only "make it harder for her to climb back off her limb." Chester Bowles wrote the prime minister what Rostow characterized as a "scorching letter," while the State Department expressed the administration's irritation to officials of the Indian embassy, but Johnson remained above these expressions of disappointment. The president went so far in the other direction as to write Gandhi a letter congratulating her for courage in installing needed economic reforms.[68]

Though Gandhi's attack may have delayed a decision by Johnson to push for India aid, of greater importance was the bad weather that afflicted the American wheat belt during the spring and summer of 1966. Predictions of a thin harvest drove wheat prices sharply upward. The price rise not only elevated the cost of an India-aid bill but also heightened the administration's sensitivity to inflation. This sensitivity translated into a reluctance to take any steps—such as sending large amounts of American grain to India—that might further raise grain prices.

In addition, the autumn of 1966 brought congressional elections. Foreign aid has never been a vote winner in the United States, except for special-case Israel, and at a time when incumbents had to explain how the country was going to pay for the Great Society and the Vietnam War, foreign aid looked like a vote loser. Gandhi's embrace of Soviet leaders in Moscow made it doubly so. All things considered, 1966 was a bad political year for India aid.

This was unfortunate for India, but not as unfortunate as a second straight failure of the monsoon. Instead of the 1966 harvest bringing an end to India's acute food shortage, it threatened to extend the shortage for another year. The country had managed to stave off starvation until now. Whether it could do so for another twelve months was by no means clear.

All of Johnson's important advisers urged him to release new aid to India. In August, Dean Rusk, Orville Freeman, and William Gaud, the director of the Agency for International Development, jointly recom-

mended approval of a shipment of two million tons of grain. Walt Ros-
tow, whose interest in Indian development had attracted him to the
Kennedy camp during the 1950s, argued strongly for approval of the
Rusk-Freeman-Gaud proposal.[69]

Johnson flatly refused. "We must hold on to all the wheat we can,"
the president said in a note to Rostow. "Send nothing unless we break
an iron bound agreement by not sending."[70]

The firmness of Johnson's refusal surprised his aides. Harold Saunders went so far as to advocate that the White House staff keep the
departments from learning of "the President's startlingly strong negative reaction." Saunders had spent weeks trying to smooth out differences between the various branches of the bureaucracy. At one point a
recommendation hung up while AID and the Department of Agriculture
fought over whose stationery to use. Saunders shuddered at what the
president's response would do to the aid process. "It would paralyze
State/AID and give USDA a roadblock to throw in front of every PL-480
program for the rest of the year," he said.[71]

Through the fall of 1966, as the food situation in India grew ever
tighter, Johnson's advisers struggled to change the president's mind.
Rostow proposed a stretched-out version of the Rusk-Freeman-Gaud
plan, contending that the Indian government was moving as quickly as
could be expected to implement the reforms the president had specified. "There have been some gaps in the Indians' performance," Rostow
conceded, "but overall they've made the right decisions." The national
security adviser analyzed price trends on the world wheat market and
declared that two million tons for India would have no more than a
negligible effect on the cost of a loaf of bread in American supermarkets.
He added that food prices in India had risen markedly, squeezing Indian
peasants and creating a dangerously volatile situation. Orville Freeman
relayed an eyewitness account from India that the situation there was
"genuinely critical." Freeman and Dean Rusk repeated their recommendation for a two-million-ton shipment, arguing that American
interests could not stand massive unrest in India. Chester Bowles
complained that the administration's policy was becoming "self-
defeating."[72]

Johnson's aides closely monitored the president's moods, trying to
catch him at a propitious moment. Saunders, passing a remodified proposal to Rostow, remarked, "This is our fourth try on Indian PL-480, so
I hope you can pick a time to send this upstairs when the President has
not been through the wringer on some aspect of domestic inflation."
After Johnson held the plan for several days without reply, Saunder's
associate Howard Wriggins wrote Rostow, "We're getting pretty close to
the line, and I am wondering whether we shouldn't probe gently upstairs
just to make sure your last memo isn't lost. In the event he has a basic
policy difficulty, we *must* talk about it promptly."[73]

Johnson continued to refuse. In September he jotted to Rostow: "Keep an eye on wheat prices and come back in a week." Rostow did, reporting stable prices and submitting another request just as the president was leaving town. "I am tempted to wait until I return," Johnson said. He did wait, but when he got back he was no more amenable than before. "I still think we ought to hold off."[74]

Part of the reason for the president's resistance to the aid requests was that he thought some of the reports out of India were unduly alarmist. Harold Saunders and Howard Wriggins, two of the strongest supporters of aid, conceded that Johnson had a point. If the pipeline ran dry for a short time, they said, no one was likely to starve. Over the centuries the Indian people had developed strategies for coping with shortages. "Not even the Indian government knows how much food is tucked away in the corners of that vast nation." The two aides did contend, however, that the administration's slowness on aid was damaging American credibility.[75]

Johnson's slowness was designed to pressure three groups: the Indian government, the governments of other potential donor countries, and Congress. Pressure on the Indian government would keep it on the path to economic reform. Pressure on other governments would dispose them to pitch in on the effort to prevent famine and generate Indian development. Pressure on Congress would produce the votes necessary to pass an aid bill without using up the president's political capital.

The pressure built with each passing week. By December 1966 Johnson's strategy was plainly yielding results. India continued its reforming ways, as evidenced in negotiations with the World Bank and in discussions with American officials. Gandhi appeared irreversibly committed to emphasizing agriculture and fostering an atmosphere conducive to foreign investment in projects beneficial to farmers and others, as well as to limiting population growth through family-planning programs. Multilateral support was growing for India aid. Canada had promised increased shipments of grain to India, and Australia was nearing a similar pledge. Administration officials expected to be able to persuade grainless countries to make cash contributions. Most important from Johnson's perspective, shortly after the November elections Congress appointed an investigating team to travel to India and deliver a firsthand report.[76]

Johnson had little doubt what the congressional investigators would say, but it was vital to his strategy that *they* deliver the news rather than accept the administration's word. As Johnson expected, Congressmen Robert Dole and W. Robert Poage and Senator Jack Miller described India's grave need for aid, and they endorsed India's capacity to put American assistance to good use. The three legislators recommended that the administration immediately release 1.8 million tons of grain and indicated that they would support a more comprehensive aid package once the new Congress convened.[77]

With this report the aid jam began to break up. Johnson, even now wary of being tagged a spendthrift on foreigners of questionable politics, approved nine hundred thousand tons to be shipped at once, in ships he had quietly directed Rostow to have ready. The rest would await congressional action. Johnson announced American participation in an India-aid consortium and consented to take the lead in setting it up. He appointed Eugene Rostow as special negotiator to the consortium, with instructions to exercise ingenuity in arranging contributions. As an example, the president agreed to let West Germany count purchases of American grain for India as part of its troop-offset payment to the United States. He asked George Woods, president of the World Bank, to make India a priority.[78]

At the beginning of February 1967, Johnson placed before Congress a comprehensive package of aid. To supplement grain already shipped, he authorized an immediate allocation of two million tons. Beyond this he requested the legislature to appropriate $190 million to purchase three million tons, to be provided on a matching basis with contributions, in kind or cash, from other countries. He requested an additional $25 million for supplementary relief supplies of various sorts.

The careful groundwork Johnson had done now paid off. His aid bill ran into minimal shenanigans in committee, although one cotton-state congressman attempted to attach a stipulation that India decrease the acreage planted to cotton, the better to grow grain—and prop up world cotton prices. The president had to modify the matching provision slightly, accepting less than a one-to-one ratio (which he had cannily refused to write into the bill) when other donors were slow to respond. But with the strong support of legislators Dole, Poage, and Miller, the measure moved swiftly to the floor of both the House and Senate, and passed easily. On April 1 Johnson signed it into law.

Although it had been a nail-biter for most of 1966, the India-aid story had a happy ending in 1967 and after. Famine was averted, despite the distraction of the June War in the Middle East, which closed the Suez Canal, trapping one luckless grain ship for years and diverting the rest of the food fleet. The Indian government's reforms, boosted by a good 1967 monsoon, launched India toward self-sufficiency in food, which it essentially achieved during the next decade.

Perhaps predictably, the happy ending did not extend to U.S.-Indian relations. While Johnson looked with not unjustified satisfaction on his role in encouraging the Indian government to adopt painful but needed measures to end India's dependence on foreign food, Indians remembered the pain and associated it with the United States. Gandhi wouldn't have been as successful a politician as she was if she hadn't allowed the president to bear the blame.

Six | # Bloody Good Luck

ON A VISIT TO Pakistan during a time when Washington was having more trouble than usual with Ayub, George Ball asked the general-president a question unrelated to his current mission. The question involved Pakistan's diplomacy, but not its diplomacy toward the United States or India or China or the Soviet Union. The question instead involved Pakistan's dealings with Indonesia. It had struck Ball, among others, that Indonesia and Pakistan had much in common. Indonesia and Pakistan comprised approximately one hundred million inhabitants each, placing the two countries among the largest in the world. Both Indonesia and Pakistan constitutionally embraced Islam, distinguishing them from the other countries with such large populations. Both Indonesia and Pakistan were non-Arab, separating them from most of the Islamic states. In light of these commonly distinctive characteristics, Ball wondered why Indonesia and Pakistan did not undertake common diplomatic initiatives.

"Do you really want to know, Mr. Secretary?" Ayub replied. Ball said he did. "The answer is very simple," Ayub continued. "Sukarno is such a shit."[1]

Lyndon Johnson could not have agreed more. Had Fidel Castro not been so much closer and so politically visible in Washington, Sukarno might have taken first place on the administration's bogey list. More

155

than most Third World leaders, the Indonesian president made a habit of provoking the United States. Anti-American diatribes formed a regular feature at Sukarno's political rallies, and if he exerted himself to curb violence against American offices and American property in Indonesia, he disguised the effort well. Although not a communist himself, he allowed the Indonesian Communist Party, or PKI, great and growing influence in Indonesian politics, and he spoke glowingly of Indonesia's close ties to Beijing.

To be sure, Sukarno had reason for disliking the United States. During the late 1940, American aid to the Netherlands had helped the Dutch fight the Indonesian nationalists Sukarno led. After independence in 1949 the Truman administration frowned on the neutralism Sukarno promoted as an alternative to the bloc-thinking of the Cold War. The Eisenhower administration carried the frowning further, underwriting a 1958 rebellion against Sukarno's government. As it would against Castro a short while later, the CIA organized a small army of antiregime dissidents who used American financial and logistical support to mount an uprising. Like the Bay of Pigs operation, the Sumatra rebellion failed, although less noisily. But the American hand was not hidden from Sukarno, who had a long memory for this sort of thing.

Kennedy adopted a conciliatory policy toward Sukarno, not unlike the approach he adopted toward Nehru and Nasser, two other leading neutralists. Kennedy brought Sukarno to Washington and augmented American aid, believing that generosity might succeed where bellicosity had failed. Sukarno took Kennedy's money but not his advice. In 1963 the Indonesian president declared psychological war on newly federated Malaysia, which he deemed a stooge for continuing Western imperialism. This *konfrontasi* threatened to undo the Malaysian federation and perhaps embroil Malaysia's allies Britain and Australia in a broader conflict. Such a conflict conceivably could have sucked in the United States, which was tied by numerous links to Britain and by the ANZUS pact to Australia.

The confrontation against Malaysia persisted undiminished when Johnson assumed the presidency. So did Sukarno's preference for the company of the Chinese communists. This preference annoyed Johnson, but it particularly provoked members of Congress who wondered why the United States was providing aid to such an egregious fellow traveler. Almost before Johnson figured out where the telephones in the Oval Office were, and well before he installed his special three-screen television—so he could keep track of what all the network news programs were saying about him—he discovered on his desk a congressional resolution barring new aid to Indonesia unless said aid served the American national interest, as determined by the president. This legislative maneuver allowed the approving lawmakers to claim credit for voting against unnecessary aid, while leaving Johnson to bear the blame for backing Sukarno, if the president decided in favor of aid, or for driving Sukarno even closer to Beijing, if he decided against.

Dean Rusk counseled the president to move carefully. The secretary summarized the background to present American policy toward Indonesia. Following Sukarno's declaration of the *konfrontasi*, Rusk told Johnson, the United States had warned Sukarno that a direct attack on Malaysia would trigger action by the United Nations, with the United States aligned against Indonesia. In addition Washington had cut back military aid and military training and had suspended negotiations regarding economic aid.

Pointing ahead, Rusk advised Johnson against further sanctions. Sanctions had not worked against Sukarno in the past, as the continuing confrontation with Malaysia demonstrated. The secretary did not think they would work in the future, and they might well move Sukarno to retaliate by expropriation or other methods. American oil companies were especially vulnerable, as were American offices in Indonesia. "In all probability Sukarno would seize the $500 million American oil properties, encourage Communist hoodlums to burn our Embassy, and break diplomatic relations," Rusk said. Sanctions might accomplish precisely the opposite of their aim, causing Sukarno to escalate his contest against Malaysia, leading to another big war in the area of Southeast Asia, which the United States definitely did not need. Rusk conceded that the dire consequences he was forecasting might materialize even in the absence of American sanctions. But if they did, the world should know where blame lay. "We should see that it is Sukarno that gets the full onus." Rusk advocated allowing economic assistance already appropriated to be delivered, and he recommended continuing as much aid as possible to the Indonesian military, the most likely source of eventual opposition to Sukarno.[2]

The White House staff seconded Rusk's recommendation. Michael Forrestal, McGeorge Bundy's East Asia specialist, thought it particularly important to maintain ties with the Indonesian army. The army's annoyance with Sukarno was no secret, although at present the generals saw no alternative to him. A cutoff of American military aid might force the army to rally to Sukarno's side against the pushy Americans. Robert Komer amplified Rusk's point about keeping Indonesia quiet while Vietnam was getting loud. "A showdown with Indonesia at a time when we want to step out more in Vietnam would be quite a complication," Komer declared.[3]

The CIA assessed Sukarno's prospects vis-à-vis the other contenders for power in Indonesia. The intelligence agency explained that though Sukarno lacked an institutional political base he could call his own, he demonstrated remarkable adroitness in playing against each other the two major sources of institutional power in Indonesia, the PKI and the army. The army tolerated Sukarno because army leaders judged him essential for maintaining national unity. The PKI relied on Sukarno to protect party members from the army. For the time being neither the army nor the PKI appeared inclined to challenge the status quo. The acquiescence of the two rival groups allowed Sukarno to exercise greater power than ever before. If anything seemed likely to bring Sukarno down, it was ill health. Rumors described a variety of maladies, some life threatening. But Sukarno's doctors were not talking, and in public appearances the president seemed hale.[4]

Johnson, sifting through this information and advice, raised the Indonesia issue at a January 1964 White House meeting. Dean Rusk repeated his warning that Sukarno would react to a cutoff of American aid by confiscating American property in Indonesia. The Indonesian president might also do worse. "In the case of a showdown he might ask help from China and even Russia," Rusk said. The secretary of state acknowledged the narrowness of the line the administration had to walk. "We want to keep the United States in a position to influence Sukarno, but we must keep our good relations with Congress and not allow congressmen to think we are disregarding the legal requirement they imposed upon us when the foreign assistance act was amended." Rusk suggested deferring a presidential finding on aid. "It would be bad to act now before the situation is ripe. The stakes are very high. More is involved in Indonesia, with its 100 million people, than is at stake in Vietnam." The question facing the president, Rusk said, was "whether we decide to stay at the table and play a little longer rather than leave the table now."

CIA director John McCone concurred that an aid embargo would probably backfire. It would not induce Sukarno to halt his confrontation with Malaysia, and it might well have "very serious consequences" for American interests in Indonesia. Averell Harriman picked up on this theme, remarking that if Sukarno nationalized American oil holdings and continued on his leftward diplomatic course, Beijing might end up in effective control of the oil operations. Treasury secretary Douglas Dillon advocated continuing the "smallest amount of aid possible" as insurance against both expropriation and congressional wrath. Joint Chiefs chairman Wheeler stressed the need to keep in contact with the Indonesian military.

McGeorge Bundy agreed that continuing aid was necessary, but he appended the thought that the president should send a "tough man" to Jakarta to tell Sukarno to shape up or write off future assistance. The confrontation must end, or American aid would. For the purpose Bundy recommended Robert Kennedy, who had stored up a "reservoir of good will" during a 1962 trip to Indonesia.

Kennedy tried to beg off. The attorney general did not deny that he enjoyed a certain favor in Indonesia, but he said he did "not look forward" to another round with Sukarno.

Johnson decided not to decide, not now anyway. The president said he wanted to talk with members of the pertinent congressional committees to gauge the depth of legislative displeasure with Sukarno. At present, Johnson said, he would find it "very difficult" to state that aid to Indonesia served the American national interest. He did not want to place himself in the position of being charged with a "lack of faith toward Congress." Perhaps in discussions with legislative leaders some arrangement could be made to avoid a cutoff of aid to Indonesia. In the meantime he would request an opinion from the attorney general on how soon he needed to state a finding on aid to comply with the law. And speaking of the attorney general, Johnson said he liked the idea of a special emissary to Sukarno.[5]

Robert Kennedy well suited Johnson's purposes as an envoy to Sukarno. During the trip Bundy mentioned, Kennedy had traveled to Jakarta to facilitate negotiations between the Netherlands and Indonesia over West Irian. He had indeed made a favorable impression on the Indonesians. One student queried Kennedy regarding an aspect of American history that provided a parallel to Indonesia's current situation. In 1846, the student pointed out, the United States had used force to settle its dispute with Mexico over Texas. How then could the United States oppose the use of force in Indonesia's struggle to gain West Irian? Kennedy responded that he was not proud of America's behavior at the time of the Mexican War. "I don't think we were justified in getting involved in it," he declared. The reply satisfied his interlocutor, although it raised some hackles among Texans who read of the attorney general's remarks. After his return to Washington, Kennedy met with his president brother, who suggested, perhaps only half in jest, that in the future the attorney general should clear all his Texas speeches with Vice President Johnson.[6]

Johnson disliked Bobby Kennedy, Mexican War or no Mexican War, and this dislike may have figured in his decision to send Kennedy to see Sukarno. If Kennedy succeeded in moderating Sukarno's behavior, good for the United States. If Kennedy's mission failed, as seemed more likely, Johnson would not mourn Kennedy's association with the failure. In either case Sukarno was a sufficiently controversial figure in American politics that Johnson was happy to have Kennedy be the one photographed arm in arm with the Indonesian president.

Sukarno met Kennedy halfway, geographically if not politically. Sukarno flew to Tokyo, where he greeted Kennedy: "Did you come here to threaten me?" He accompanied this challenge with a smile, but the smile did not hide his seriousness. Kennedy answered, "No, I've come to help get you out of trouble." At this both men laughed. In discussions in Tokyo and subsequently in Jakarta, Kennedy got Sukarno to agree to a cease-fire in the guerrilla combat the dispute with Malaysia had become. At the end of January the two men announced the cease-fire to a joint press conference.

But Kennedy had no sooner left Jakarta than Sukarno reneged. The Indonesian president declared that he still intended to "crush Malaysia." He refused to withdraw Indonesian irregulars from Malaysian territory prior to negotiations with the Malaysian government. Because the Malaysians refused to talk until the Indonesians left, the promised cease-fire never took hold.[7]

The failure of the Kennedy mission came as no great surprise to anyone, although Sukarno's sudden change of mind did embarrass Kennedy. Nor did the mission's failure alter Johnson's desire to put off a decision regarding aid to Indonesia as long as possible. Shortly after Kennedy returned to the United States, Johnson approved a national security action memorandum declaring that he found neither for nor against future aid to Indonesia. He would make a finding only after further consideration of the matter. In the meantime existing aid programs would continue.[8]

Johnson invited a group of legislative leaders to the White House for an explanation of the administration's policy. George Ball put the issue succinctly. "Our purpose is to try to keep Sukarno from turning to the communists for support," Ball said. "We do not approve of Sukarno but we are trying to get along with him to the extent of maintaining some influence in the hope that we can prevent him from using force to destroy Malaysia." Johnson added that the administration was planning no new aid. Only aid already in the pipeline would continue. "If we cut off all assistance," Johnson declared, "Sukarno will probably turn to the Russians."[9]

In choosing to keep aid flowing, Johnson was playing to three audiences. The first sat on Capitol Hill. Congressional irritation with Indonesia varied from week to week, depending on what outrage Sukarno had committed lately and what else was on the legislative calendar. Postponing a definitive decision regarding Indonesia aid bought Johnson time. The president did not want to cut American ties to Indonesia, for the reasons Rusk and Ball had described. He hoped if he put the matter off, Sukarno's congressional critics would find other topics to distract them.

The second audience was Sukarno. Johnson had difficulty reading the Indonesian president. Unlike Johnson, who fell into a frozen reserve before large audiences—in marked contrast to his one-on-one irresistibility—Sukarno was a stem-winder extraordinaire, a speechmaker rivaled only by Nasser and Castro in his ability to rivet huge crowds and impart passion to the masses. A Western observer present at the inauguration of Indonesian independence watched in amazement as Sukarno mesmerized a throng of a quarter million for hours under a broiling sun.

"Could the Dutch ever have held this, in the face of that?" the observer asked.[10]

Like Nasser and Castro, Sukarno sometimes got carried away by his own speaking ability. In private he was more subdued and less forceful— just the opposite of Johnson—and he often seemed more reasonable. Johnson, in delaying a decision regarding aid, hoped to take advantage of one of Sukarno's quieter moments, when he was not before a crowd. The American ambassador to Indonesia, Howard Jones, had developed a friendly working relationship with Sukarno during five years at the Jakarta embassy. Perhaps Jones could persuade Sukarno that a rupture of relations would serve neither Indonesia nor the United States.

The third audience was the Indonesian military. Beginning in the 1950s the Pentagon had brought more than two thousand Indonesian officers to the United States for training. American military assistance to Indonesia ratified the American connection once the officers returned home. The Indonesian army was reliably anticommunist, and although for now the army chose not to move against the PKI, given some encouragement it might do so, especially if the PKI threatened a coup.

Johnson intended the extension of American aid programs to signal a desire to stay in touch with the generals. Perhaps the aid would nudge the army to action. Even if it did not, it would remind the generals that Washington shared many of their views and would be favorably inclined toward any efforts they might make to keep Sukarno or the PKI from taking Indonesia communist. American officials had little doubt that the latent conflict between the PKI and the army would break into the open sooner or later. Sukarno would not live forever. If the rumors about his health were true, he might not live for long. After his death, if not before, someone would bid for power. Keeping aid going was a way for the United States to prepare for the day. As Rusk explained, American aid served "the useful purpose of strengthening anti-Communist elements in Indonesia in the continuing and coming struggle with the PKI."[11]

II

American officials attempted to advance the date of the struggle's climax. Failing this, they hoped at least to raise in Sukarno's mind the possibility that the climax was approaching and that it might sweep him away. In the face of such a prospect, Washington thought, Sukarno might modify his anti-Americanism.

In March 1964 Rusk instructed Howard Jones to make discreet overtures to Indonesian military leaders. In an understatement, Rusk described Sukarno's reaction to American initiatives to date as "disappointing." The problem was not simply Sukarno's perverseness. The problem was also "that those who are in a position to influence Sukarno toward rational foreign and economic policies, particularly the military, are not doing so." The generals seemed to lack understanding of where Indonesian policy was leading. The secretary directed Jones to engage in "educational" activities aimed at top-level military men, although he cautioned the ambassador against anything that "would get back to Sukarno." Rusk wrote, "The Department believes we should now try to build up pressures on Sukarno from Indonesian military sources in favor of a rational settlement with Malaysia and decent relations with the free world." Rusk added, "The Department realizes that carrying out the foregoing will be a delicate task, but it believes that the time for such an effort has arrived."[12]

Jones prepared for the specified effort at once. He visited a former Indonesian ambassador to the United States, an individual he had known for years and who held no particular brief for Sukarno. Jones asked his friend a hypothetical question. What would happen if Sukarno were suddenly removed from the scene, by illness perhaps? The Indonesian diplomat replied that the country's politics would polarize around General A. H. Nasution, the Indonesian defense minister, who had spearheaded an attempt to push Sukarno aside twelve years before, and D. N. Aidit, the leader of the PKI. Jones's friend did not venture to predict who would win the struggle, although his characterization of Nasution as the "strongest man in the country" suggested he would not be betting on the communists.[13]

Shortly after this interview, Jones visited Nasution. Without quite asking whether the defense minister was plotting a coup, Jones probed him regarding the army's intentions. The Indonesian economy, depressed by low commodity prices and governmental mismanagement and recently strained by Sukarno's Malaysia adventure, was sliding downhill rapidly. Jones asked whether the army welcomed economic disintegration on the reasoning that this would provoke the PKI to try for power, thereby giving the army an excuse to attack the PKI. "He denied this flatly," Jones reported. Jones tried a couple of other approaches, seeing if he could draw Nasution out. The general refused to be drawn. "He avoided like the plague any discussion of a possible military takeover, even though this hovered in the air throughout the talk, and at no time did he pick up obvious hints of U.S. support in time of crisis." Yet Jones was not discouraged. "I intend to continue this type of conversation with other military leaders," the ambassador said.[14]

Several days later Jones spoke with General Achmad Yani, the army chief of staff. Yani said the United States must bear in mind that Sukarno was not Indonesia. However irritating the Indonesian president might be, Washington should remember it had friends in the Indonesian military. He and his fellow officers hoped President Johnson would not cut them off. Especially during this trying hour, they did not wish to become "isolated" from the United States.[15]

Nasution agreed entirely, as Jones discovered on a second visit. Nasution said the army's ties to the United States were "vital" to Indonesia's future. He assured Jones that the Indonesian military was strongly pro-American and as anticommunist as ever. He deemed Sukarno's confrontation with Malaysia nearly as misguided as the Americans did. Unfortunately he could do little to change the policy. For now the army had to follow orders. Jones detected significance in the temporal qualification in this remark. "He appeared to be saying that the time would come when the situation might be different." The ambassador followed up. He inquired as to how different the situation might be in the future. Nasution said he did not think the PKI would break with Sukarno and grab for power soon. If it did, though, the army would be ready. Referring to a violent 1948 antileftist purge, Nasution promised, "Madiun would be mild compared with an army crackdown today." To demonstrate his cooperative nature, Nasution allowed Jones to draft a letter from the Indonesian defense ministry to the American government requesting that Washington not cut off aid.[16]

Jones was encouraged by Nasution's resolve. The ambassador believed that the resolve would translate into pressure on Sukarno and that eventually the pressure would work to America's benefit. Yet the outcome was no sure thing. Sukarno had a knack for confounding prognosticators. Jones declined to prognosticate, saying, "It is too early to predict which way this cat will jump."[17]

The cat jumped into the middle of a fresh controversy at the end of March 1964. At the dedication of a new building in Jakarta, Sukarno took the opportunity to ridicule some American predictions that Indonesia's economy bordered on ruin. How could any country with the riches Indonesia possessed fail?, he asked rhetorically. Warming to his audience, Sukarno denounced demands by American legislators and editorialists that American aid be conditioned on his calling off the confrontation with Malaysia. "We receive aid from many countries. We are grateful for such aid. But we will never accept aid with political strings attached. When any nation offers us aid with political strings attached, then I will tell them"—here he switched to English for the benefit of American headline writers—"'Go to hell with your aid!'"[18]

Jones recalled afterward, "He had really done it this time." He certainly had. Rusk tried to make light of Sukarno's remark, telling reporters the administration did not interpret the go-to-hell message "as an intergovernmental communication." But Congress, which had almost forgotten Sukarno in the rush of other issues, erupted in protest at the effrontery of the Indonesian megalomaniac. Democrat Birch Bayh of Indiana told his Senate colleagues that the time had come for America to stop throwing good money after bad. "I do not believe in continuing to ask the American people for their tax dollars to support a man who is arrogant, insulting, incompetent, and unstable—a man who, by his own foul admission, despises the very people whose energy and hard work make it possible for many of his people to eat and to ward off the ravages of malaria." Bayh continued, "It is absolutely imperative for the Senate to determine that when our foreign dollars are spent, they are spent for the cause of perpetuating freedom, and not spent as they have been spent in the past years in Indonesia." Republican congressman William Broomfield charged Sukarno with transforming Indonesia "into a living hell." Broomfield directed his fire at the Johnson administration as well, blaming the White House for "mollycoddling this minor-league Hitler." Representative Harold Ryan took his complaints straight to the Oval Office. In a letter to Johnson, the Michigan Democrat asserted that this most recent manifestation of Sukarno's "insulting attitude" should be "the last straw" that forced a cutoff of American aid. Ryan blasted Sukarno as "a grossly incompetent and completely immoral, highly insulting person." To substantiate this judgment, Ryan confided to Johnson that once on a visit to Michigan Sukarno had made the "astonishing request that our Detroit Police Department supply him with women for immoral purposes."[19]

Sukarno's outburst and the reaction it evoked on Capitol Hill forced Johnson's hand. Despite the continued admonitions of Rusk and Jones that aid cuts would weaken American influence with the Indonesian military, which increasingly seemed the final barrier between Indonesia and communism, Johnson decided he had to lessen his political exposure. The American presidential election was only seven months away, and there was no telling what Sukarno would do in the meantime. Johnson refused to take any more responsibility for Indonesia than was absolutely necessary. At the end of March the president ordered aid to Indonesia slashed, to the lowest level in years.

III

The announcement on aid temporarily solved Johnson's Indonesia problem in Washington, but it did nothing for his Indonesia problem in Jakarta. If anything, it made matters worse. Sukarno grew increasingly fractious and unpredictable by the week. During the spring of 1964 the CIA sent Johnson intelligence estimates describing unusual behavior on the part of the Indonesian president. Agency informants close to Sukarno reported that the Malaysia business and the weakness of the Indonesian economy, not to mention the precariousness of Sukarno's position between the army and the PKI, were taxing Sukarno's health more than ever. Marital troubles added to his woes. The result was that he seemed to be losing his grip, emotionally if not yet politically. Insiders described, as the CIA relayed the accounts, "an uncharacteristic concern and need for personal funds." Sukarno intimates expressed worry regarding his "rationality and ability to receive and deal with basic facts." Sukarno's friends could not remedy the political causes of the stress afflicting him, nor could they cure his physical ailments, but they were said to be plotting to kill one of his wives, the one alleged to be the source of his household strife.[20]

Sukarno's tribulations did not prevent his undertaking further annoying actions. He encouraged a PKI-organized boycott of American films, and when PKI members attacked libraries of the United States Information Service in Jakarta and elsewhere, the Indonesian government made no effort to stop the violence. Instead the government announced that it could "understand the feeling" of the individuals responsible. Sukarno personally greeted a delegation from North Korea with conspicuous cordiality. In July he sent General Yani and the Indonesian foreign minister to Moscow, evidently to seek economic and military aid. Two weeks later the CIA reported a successful—from Sukarno's perspective—conclusion to the mission. The CIA asserted that Yani and the foreign minister had come back with a Soviet commitment to furnish MiG-21s to the Indonesian air force, with delivery slated to start early in 1965.[21]

Sukarno's irritating climaxed in a speech he gave on Indonesia's independence day in August 1964. Taking as his theme "the year of living dangerously," he recited a list of Indonesian grievances against the United States, including the CIA's role in the 1958 revolt and a recent remark by Johnson supporting Malaysia against Indonesia. The United States, Johnson had said, affirmed Malaysia's right "to maintain her security, preserve her sovereignty and continue her development in peace and harmony." Sukarno decried this blatant unneutrality and illegitimate meddling. As a rejoinder he announced the landing of some thirty thousand Indonesian troops on the Malay peninsula. In the part of his speech touching domestic questions, he aligned himself more closely with the PKI than ever before.[22]

When Francis Galbraith, filling in at the American embassy for vacationing Jones, reported Sukarno's speech to Washington, the chargé d'affaires commented, "Despite its many blatant contradictions, errors of fact and ridiculous statements, one thing stands out: Sukarno declares Indonesia in the camp of the Asian communist countries and opposed to the United States—opposed not only on issues of the day like Vietnam and Malaysia, but fundamentally opposed to our thought, our influence and our leadership." Galbraith accounted Sukarno's endorsement of much of the PKI's agenda the "most diabolical part" of the speech. Galbraith concluded, "It would be fatuous to pretend that the speech is other principally than a declaration of enmity toward us."[23]

The CIA read the situation similarly. The agency sent Johnson a memo asserting that Sukarno's speech indicated that the Indonesian president had chosen "to stand internationally with the anti-Western Asian world." Sukarno's handling of the situation within Indonesia afforded even less cause for hope. "He has allowed too much influence to slip into Communist hands," the agency's analysts warned. "He is well on his way to becoming a captive of the Communists."[24]

The independence day speech and the escalation of the fight against Malaysia produced a new round of protests in the United States. For Johnson the complaints could hardly have arrived at a less opportune moment. Republican presidential nominee Barry Goldwater was portraying Johnson as insufficiently attentive to the communist threat in Southeast Asia. The president was countering by painting Goldwater as an extremist whose alarmism might start a third world war. Johnson's strategy rested on the assumption that communism was not going to engulf Southeast Asia in the next few months. Sukarno made it appear that the largest country in the region had just about slipped behind the Bamboo Curtain, and by escalating the fight against Malaysia he suggested that he intended for communism to spread. Though he could not have cared less about the American election, his actions came closer to proving Goldwater right than vindicating Johnson.

Sukarno's actions also threw red meat to congressional critics of Johnson's Indonesia policy. At the beginning of August the Senate had attached to a foreign aid authorization bill an amendment canceling all remaining aid to Indonesia and barring training of Indonesian military officers in the United States. The vote on the Tower amendment— named after Texas Republican John Tower—was 62–28, indicating the breadth of opposition to Sukarno. And this vote came *before* Sukarno's procommunist speech. The amendment had yet to pass the House of Representatives, where Johnson hoped to have better luck keeping hands off the foreign aid bill. But the lopsided count in the Senate reminded Johnson once more that Indonesia was political bad news.

On the other hand, the strategic reasons for continuing at least some aid remained strong. In certain respects the reasons were stronger than ever. Perhaps Sukarno had placed himself beyond reclamation. Perhaps he had thrown his lot in with the PKI. Perhaps American aid had no influence on him. But by cozying up to the communists he might be hastening the day the army would move to save Indonesia from going the way of China. The Johnson administration needed to encourage the generals, especially now.

Robert Komer summarized the dilemma facing Johnson. In a memo to the president, Komer wrote of Sukarno: "We've strung him along for years (with our eyes open), on the basic premise that if he swung too far left we'd lose the third largest country in Asia—whose strategic location and 100 million people make it a far greater prize than Vietnam." Continued stringing, though, entailed serious political difficulties. The Tower amendment "not only puts you on the spot but moves us dangerously close to a final break with Indonesia." Komer advocated hunkering down and hoping for the best. "What's essential is not to force on you the impossible choice of either defying the will of Congress in an unpopular cause or letting the break with Indonesia move further to the point of no return.''[25]

Fortunately for Johnson, the Tower amendment got caught in the eddies that swirled out from the Tonkin Gulf affair of the same period. With Congress almost without dissent placing the conduct of the war in Vietnam in Johnson's hands, Sukarno's antagonists in the legislature had a hard time contending that the president could not be trusted to manage some small aid programs for Indonesia.

Johnson's advisers, breathing a sigh of narrow escape, unanimously advised the president to do what he could to maintain a connection to Indonesia. Differences arose, however, regarding the optimal form of the connection. The State Department considered the American military aid program expendable, thinking that the president would have an easier time defending money spent on malaria-eradication campaigns and the like. The Defense Department considered military aid and so-called civic-action, or anti-leftist rural pacification, programs essential. Military aid, the Pentagon argued, promised more cooperation for the dollar, since it went directly to the Indonesian army, the group the United States was relying on to rescue Indonesia. American defense officials suggested that the American military attachés in Jakarta visit Nasution and Yani for what James Thomson, the White House aide who interpreted the Pentagon's view for Johnson, described as ''a candid where-the-hell-do-we-go-from-here session.'' Thomson continued, ''If Nasution & Co. were to ask us to lie low for a while, it would be quite possible to taper off on military aid while continuing the civic-action programs with considerably reduced staff.'' Thomson thought the Defense Department had the better of the argument. Speaking of assistance to the Indonesian military, Thomson remarked, ''Any fast motion toward a cutoff would be a foolish waste of fifteen years' investment. Far better to play it cool, as long as the issue is reasonably quiescent in this country, and to make a fast pitch to our real pals, the Indonesian military.''[26]

After some interdepartmental bargaining, Dean Rusk recommended that Johnson split the difference between the State and Defense Departments. Rusk advocated continuation of civic-action programs and training for police and internal-security forces, while urging suspension or deferral of deliveries of military equipment. Certain economic and technical-assistance programs should be extended. Regarding training of regular military officers, the administration should ''work toward a quiet mutual agreement'' with the Indonesian military.[27]

McGeorge Bundy judged Rusk's proposal a fair compromise and urged Johnson to accept it. The national security adviser conceded that Sukarno was ''as unreliable and dangerous as he can be.'' He acknowledged that the administration's footing in dealing with Sukarno was treacherous. But he thought the situation warranted risk-taking. ''The fact that we're on a slippery slope makes it all the more important not to burn all our bridges to Indonesia," Bundy said. He cited four reasons: "(1) With Vietnam and Laos already on our Southeast Asia plate, we can ill afford a major crisis with Indonesia just now; (2) we ought to keep a few links, however tenuous, to the Indo military, still the chief hope of blocking a Communist takeover; (3) there's still a slim chance of Sukarno drawing back from a full-fledged push on Malaysia, and we want to keep dangling the prospect of renewed aid; and (4) we do not want to be the ones who trigger a major attack on U.S. investments there."[28]

Johnson only rarely overruled Rusk and Bundy together, and he did not now. At the beginning of September he approved Rusk's proposal.[29]

IV

This decision got Johnson safely past the 1964 election and into the White House on his own hook. The president's smashing victory at the polls put other thoughts into the heads of any who had intended to hector him on the Indonesia issue.

Sukarno likewise seemed to have his political ducks in a row. The week before Johnson's 1965 inauguration the CIA delivered a report indicating that Sukarno's personal difficulties had abated and that once more he was in command of Indonesian affairs. Currently he was boasting that he could crack the whip over the communists whenever he chose. A source close to the Indonesian president quoted him as saying, "Someday I will take over the PKI, but not now." Sukarno was also predicting deeper involvement by the United States in Vietnam. Eventually the Americans would go to war with China. When they did, Washington would pay dearly to woo Indonesia away from Beijing.[30]

This intelligence hardly heartened the Johnson administration. The administration could never quite figure out whether it wanted Sukarno stronger or weaker. A strong Sukarno would be better positioned to call off the confrontation with Malaysia, notwithstanding leftist criticism. A weak Sukarno would be more susceptible to pressure from the army, not to mention pressure from the United States.

The State Department saw only gloom in the Indonesian situation. Sukarno had cowed the army, Rusk said. Whether or not he possessed the power to defy the PKI, he did not seem to possess the will. Rusk remarked, "Because of the way Sukarno dominates all others in this country, we now believe the evidence is fairly convincing that effective internal anticommunist action independent of Sukarno will not develop during Sukarno's lifetime." The secretary of state added, "We are therefore faced with the cold possibility that before long this key strategic nation may be for all practical purposes a communist dictatorship and that when events have progressed that far they will be irreversible." The noncommunist forces in Indonesia were in disarray. The army saw no alternative to Sukarno. The Indonesian president was "the virtual sole remaining key to stabilizing the situation and averting a further leftward stampede." As keys went he could hardly be less promising.[31]

At the end of January 1965 the CIA sent Johnson a summary of its predictions for Indonesia for the coming year. As before, Sukarno's health remained an uncertain factor in the equation. The CIA did not know whether to hope Sukarno lived or hope he died. If he died in the near future he would leave the country in a sad condition. "His bequest to Indonesia would be international outlawry, economic near-chaos, and weakened resistance to Communist domination." The other side of the coin looked no better. "If Sukarno lives on for some time to come, the chances of the Communist Party (PKI) to assume power will probably continue to improve." The noncommunist forces in the country displayed a dismaying lack of "backbone, effectiveness and unity." Until they decided to stand up to the PKI the prognosis for Indonesia would continue grim.[32]

Johnson decided something had to be done. Briefly he considered inviting Sukarno to Washington. He tried out his idea on British prime minister Wilson, whose government's close ties to Malaysia made it a critically interested observer of Sukarno. "Sukarno's personal vanity is maddening," Johnson wrote Wilson. "But it may be a possible handle that might be turned to use." Johnson said he thought an invitation to Sukarno to come to the United States might serve to "drive a small wedge between Sukarno and the PKI." Johnson assured Wilson that an invitation would in no way indicate American support for Sukarno in Indonesia's contest with Malaysia.[33]

Wilson did not think much of Johnson's feting Sukarno in Washington, regardless of the president's disclaimers. On second judgment Johnson had to agree that a visit by the unpredictable Indonesian leader might be unwise. Instead he chose to send another special envoy to Jakarta to talk with Sukarno. For the task he called on Ellsworth Bunker.

Bunker had helped negotiate a resolution to the West Irian dispute during Kennedy's tenure, and for a time under Johnson the White House thought to send him to Jakarta as a permanent replacement for Howard Jones. Jones was nearing retirement, and after seven years in Indonesia his usefulness was running low. The ambassador's friendly relationship with Sukarno allowed him access to the Indonesian president even when Sukarno was raging against all else American, but the same friendliness made it difficult for Jones to convey Washington's displeasure. While Bunker was not unfriendly, neither was he the ingratiating sort. He had a well-deserved reputation for toughness in tough situations. Robert Komer—no softie—judged Bunker the administration's best man for Southeast Asia. In February 1965 Komer argued to McGeorge Bundy that Jakarta was the most sensitive and important diplomatic post in Southeast Asia—Vietnam not excepted—and he concluded that Bunker stood a better chance of halting the decline in American relations with Indonesia than anyone else. Johnson agreed to send Bunker to Jakarta, but only temporarily, as a special envoy.[34]

Bunker spent two weeks during April 1965 with Sukarno and other top Indonesian officials. He devoted considerable effort to explaining how American activities in Vietnam did not represent a continuation of Western imperialism in Southeast Asia. Sukarno was not convinced. Bunker attempted to demonstrate to Sukarno that the United States had not taken the side of Malaysia in the Indonesia-Malaysia quarrel. Sukarno did not accept this either. American policy toward Malaysia, Sukarno said, was a "slap in Indonesia's face." Nor did Sukarno's associates appear moved. Writing to Washington, Bunker described a "well-orchestrated chorus" of complaint among Indonesian officials at America's policy regarding Indonesia's troubles with Malaysia. Bunker assured Sukarno that President Johnson desired good relations with Indonesia. President Johnson wanted to know whether President Sukarno reciprocated this feeling. Sukarno refused to commit himself.[35]

An escalation of anti-American activity backdropped Bunker's visit. Indonesian leftists, obviously with Sukarno's approval, attacked the American embassy and American businesses in Jakarta and beyond, and harassed personnel of the Peace Corps and the Ford Foundation.

Bunker assessed these attacks and Sukarno's rebuff in a report to Johnson upon his return to Washington. He predicted that Indonesian-American relations were "unlikely to improve in the near future," by which he meant as long as Sukarno remained in power. Relations with Indonesia had foundered on "Sukarno's mystical belief in his own destiny." Sukarno thought he had a unique mission to lead Indonesia to unity and greatness, regardless of the complications this created with other countries. Sukarno's sense of exceptionalism was nothing new, but his declining health gave it an urgency it had not previously had.

Bunker thought Washington would waste its time trying to deal with Sukarno. Better to step back and wait for the day Sukarno departed the scene. "U.S. visibility should be reduced so that those opposed to the communists and extremists may be free to handle a confrontation which they believe will come, without the incubus of being attacked as defenders of the neo-colonialists and imperialists." Events in Indonesia were beyond American control. "Indonesia essentially will have to save itself." Washington should concentrate on letting the saving, if such were in Indonesia's future, proceed. "U.S. policy should be directed toward creating conditions which will give the elements of potential strength the most favorable conditions for confrontation." Bunker advised a reduction in American activities in Indonesia. "They divert attention from the main aspects of the sharpening internal power struggle between the communists and noncommunists."

Like other American officials Bunker placed most of his hopes on the army. "The Army has lost political influence since its heyday in the early 1960s but still retains the physical power to overcome any rival political force," he said. Bunker had doubts regarding how much of the army's opposition to the PKI demonstrated disapproval of Marxism-Leninism as a method of social organization and how much simply considerations of power. He suspected that the latter motivated most of the officers. "The bulk of them are strongly anti-PKI, less from ideological considerations than from an awareness that the PKI represents the chief threat to their own position and privileges." But from whatever cause, the army could be counted on to oppose the PKI, which was considerably more than could be said of Sukarno.[36]

The State Department seconded Bunker's recommendation for a reduction in the American presence in Indonesia. George Ball rejected counterarguments that pulling back meant surrendering the country to the communists. "The reduction or even removal of our presence would not mean turning the country over to the communists," Ball wrote Johnson. "On the contrary, it is more likely to mean a sharpened confrontation between the Communist Party and anti-Communists in the country." Yet Ball did not hold out any hope that the sharpening would soon reach the point of decision. "So long as Sukarno is alive, the prospects of an army revolt against him are slim."[37]

At just the time of Bunker's report and Ball's supporting memo, Johnson had much on his mind. The Great Society was moving rapidly through Congress. The war in Vietnam was accelerating, with an air offensive against North Vietnam under way and expanded troop involvement in the offing. Most pressingly, the day Bunker's report crossed Johnson's desk the Dominican insurrection broke out. For a month Johnson had scarcely a moment to spare for Indonesia.

Johnson's distractions predisposed him to accept Bunker's and Ball's advice. He probably would have gone along even without the distractions. Sukarno evidently did not want Americans in Indonesia, and as Johnson would demonstrate in response to de Gaulle's disinvitation regarding France, the president did not desire for the United States to stay where it was not wanted. He approved the reductions Bunker recommended. During the next several months the number of official Americans in Indonesia fell to a few dozen, from a high of several hundred. Aboveboard American aid to Indonesia essentially ceased, although a trickle of off-the-record assistance seems to have continued.[38]

At the beginning of the summer of 1965 McGeorge Bundy summarized the administration's policy toward Indonesia as "cool and correct." Bundy said, "We are keeping the door open to friendly relations, but we have removed the Peace Corps and other targets of Communist agitation. We are really playing for the breaks in a situation in which the Communists are gaining in influence, but the prospect of a reaction is strong."[39]

On this last point Bundy was more hopeful than others in the administration. With Indonesia's independence day approaching, Washington braced for a further blast of Sukarnoist diatribe. The Indonesian president had intimidated all opposition. Howard Jones described a rally of three hundred thousand persons in Jakarta at which Sukarno's ranting called to mind "Hitler's strutting pomposity at Nazi totalitarian rallies in the 1930s." Jones continued, "The saddest vignette of the morning was Defense Minister/coordinator General Nasution screaming 'Long live Bung Karno' with the rest of the ministerial sycophants, and receiving in return a hand gesture from Sukarno as though the latter were shooting at him with a water pistol." A short while later Sukarno delivered another speech in which, according to Jones, his "ideological affinity to the communists, while not exactly new, was remarkable in its openness and extent."[40]

For reasons best known to himself, Sukarno chose not to use the August 17 festivities as a forum for particularly lambasting the United States. Marshall Green, who replaced Jones when the ambassador took his retirement at the end of May, described the address as thoroughly unsurprising. Sukarno denounced Johnson's decision to send large numbers of combat troops to Vietnam and to continue the bombing campaign, but plenty of other world leaders were doing the same.[41]

Whatever the source of Sukarno's relative restraint, administration officials did not expect it to last, and they prepared for its end. Rusk placed a ban on new personnel assignments to Indonesia and ordered American dependents brought from outlying regions to Jakarta for quick evacuation if necessary. In mid-September the secretary of state directed Marshall Green to destroy most of the embassy's classified files, keeping only thirty days' worth of working records.[42]

V

The storm began at the end of September, but its nature caught the Johnson administration by surprise. On the last night of the month a group of junior officers sympathetic to the PKI attempted to decapitate the army high command. The colonels nearly succeeded, assassinating Yani and five other top generals. Nasution barely escaped, his flight made possible by a brave and unfortunate aide who donned Nasution's jacket and cap and in the dark was mistaken, mortally, for the defense minister.

The conspirators overlooked General Suharto, the chief of Indonesia's strategic command. They never explained their oversight—not that they were given much chance. Suharto had his own answer. "They didn't think my command that important," the general said. The error proved the rebels' undoing, for within hours of the multiple assassinations Suharto organized a counterattack.[43]

The assault and counterassault thoroughly confused American officials. Ambassador Green thought Sukarno might have died and the fight for the succession begun. On the other hand, the turmoil might mark some especially deep ploy by the devious Indonesian president. In either case the silence of the PKI seemed strange. "It is disturbing that the Communists are so complacent," Green wrote on the morning of October 1.[44]

The communists had plenty of reason to keep still, as events soon demonstrated. Suharto moved tens of thousands of loyal troops—loyal to him, at any rate—into the capital and began running the communists to ground. Although American officials didn't know it, they were witnessing the start of a breathtakingly broad purge of Indonesian communists. Before the bloodletting ceased, somewhere between two hundred thousand and five hundred thousand communists would die, with the estimates varying according to the politics of the estimators and the definition of *communist* applied.

Johnson first got the news from the White House briefing officer just before midnight, Washington time, on September 30. By the next morning a special Indonesia task group had formed, and the CIA began pulling together information on Suharto. The initial biographical sketch was sketchy indeed, partly because Suharto was one of the few leading Indonesian officers not trained in the United States—which may have contributed to the colonels' missing him. The CIA knew little about his background. Of his politics the agency could offer only: "Suharto is considered to be anti-Communist." Yet American officials could reliably conclude almost nothing more. George Ball remarked on October 1 that the situation in Jakarta was "still as muddy as it could be."[45]

As Washington realized just how anticommunist Suharto was, American leaders kept their fingers crossed that the army would capitalize on its opportunity to draw Indonesia back from the brink. Sukarno had resurfaced and appeared to be trying to recoup his position, although Marshall Green, after listening to a pronouncement by the Indonesian president, said he sounded like a "man in trouble." Green added, "The key question is whether the Suharto forces will have the courage to go forward against the PKI despite whatever moves and statements Sukarno may make to try to stop them." The next day, October 4, the Indonesia task group in Washington reported, "The army has not yet reached a decision on whether to maintain its drive for complete victory over the PKI." Two days later the task group offered hopefully, "The army appears now to have determined to move vigorously against the PKI."[46]

The obvious question for the Johnson administration was what it could do to encourage the army to keep up the good work. Green in Jakarta thought that for the moment Washington could not do much. The ambassador allowed a modest role for American propaganda designed to "spread the story of the PKI's guilt, treachery and brutality," if the American provenance of such propaganda could be disguised. But otherwise the United States should stay aloof. Premature intervention would simply spoil things. "Events will largely follow their own course as determined by basic forces far beyond our capability to control," Green asserted.[47]

Administration officials in Washington concurred. George Ball believed that the administration would best serve American interests by keeping far removed from the situation in Indonesia. "The army clearly needs no material assistance from us at this point," the undersecretary of state said. Nor did the generals require American moral support. Ball commented that years of American military training and millions of dollars of American military aid "should have established clearly in the minds of Army leaders that the U.S. stands behind them if they should need help." Ball thought the United States should stay away from even propaganda. He rejected Green's suggestion for an anti-PKI information campaign. Further, he told the ambassador to use "extreme caution in contacts with the army" lest the contacts play into the hands of Sukarno and the radicals.[48]

The CIA pointed out that although Suharto was the leading figure in Jakarta for now, Sukarno might manage a comeback. Much of the emotion against the PKI reflected the party's perceived—though yet unproven—connection to the murders of September 30. Sukarno seemed to have distanced himself from the conspiracy, and once emotions died down Indonesians would have to choose between Suharto and Sukarno. Suharto "temporarily will retain a political ascendancy," the CIA predicted. But the ultimate outcome remained a puzzler.[49]

Rusk indicated an additional uncertainty lending to caution on the administration's part. "We are not at all clear as to who is calling the shots within the military," the secretary commented on October 13. Although Suharto was the most visible of the military leaders, perhaps Nasution or some other general controlled the situation from behind the scenes. Yet Rusk recognized the risks in moving too slowly. "If the army's willingness to follow through against the PKI is in any way contingent on or subject to influence by the United States, we do not want to miss the opportunity to consider U.S. action." The secretary told Green to inform Suharto that the United States was "as always, sympathetic to the army's desire to eliminate Communist influence," but that it could not commit itself at this stage.[50]

After the initial show of force, the army demonstrated that it too aimed to move carefully. Sukarno, slippery as ever, was maneuvering to regain control. Without the assistance of the PKI, most of whose surviving members had dropped out of sight, the Indonesian president lacked much of his previous power, yet even so he could still hope to mobilize the masses. Besides, Sukarno retained a legitimacy Suharto and the other generals could use if somehow they could appropriate it for themselves.

At the end of October the CIA summarized for Johnson what was shaping up as a standoff: "Neither Sukarno nor the army wants civil war or serious internal disturbances. Each appears fearful of provoking it if it presses the other too far." Green offered a similar assessment. Focusing on Suharto, the ambassador remarked, "In typically Indonesian, if not Javanese, fashion, Suharto's strategy calls for extremes of patience and a slow-moving time framework." Green added, "It will require an equally patient approach on the part of the U.S. if at any point we are to mesh our efforts with the army's."[51]

Two weeks later the CIA attempted a bit more illumination. In a memo to the White House the intelligence agency described the immediate objective of the Indonesian army as "the destruction of the PKI as an effective political and military force." Over the longer term the army would seek to neutralize Sukarno, although the generals continued to value him as a "national symbol." During some indeterminate period the army would probably follow a publicly anti-Western line, since any abrupt departure from Sukarno's policies would render the army suspect in the eyes of the scores of millions of Indonesians who still revered the only president Indonesia had ever had. A number of the generals seemed reluctant to get too involved in politics. The generals' line on the purge of the PKI was that they opposed the PKI but not communism. The authors of the CIA memo remarked that this reasoning involved "some Javanese subtlety." As for Sukarno himself: "Sukarno is in a political corner, but the game is not over."[52]

VI

To this point Johnson had not had to make any significant decisions regarding Indonesia since the events of September 30. As Bunker and Green had predicted, events in Indonesia moved in a direction and at a pace determined by Indonesians. Just as the administration had earlier recognized that it might have to watch Indonesia go over the edge, unable to do anything to prevent the fall, during this period it realized it could only watch as Indonesia pulled back from the precipice, equally unable to speed the rescue.

At the beginning of December, though, somewhat greater scope for American action appeared. On December 1 Green received a visit from two of Suharto's associates. As the ambassador related the conversation, "Both emphasized that the right horse was now winning, and the U.S. should bet heavily on it." The two officers mentioned the 1958 rebellion, which the United States had backed only feebly, as a poor precedent. "Their bluntest remark was the question: How much is it worth to the United States that the PKI be smashed?" They went on to liken Indonesia to "a ship with a new captain at the helm, anxious to sail a proper course but unable to leave harbor without fuel." In perhaps their most effective comparison, given Washington's current preoccupations, they reminded Green how much the United States was spending on Vietnam. By comparison, an insurance policy for Indonesia would come cheap.[53]

Johnson was not ready to resume major aid to Indonesia yet. During October and November a small amount of covert American aid seems to have gone to the Indonesian military, in the form of hand radios. But Indonesia remained a sufficiently sensitive issue in the United States, and the complexion of a post-Sukarno government, if indeed such was taking shape, remained sufficiently uncertain, that Johnson chose to keep clear. American aid would not make the difference between an army victory and an army defeat. That issue the Indonesians would have to decide. American aid *would* raise potential problems for Johnson at home. These he preferred to avoid.[54]

Dean Rusk thought the generals were getting pushy for a group that only recently had been an endangered species. They seemed to believe American aid would flow as soon as they indicated they wanted it. "A cargo-cult mentality has developed," Rusk remarked. "They appear to see fully loaded ships ready to arrive when they consent to aid and press the button." Rusk could have lived with the generals' pushiness, but he didn't like their suggestion that American aid be kept secret. He understood their reasoning: that a turn to the United States would call their nationalist bona fides into question and afford Sukarno an opening for a political counterattack. Yet from the administration's standpoint, secret aid was too risky.[55]

Johnson was even less inclined than Rusk to go the secret route. After the Dominican intervention of the previous spring and especially after Fulbright's September defection, the president faced a growing credibility problem. With the press greeting everything the administration said with skepticism, keeping secret aid secret would be impossible. For the president to be caught cheating on Indonesia could be ruinous— and not only for policy toward Indonesia. Steadfastness in Vietnam depended on the American public's faith in the president, in particular on the public's willingness to believe his assurances that the war was winnable. When credibility vanished, the prospects for victory in Vietnam would dangerously diminish. Incredibility's effect on Johnson's political prospects would be stronger still.

Besides, the Indonesian generals were doing well enough on their own. From Jakarta, Marshall Green cabled, "The elimination of the communists continues apace." George Ball noted that the army's campaign to destroy the PKI was "moving fairly swiftly and smoothly." Bundy commented to Johnson, "At the moment the political situation in Djakarta is more promising for us than at any time since 1963." Early in January 1966 the CIA declared, "The era of Sukarno's dominance has ended." The agency continued, "In the last three months his prestige has been seriously eroded; he is less the father and political hub of the country and more the petulant old man." Regarding the struggle between the army and the PKI, the CIA asserted, "The army has virtually destroyed the PKI." Although the PKI had not been legally proscribed, such action would be "largely a formality."[56]

VII

Officials of the Johnson administration were right about the PKI being destroyed, but they pronounced Sukarno's demise prematurely. Suharto continued reluctant to attack the Indonesian president directly, and during late January and early February Sukarno showed signs of rebounding. Sukarno's leftist sympathizers held anti-American demonstrations. At one point in February a radical group invaded the compound of the American embassy. Robert Komer told Johnson, "In the last 4–5 months some 100,000 Communists have been killed. But the outcome in this key country of 100 million is still uncertain." Marshall Green arrived from Jakarta to brief the president personally. Green said of Sukarno, "He is clever and persuasive and still seems to have extraordinary physical reserves." The army was on top, Green said, yet Sukarno was hard to pin. The CIA described the situation as one of "paralysis." The intelligence agency thought it significant and not reassuring that "Sukarno appears to operate on the premise that time is on his side."[57]

Green thought that while matters hung in the balance the administration ought to maintain its low visibility. Rusk agreed. Once more the generals had indicated their desire for covert American aid. Repairing the economic damage of the Sukarno era was proceeding more slowly than expected, and Suharto and the other generals believed they needed to show some improvement if they hoped to gain the support of the people. They asked the United States to quietly underwrite the purchase of rice from Thailand. Rusk rejected the request. The American government, the secretary granted, was "sympathetic" to the army's problems. But he saw no way aid would remain quiet, and unless the army made the request public he could not recommend that the president approve it. "While we would like to be helpful, there is not much we can do under present circumstances."[58]

Robert Komer, now interim national security adviser, criticized with characteristic forthrightness what he called the "small-bore thinking" of the State Department. The diplomats excelled at finding ways to justify inaction. Komer concurred that the time for aid to Indonesia was not quite at hand. "Let's keep them coming to us a while longer," he said. But the administration must ready plans. "Once we see an exploitable opportunity we ought to be prepared to move in deftly." The administration must take care not to "miss the boat" on Indonesia "when we may at last have Sukarno on the run."[59]

Sukarno's comeback fizzled, saving Johnson a difficult decision. Suharto set his sappers on Sukarno's position, leaving the Indonesian president in office but depriving him of power. The general forced Sukarno to accept a cabinet controlled by the army and other conservatives. By the middle of March, American officials were nearly ready to declare Sukarno defeated. "It is hard to overestimate the potential significance of the army's apparent victory over Sukarno (even though the latter remains as a figurehead)," Komer explained to Johnson. "Indonesia has more people—and probably more resources—than all of mainland Southeast Asia. It was well on the way to becoming another expansionist Communist state, which would have critically menaced the rear of the whole Western position in mainland Southeast Asia. Now, though the unforeseen can always happen, this trend has been sharply reversed." The CIA declared, "The struggle between Sukarno and the army appears all but resolved."[60]

Suharto felt sufficiently strong to request overt American aid. At the end of March he sent the State Department an application for fifty thousand tons of PL-480 rice. Rusk forwarded the application to Johnson with a recommendation for approval. The president, remarking a recent improvement in Indonesia's relations with Malaysia and a shift toward policies likely to strengthen the Indonesian economy, accepted the recommendation and okayed the rice.[61]

At the same time, those at the State Department whom Komer had called small-borers began drawing up broad-gauged blueprints for Indonesian reconstruction. The department's Policy Planning Council produced a 130-page document entitled "A Contingency Plan for the Rescue, Stabilization and Rehabilitation of the Indonesian Economy." The paper presented a full-featured program of economic and military aid designed to reinforce the positive actions the Suharto government had taken thus far and to facilitate Indonesia's economic modernization.[62]

Rusk endorsed the idea of large-scale aid. Indonesia certainly needed help. "The basic fact is that Indonesia's economy is in ruins and its Central Bank is bankrupt," the secretary wrote. Rusk sent Johnson a brief account of developments in Indonesia since the assault on the army leadership, describing the country's turnaround in dramatic terms: "Between October 1 and the middle of March of this year the Communist Party was virtually eliminated as an effective political organization. Perhaps as many as 300,000 Indonesians were killed—the great bulk of whom we believe were in fact associated with the Communist apparatus." The new regime had completely suppressed manifestations of anti-Americanism in Indonesia. American property in the islands was safe.

Yet Rusk hesitated for the administration to leap too quickly back into Indonesian affairs. The rice deal had registered America's desire that the Suharto government survive and prosper. The State Department currently was working on a concessionary sale of seventy-five thousand tons of cotton to Indonesia. To undertake to underwrite the Indonesian economy, however, was indiscreetly and probably unwisely ambitious, at least at present.[63]

It was also something Johnson was unlikely to accept. During the spring of 1966 Johnson was keeping the tether tight on aid to India, fearing a congressional backlash on foreign assistance and trying to enforce reform on Indira Gandhi's government. To request a major program of aid for Indonesia, even the new Indonesia of Suharto, would contradict the careful image Johnson was trying to convey.

Suharto rendered assistance easier in August 1966 when his government formally canceled the confrontation with Malaysia. A short while later Indonesia returned to the United Nations, from which Sukarno had withdrawn his delegation early in 1965.

At a White House meeting in the late summer of 1966, Johnson reviewed American policy toward Indonesia. The president opened the session by noting that less than one year earlier Indonesia had seemed well on the road to becoming "an out-and-out Communist state." The situation since then had changed entirely. Johnson asked CIA director Richard Helms whether there existed any prospect of a Sukarno comeback. Helms responded that the army had the situation thoroughly in hand.

Rusk was not so sure. The secretary perceived an "outside chance" Sukarno might cause further trouble. Consequently he continued to recommend caution in providing aid. The secretary said the administration had moved deliberately thus far, and success had followed. Hasty American involvement could be "the kiss of death" for Suharto. Rusk advocated working through international aid channels in helping Indonesia. This would not only save the United States money but make the aid more acceptable politically in Indonesia.[64]

At this meeting Johnson decided in principle to declare American aid to Indonesia in the American national interest—the condition Congress still required for significant assistance. But before finalizing the decision he wanted to check with congressional leaders. The president sent representatives of the State Department to meet with important lawmakers to determine whether Indonesia remained connected to any raw nerves. It did not, as Rusk reported. In September, Johnson declared aid to Indonesia "essential" to American interests.[65]

VIII

This decision opened the way to a renewal of large-scale American assistance. But the renewal started gradually. Johnson approved the cotton deal the State Department had been negotiating, and with it various smaller packages. Through the end of 1966 American aid since Suharto's emergence totaled about $20 million, mostly discounted loans for American agricultural commodities.

At the beginning of 1967 no one questioned that Suharto held the reins in Jakarta. As a measure of his self-confidence the general assumed the title of acting president, pushing Sukarno further into the shadows. Some uncertainty remained regarding the policies Suharto intended to follow. The CIA noted with concern that the Indonesian government still relied on the Soviet Union and other countries of the communist bloc for military assistance. Even if the United States started shipping weapons to Indonesia in generous quantities, the CIA said, the dependence of Indonesia on the Soviets for servicing systems already in place would continue for some time.[66]

Johnson refused to get especially exercised about the Soviet connection. The president did not blame Suharto for the sins of Sukarno, specifically not for the fact that with American military aid severely curtailed since 1963 the Indonesians had had little choice but to look East for arms and ammunition. When Suharto's foreign minister Adam Malik—whom Walt Rostow described as "a shrewd Asian whose judgment is just about like ours" on important issues—assured Marshall Green that Indonesia was no longer getting Soviet weapons and had sent home many Soviet technicians, Johnson was even less inclined to hold Russian aid against the new government.[67]

In February 1967 Rostow sent Johnson a request for another batch of assistance. As per Rusk's recommendation and Johnson's instructions, the new aid would be part of a multilateral package backed by the International Monetary Fund. The United States would pay one-third of a total of approximately $200 million. In forwarding the request Rostow argued, "The Indonesia leadership has been fighting an uphill battle to undo the damage of Sukarno's years of misrule. They have worked closely with the IMF in laying out their plans for the future. Our specialists consider their plans to be realistic. But they do need help, from us and from others."

Johnson again insisted on touching congressional base. In the margin of Rostow's memo the president scribbled, "Check out House and Senate leadership. Also Foreign Relations Committee and Foreign Affairs. Top 3 on each side. Report reactions." Two days later, not having heard back, Johnson reminded Rostow, "I want 2–3 State Department people to go on the Hill and check it out carefully."[68]

Twenty-four hours after this reminder the State Department and the White House staff sent Johnson their reports. Undersecretary of State Nicholas Katzenbach said he and his assistants had spoken with Senators Mansfield, Dirksen, Fulbright, and Hickenlooper and Representatives McCormack, Albert, Ford, Morgan, Bolton, Zablocki, and Broomfield. "None had any objection or unfavorable reaction," Katzenbach told Johnson. "There were many affirmative statements of full support." White House aide William Jorden described similar positive responses.[69]

Johnson, now fully confident of his political footing, signed off on the aid proposal. He annotated Jorden's memo: "OK on assumption Japanese and Europeans go ⅔ to match our ⅓. We will go on that basis."[70]

During the succeeding months the president received additional requests for Indonesia aid. Continued consolidation by Suharto encouraged Johnson to continue to respond favorably. In March, Robert McNamara surveyed the situation. "A year and a half ago Indonesia posed an ominous threat to the United States and the Free World," the defense secretary said. "Today the prospect is dramatically altered for the better. General Suharto's government is steering Indonesia back toward a posture that promises peace and stability in Southeast Asia."[71]

The situation was equally promising in Congress. As Sukarno's position became irretrievable and Suharto expanded economic reforms, American legislators almost outshouted themselves calling for more American aid. At one meeting between AID officials and Senate leaders, a previously outspoken opponent of assistance to Indonesia complained that the Johnson administration was not doing enough to help Suharto. Rostow related the exchange to Johnson with the comment, "This belongs in the man-bites-dog category."[72]

Such a response was just what Johnson had been waiting for. As with India, regarding which he had insisted that congressional support for aid build to the point where he seemed to have no choice but to accept it, so with Indonesia. By now convinced that an investment in Indonesia's future would not come out of his supply of political capital, Johnson threw his weight behind increased aid. At a meeting in August 1967, shortly after a cabinet-room session at which legislative leaders once more urged increased aid for Indonesia, Johnson declared, "Here is a country which has rejected communism and is pulling itself up by its bootstraps." The United States should not miss the opportunity to demonstrate approval for what Indonesia was doing. "We can make it a showcase for all the world." The president added—especially meaningfully during a summer of war in Vietnam, war in the Middle East, and riots in many of the big cities of America—"It is one of the few places in the world that has moved in our direction."[73]

The foreign-aid bureaucracy geared up for an expanded program of assistance, but now the wheels moved too slowly for Johnson. "I want to do everything I can for Indonesia—as quickly as I can," Johnson told Rostow. "Send me a program."[74]

Johnson's enthusiasm for the Suharto regime not surprisingly led to some attempts to appropriate responsibility for the general's rise to power. Robert McNamara made the Pentagon's case when he asserted that the American program of military aid to Indonesia over the years had had a significant effect in predisposing Indonesia's military officers to look favorably on the United States. In particular, the defense secretary said, the Pentagon's policy of bringing promising Indonesian officers to the United States for training, followed by Washington's efforts to maintain ties with the military even during the final bleak stretch under Sukarno, had been "very significant factors in determining the favorable orientation of the new Indonesian political elite." Moreover, American material and moral support of the army had "encouraged it to move against the PKI when the opportunity was presented."[75]

Doubtless there was a certain truth in what McNamara said, but the implication of substantial American responsibility for Sukarno's replacement by Suharto was unwarranted. Even the CIA admitted as much. In the spring of 1966 Rostow had directed the intelligence agency to examine the effects of past American policies on events in Indonesia. Rostow was especially interested in knowing the effect of the tough American line against communism in South Vietnam on the Indonesian army's ascent. The CIA found no connection. "We have searched in vain," Richard Helms wrote, "for evidence that the U.S. display of determination in Vietnam directly influenced the outcome of the Indonesian crisis in any significant way." Vietnam aside, Helms discounted the importance of American actions in shaping developments in Indonesia. What had happened there had originated there. The Indonesian army's rise to power had "evolved purely from a complex and long-standing domestic political situation."[76]

Johnson was less concerned with what had led to the reversal of the situation in Indonesia than with the fact that it had occurred, and he wished to see that the new arrangements persisted. When Rostow sent the president the aid program he requested, Johnson examined it, then dispatched Hubert Humphrey to Jakarta with the good news that a sizable fresh package of American assistance was on the way. Humphrey brought back glowing reports of the Indonesians' progress. "They are enthusiastically trying to restore their economy," Humphrey said. The government was sound and responsible. "Suharto is an honest, hardworking man." Suharto and his colleagues looked to the United States for help. "The Indonesians really want our friendship."[77]

They got it, and got the American money—eventually running into hundreds of millions of dollars—such friendship entailed. Johnson judged the exchange more than fair. Suharto not only had saved Indonesia from communism, but by various other actions, including sponsoring and furnishing headquarters for the Association of Southeast Asian Nations, chartered in 1967, he helped stabilize the Southeast Asian region, which was exactly what Johnson was trying to accomplish in Vietnam. Indonesia might be a bit less than a "showcase for all the world"— in the area of civil liberties it still required considerable work—but Johnson was not complaining.

Seven | **Six Days in June**

WHEN JOHNSON became president at the end of 1963, the dispute between Israel and its Arab neighbors was fifteen years old—slightly older than the Atlantic alliance. In its decade and a half, the dispute had provoked one and a half wars: the one being the war of the 1948 Palestine partition, the half being the Suez War of 1956, for which the troubles between Arabs and Israelis deserved no more than partial responsibility, with the rest falling on Britain and France. During the first ten of the fifteen years of conflict the United States had basically stayed away. The Truman administration provided diplomatic support for partition that proved crucial to the creation of Israel, but afterward it refused to stake American prestige on the success of the Zionist experiment. Whether it would have rescued Israel had the new state faced imminent annihilation was a question the Israelis thoughtfully spared Washington.

Eisenhower showed considerably less personal interest in Israel than Truman. With most who came to American foreign policy from the international side, as opposed to those like Truman and Johnson who came from the domestic side, Eisenhower considered Israel an annoying and destabilizing implant in a neighborhood never known for stability and now, by virtue of its oil, a region of economy-shattering importance to Western security. Further, since Jewish-American politics in the 1950s was largely Democratic politics, a Republican administration could afford to pay less attention than a Democratic one to concerns close to Jewish hearts.

Yet Eisenhower was no less convinced than Truman of the need to prevent communist penetration of the Middle East. Where Truman in 1947 had announced the Truman doctrine as a measure to keep the western portion of the region in the Western camp, Eisenhower in 1957 unveiled the Eisenhower doctrine for the same purpose with respect to the area as a whole. Conceived amid the wreckage of the Suez debacle, which discredited Britain and France as guarantors of the status quo, and embodied in a congressional Middle East resolution, the Eisenhower doctrine promised American support for conservative regimes threatened by communist subversion. Critics charged Eisenhower and John Foster Dulles with irrelevance, asserting that Arab nationalism rather than communism posed the principal danger to regional things-as-they-were. The Republican administration implicitly granted the force of the objection in 1958 when, in the initial application of the doctrine, Eisenhower sent troops to Lebanon to prevent a takeover by radicals inpired by the Arab-nationalist preachments of Egypt's Gamal Abdel Nasser. Communists were nowhere to be seen, as even Eisenhower and Dulles were forced to concede. Nasser had great fun pointing out to communist-preoccupied Americans that his government put communists in jail, while Israel's—let alone America's—did not. He also got a laugh out of the fact that the American marines oversaw the installation of a Lebanese government he had suggested months before.

As matters turned out, a chief legacy of the Eisenhower doctrine, which was quietly shelved after the Lebanon affair, was the example it provided Lyndon Johnson in preemptive co-option of Congress. When the 1957 Middle East resolution came before the Senate, Hubert Humphrey asserted that it amounted to a "predated declaration of war." Humphrey meant his characterization critically. Johnson, however, recognized the shrewdness of Eisenhower's maneuver and copied it in 1964 regarding Vietnam.[1]

John Kennedy attempted the impossible in the Middle East: good relations with all parties. In Congress, Kennedy had established a record of support for Israel—perhaps partly to distance himself from the pro-appeasement and hence presumptively anti-Semitic sentiments of his father, Joseph, Franklin Roosevelt's ambassador to Britain during the late 1930s. In the 1960 campaign the Democratic nominee made much of the Eisenhower administration's neglect of Israel and pledged to do better. The Jewish vote came in heavily Democratic, and in a contest in which every vote—and maybe more, in some Illinois precincts—counted, Jewish support contributed materially to pushing Richard Nixon aside and Kennedy into the White House.

Kennedy took pains to ensure a full hearing in his administration for

the Zionist viewpoint. He brought Abraham Ribicoff and Arthur Goldberg, two committed Zionists, into his cabinet. To balance the professionally Arabist perspective of the State Department, he gave executive
counsel Myer Feldman a second job as de facto White House desk officer
for Israel. And he listened closely to the advice of Johnson, a committed
friend of the Zionist state.

On the advice of this group, Kennedy decided in September 1962 to
approve the delivery of Hawk surface-to-air missiles to Israel. The decision suggested a significant change in American policy toward the Middle East. Since 1950, when the United States had joined with Britain and
France in the so-called Tripartite Declaration to limit arms sales to the
disputants in the Arab-Israeli conflict (and guarantee frontiers), the Truman and Eisenhower administrations had discouraged a weapons race
in the area. The Soviets, however, took no self-denying pledge, instead
shipping arms first to Egypt and then to Iraq—not least because Britain
and France had managed to find loopholes in the joint Western policy.
Kennedy's Zionist advisers argued that Israel required American arms
to restore balance to the region. They contended that the Hawks, being
antiaircraft missiles, were useful only for defense, and that approval did
not commit the United States to further arming of Israel. Kennedy accepted the argument, just in time for the 1962 congressional elections.

Meanwhile Kennedy sought to establish closer ties to Egypt and Nasser. In this respect he continued a process of reconciliation begun during Eisenhower's last two years in office. A revolution in Iraq in 1958—
the trigger for the American intervention in Lebanon—brought to
power a group that proved violently antipathetic to Nasser. Before long
the Fertile Crescent witnessed a fresh outbreak of the age-old rivalry
between the valleys of the Nile and the Tigris-Euphrates. The Iraqi regime looked East for a sponsor, prompting Nasser to look West. Eisenhower responded positively but cautiously.

Kennedy went further. Kennedy's cultivation of Nasser dovetailed
with his overall policy of friendliness to important Third World leaders,
including India's Nehru and Indonesia's Sukarno. Kennedy increased
American aid to Egypt, particularly PL-480 grain. He spoke sympathetically of the aspirations of Arab nationalists. When possible he supported
Egypt's interests diplomatically. In April 1962 he went so far as to allow
Adlai Stevenson at the United Nations to vote in favor of a Security
Council resolution condemning an Israeli raid on Syria—although he
attempted to distance himself personally from the vote when pro-Israel
groups protested.[2]

Kennedy's efforts at demonstrating goodwill for all did not founder
on the Arab-Israeli reef, which for the three years of his presidency remained fairly well submerged. Instead his policy ran onto an inter-Arab
shoal at the southern end of the Red Sea. In September 1962 the imam
of Yemen died, leading pro-Nasser elements in the army there to declare

the Rassite dynasty dissolved and a republic established. Royalist backers of the imam's heir denounced the coup and took to the outback. The dispute acquired international ramifications when Saudi Arabia and Jordan—two staunch but worried monarchies—took the side of the royalists while Nasser backed the republicans.

Kennedy had to decide whether to recognize the new government, which would please Nasser and might further the cause of individual rights in Yemen, or not, which would reassure the Saudis and Jordanians. After three months' delay the president recognized the republicans. "We must keep our ties to Nasser and other neutralists even if we do not like many things they do," Kennedy remarked privately, "because if we lose them the balance of power could swing against us."[3]

Recognition might have placated Nasser had Kennedy not proceeded to placate the Saudis by sending them eight F-100 fighter-bombers and a contingent of Green Berets as military advisers. Prince Faisal (soon King Faisal) accepted the American help and remained on amicable terms with the United States, to the great relief of American oil companies and other partisans of a conservative status quo. Nasser, angered by this action, accused Kennedy of double-dealing and stepped up efforts to topple Faisal and Jordan's King Hussein.

II

Nasser was still angry when Johnson became president. He seemed likely to grow angrier when Johnson received a proposal to authorize the delivery of several hundred American tanks to Israel. As in the case of the Hawk missiles, the pro-Israel lobbyists in Washington contended that Israel required the tanks to keep pace with the radical Arabs, whom Moscow continued to arm. The tank brigade threw in the additional argument that American military assistance to Saudi Arabia, while designed to secure the oil kingdom against Nasser, potentially threatened Israel as well.

In February 1964 the CIA assessed the probable impact of approval of the tank deal. Reviewing the recent past, the agency explained, "After an initial period of concern Arab leaders had come to consider the late President Kennedy as a man who understood their problems and was at least reasonably impartial on Arab-Israeli questions. The accession of President Johnson has revived Arab fears that a Democratic administration in Washington is bound to favor Israel, especially in an election year." Shipping the tanks would represent "a fundamental break with the fifteen-year old U.S. policy of not being a major supplier of arms, especially offensive ones, to either Israel or the Arab states."

Approval of the tank deal, the CIA continued, would produce negative reactions in virtually all the Arab countries. Arab petroleum producers would respond economically. "U.S. oil interests could expect, not nationalization or confiscation, but a sharp increase in pressure from the host countries for larger shares in oil revenue." Arab diplomats would grow even less inclined than at present to support American initiatives in the United Nations.

For Israel's part, approval would give the Israelis unfortunate ideas and added leverage. When the Arabs turned to Moscow for support, as they assuredly would, Israel would try to play the anticommunist card. "The dominant policy makers in Tel Aviv would welcome the opportunity to use the Cold War as a device to assure themselves of closer U.S.-Israeli ties, particularly in the military field." The Israeli government had waged an "intensive campaign" during the last several months to get the tanks. In the current election year they were increasing the pressure. To concede them victory would make denial of future requests very difficult.[4]

Although the State Department joined the CIA in opposing the tank deal, Johnson's White House staff argued in favor. Myer Feldman judged the logic behind sending the tanks "inexorable." Appealing to Johnson's sense of American and presidential honor, as well as to the rationale of deterrence, Feldman asserted, "In view of the commitments expressed many times by many Presidents to come to the assistance of Israel if she is attacked, our basic policy must be directed toward the prevention of any aggression. Our policy must be such that American intervention will not be necessary." Robert Komer, by no means a pushover for Israel, used a different line. Johnson had been considering postponing a decision on the tanks pending a scheduled visit by Israeli prime minister Levi Eshkol. Komer figured that the president would probably say yes eventually, and he thought he ought to go ahead and say it now. With elections approaching, Komer reasoned, the president should get credit for doing Israel a good turn lest the Republicans seize the issue and negate any benefit the administration would derive. "If we are going to supply tanks to Israel, I feel it would be foolish to wait too long before the announcement is made," Komer said.[5]

Johnson chose to wait. From his days in the Senate, Johnson had distinguished himself as a firm supporter of Israel. He admired the gritty determination of the Israelis, and he knew and valued the friendship of many American Jews for whom Israel was a personal issue. He developed a warm attachment to Ephraim Evron, the minister of the Israeli embassy in Washington. Many observers detected in Johnson a distinct temperamental and emotional affinity for Jews. White House aide Harry McPherson had an unusual explanation for this, suggesting "that some place in Lyndon Johnson's blood" there were "a great many Jewish corpuscles." McPherson went on, "I think he is part Jewish, seriously. Not merely because of his affection for a great many Jews but because of the way he behaves. He really reminds me of a six-foot-three-inch slightly corny Texas version of a rabbi or a diamond merchant on 44th

Street." Personal considerations aside, the leader of the Democrats could hardly do otherwise than back a cause so important to so important a segment of the Democratic Party. And though Johnson got along with the oil interests that figured centrally in Texas politics, he did not carry cooperation to the point of pro-Arabism.[6]

Yet Johnson as president recognized that the security interests of the United States came before those of Israel. Without diminishing his devotion to Israel's ultimate welfare, he looked skeptically at what that welfare required. In the case of the tanks he aimed to get something from Eshkol in return. Of particular concern to the administration at this time was the possibility that the Israelis were developing nuclear weapons. McGeorge Bundy, briefing the president for the prime minister's visit, thought the United States should insist on an Israeli pledge not to build nuclear weapons.

Johnson agreed, and when Eshkol arrived Johnson pressed him for a promise to allow American inspection of Israel's nuclear facilities to ensure that only peaceful activities were taking place there. Eshkol gave the pledge.

At this, Bundy urged approval of the tank deal. "The Israelis have played ball with us so far," the national security adviser said. "We owe them this much."[7]

Still Johnson hesitated. He recognized the tumult the tanks would produce among the Arabs, and he preferred to let someone else take the heat—the British, perhaps, or the Germans. He told Eshkol he would help Israel get the tanks from other sources. Only if other sources refused would the United States supply them.[8]

This arrangement lacked the diplomatic and political impact of a forthright American commitment to arm Israel. "It will not give pleasure in Israel," Bundy remarked, not needing to add that Israel's displeasure would find its way back to the United States. Even so, Johnson did not deem Israel's safety at stake, and he preferred to avoid alienating the Arabs any more than necessary. Nor did he wish to provide another fillip to the Arab-Israeli weapons race.[9]

In the end the outcome was the same. For several years Israel had received weapons from West Germany. The connection got little publicity, although Nasser and other Arab leaders probably knew of it. In 1964, not long after Eshkol left Washington, Chancellor Erhard arrived. Johnson strongly encouraged Erhard to sell Israel the tanks Israeli leaders said they needed. Erhard, not wishing to annoy an administration already irritated about the offset problem, consented. The arrangement leaked, and this time Nasser made a big deal of the Bonn-Tel Aviv connection. Erhard thereupon backed out for fear of wrecking relations with the Arabs. Eshkol proceeded to cash in Johnson's commitment. Shipments of American tanks began shortly, and through the end of 1965 more than two hundred landed in Israel.[10]

III

No one, certainly not Johnson, doubted that the delivery of the American tanks would upset Nasser. It did. But Johnson did not intend to write off Egypt entirely. Despite, and partly because of, the president's strong support for Israel, the administration made an effort to reassure Cairo that a pro-Israel policy need not be anti-Egypt. Just before Eshkol arrived in Washington in the spring of 1964 Johnson met with Mustapha Kamel, the Egyptian ambassador. Bundy summarized the purpose of the meeting with Kamel. "A private conversation with him before the Eshkol visit would be very valuable political insurance with Nasser and therefore with the Arabs generally," Bundy said. "Nasser is the only leader of first-rate importance, and massage to his Ambassador would undoubtedly be received by him as a personal compliment."[11]

Unfortunately, Nasser did not take to Johnson's massage. Aside from the fact that American policy objectively favored Egypt's principal enemy, Nasser recognized the political advantages of attacking the United States, at least verbally. When the American ambassador, Lucius Battle, brought him a message from Johnson declaring that constructive relations between the United States and Egypt required quiet diplomacy rather than public name-calling, Nasser replied that quiet diplomacy might work for status-quo America but served little purpose in revolutionary Egypt. "You have got money and atom bombs, riches and power without limit," the Egyptian president complained. "These are your means. What have I got? The main weapon of the revolution is its masses: the conviction of the masses and the mobilization of those masses." The masses demanded unambiguous declarations of the regime's purposes. "Quiet diplomacy would not suit us because I would be cut off from the support of my masses."[12]

To keep the masses mobilized, Nasser decried American policy on nearly all questions: Israel, the still-running Yemen war, Vietnam, various issues in Africa, and matters dividing the industrialized nations from the Third World generally. Nasser's policy of provocation made it nearly impossible for the Johnson administration to improve relations with Cairo. Nasser's neutralism had long since alienated Congress, and his antagonism toward Israel rendered him even more irresistible as a target of congressional criticism—and even more dangerous for an administration to deal with. Bundy remarked to Johnson that if the administration sought to foster closer ties to Cairo, it should send "a very hard-hitting private message which will aim to get it across to Nasser that there is just not any real hope for our relations unless he pays proper attention to our political problems as well as his own." For a messenger, Bundy nominated Robert Komer.[13]

Komer had just returned from a trip for Johnson to Israel, where he dealt with the Eshkol government in typical straight-to-the-jaw fashion.

At the beginning of 1965 Johnson had approved an increase in American arms aid to Jordan. King Hussein, noting the assistance Washington was bestowing on fellow conservative Faisal, had thought he deserved more than he was getting. Not entirely convincingly, but with sufficient verve to worry Washington, Hussein threatened to turn to the Soviets. Johnson agreed to meet Jordan's needs.

The hard part of the transaction was selling the idea to Israel. If the Israeli government raised a stink, Congress would catch a whiff and Johnson would catch hell. As usual, Johnson desired to keep the legislature calm and focused on passing his domestic reform bills, under which he currently was burying it. To prevent Israel from complaining, the president sent Komer and Averell Harriman to Israel. Johnson chose Komer for his no-nonsense style, and Harriman because, as George Ball commented to the president, the former New York governor had "a lot of vested capital with the New York Jewish community."[14]

The two envoys told the Israelis that the United States must preempt the Soviet Union in providing weapons for Amman. "The dangers to Israel would clearly be greater if we did not help Jordan," Harriman said. Hussein would open the door to the Russians, who would leap at the chance to cause mischief. Harriman assured Eshkol that America's commitment to Israel remained strong. President Johnson was "keenly concerned about Israel's security" and recognized Israel's need for an adequate deterrent against Arab aggression. To this end, Harriman explained, the United States was willing to increase military aid to Israel. But Israel must help the president with his political difficulties. In particular the Israeli government should place the administration's arms deal for Jordan in the "proper perspective" for the American Jewish community. Such perspective-placing, Harriman said, constituted "an essential part of our relationship."

Eshkol and Israeli foreign minister Golda Meir wanted a clearer expression of the administration's intentions. They demanded a public statement of Washington's willingness to provide weapons to Israel, and they insisted on receiving more and better arms than Jordan. They also called for American support in a water dispute with Jordan.

Harriman and Komer refused to commit the administration to public arms sales to Israel. To do so, Harriman said, would provoke the Arabs excessively. The two envoys stated that the administration backed Israel's water claims in principle, but they told Eshkol to take the matter to the United Nations. Lest Israel think about acting alone on the water question, they declared explicitly that Washington could not accept unilateral measures by Israel.

Bridging the gap between the American and Israeli positions required several days of bruising bargaining. Harriman gave up and went home, leaving the problem to Komer. Komer hectored and berated Eshkol, then eventually struck a deal. Israel agreed to advise its friends in the United States not to fuss over the Jordan arms package. The Johnson administration agreed to help Israel acquire some new military aircraft. Israel first would seek the planes from third countries, but the United States would furnish them if Israel could not find what it needed. The water question was left unresolved.[15]

As in the case of the tanks, the administration within several months found itself having to make good on its pledge to act as supplier of last resort. The Israelis decided their security required American A-4 fighter-bombers. Nothing else would do. Johnson reluctantly approved the transfer.

While the administration got ready to announce the sale, it prepared for bad times in Egypt and other Arab states. Johnson had decided against sending Komer to Egypt, guessing that not much good and too much bad would follow what might appear an effort to appease Nasser. Relations remained rocky with Cairo and the other Arab capitals. Walt Rostow warned that the jets would jolt the region. "We have made every effort to stave off an explosive Arab reaction," Rostow told Johnson at the end of April 1966. "But I want to forewarn you that, at worst, Arab reaction could endanger some of our embassies."

On the other hand, there would also be positive repercussions. "This announcement will win loud plaudits from Israelis and from their friends here," Rostow predicted. Along with continuing programs of American aid, the airplane package provided "handsome evidence of your current support for Israel."[16]

A short while later Rostow tallied up just how handsome American support for Israel was. Total aid due for delivery through fiscal 1966 amounted to some $1.1 billion. Most of this took the form of nonmilitary assistance, but the weapons weighed heavily also. In 1962 the United States had supplied Israel with five batteries of Hawk missiles worth $21.5 million. In 1965 it delivered 210 tanks worth $34 million. The A-4 agreement included forty-eight planes worth $72.1 million. In addition the United States had sent Israel large quantities of ammunition, spare parts, and communications equipment. All this, Rostow noted without apparent irony, had been delivered or scheduled for delivery "despite our standing policy not to become a major arms supplier in the Middle East."[17]

IV

Irony had long formed a staple of American approaches to the Middle East. The Truman administration had sought to organize the Arab states of the region into an anticommunist alliance while pursuing policies toward Israel guaranteed to push the Arabs toward Moscow. Eisenhower had undertaken measures to limit the influence of Nasser only to be forced to Nasser's rescue during the Suez War. American troops entered Lebanon in 1958 to prevent the spread of Nasserist radicalism, but before they left they had helped install a president and prime minister recommended by Nasser. Kennedy and Johnson contributed to the arms race they had hoped to curtail.

While Johnson may not have noticed all aspects of the irony, one additional aspect certainly did not escape his attention. By 1966 opposition to the war in Vietnam had become a major political problem for the administration. Opposition was concentrated among liberals and intellectuals, a group in which American Jews stood prominently. Johnson thought it ironic—and more than ironic: positively dumbfounding—that supporters of Israel could not see the connection between what he was trying to accomplish in Vietnam and what he had pledged to accomplish for Israel. He thought the connection worth pointing out, to the mutual benefit of Vietnam and Israel. When Zalman Shazar, the president of Israel, called at the White House during the summer of 1966, Johnson asked, "If because of the critics of our Vietnam policy we did not fulfill our commitments to the sixteen million people in Vietnam, how could we be expected to fulfill our commitments to two million Israelis?" The president lamented that some "friends of Israel in the United States" were demanding American withdrawal from Vietnam. He requested "sympathetic understanding" of the administration's efforts and hinted that Israel might help.[18]

Whether at Israel's encouragement or otherwise, certain members of the American Jewish community took the hint. Abraham Feinberg, a deep-pocketed Democratic stalwart and a Johnson liaison with American Jews, offered to organize a campaign demonstrating the connection between American credibility and Israeli security. Feinberg explained his idea to Walt Rostow, who related it to Johnson. "The theme is as follows," Rostow wrote: "The whole fate of Israel depends on the credibility of U.S. commitments. If the U.S. were to fail to meet its commitments in Viet Nam, what good would its commitments be to Israel?" Rostow appended his own comment, "I think this is a first-rate approach and I told him so." Johnson thought so too.[19]

But Feinberg's efforts produced no noteworthy conversions on the Vietnam issue, although they may have tempered some criticism. Part of the problem was that Vietnam cut the other way as well. Supporters of Israel could fear that the American commitment to Vietnam would so preoccupy the president that he would be distracted from Israel. While a great power has to demonstrate its determination, it also has to show judgment and a sense of proportion. Few Vietnam critics doubted Johnson's determination, but many doubted his judgment.

V

Those who feared American distraction need not have worried. If Johnson often appeared preoccupied by Vietnam, as indeed he often was, when the biggest Middle East crisis of his presidency occurred he refocused his faculties at once. The crisis climaxed in June 1967, but it had been building for some time. In February 1966 a coup in Syria brought an ardently socialist and militantly anti-Israel government to power. The Baathist regime launched simultaneously a diplomatic offensive toward the Soviets and a guerrilla offensive against the Zionists. The former produced increased Soviet attention to Syria and corresponding American concern. The latter produced raids on Israel—some from Syrian soil, some from Jordanian—and consequent Israeli reprisals.

A major Israeli attack on the Jordanian village of Samu in November 1966 quieted the guerrilla assaults for a time, but the new year brought renewed violence. By the beginning of April 1967 artillery exchanges across the Israel-Syria frontier in the Golan Heights had become an everyday occurrence. On April 7 Israel launched air raids against the Syrian gunners, provoking Damascus to scramble its own planes and join the air battle. The Israelis shattered the Syrian force and buzzed the Syrian capital to skywrite their scorn.

Though Syria had seceded from its short-lived union with Egypt in 1961, Nasser, still aspiring to leadership among the Arabs, felt great pressure to respond to the Israeli attacks. Nasser possessed a more realistic sense of the power balance between the Arab states and Israel than the Syrians did, and he had no compelling desire to take on the Israelis. But because opposition to the Zionists was the one big issue holding the Arab movement together, he could not allow anyone to seem more determinedly opposed than he. In November 1966 he concluded a mutual defense pact with Syria. Following the April 1967 Israeli air victory, the Syrians demanded that Egypt honor its commitment.

In a related area Nasser felt equal pressure. Arab irredentists charged the Egyptian president with cowering behind the United Nations Emergency Force, which had separated Egyptian and Israeli armies in the Sinai since the 1956 war. Syrians, Jordanians, Iraqis, and other Arabs complained that Nasser knew only how to talk, that if he was serious about Arab unity and the struggle for Palestine he would order UNEF out of Egypt and punish Israel.

Nasser bowed to the criticism. On May 16 he told the United Nations to withdraw its troops from Egyptian territory. Two days later the secretary general, Burma's U Thant, ordered compliance. On May 22 Nasser cranked matters tighter. Just as the last UNEF troops left Sharm el Sheikh on the Strait of Tiran at the entrance to the Gulf of Aqaba, the Egyptian president announced that Egypt would close the strait and thereby the gulf to Israeli traffic. Meanwhile he increased the number of Egyptian troops near Israel's southwestern border.

To Israel, Nasser's actions represented an intolerable threat. The eviction of UNEF and the massing of troops appeared to presage an attack on Israel's frontier, while the closing of the Strait of Tiran deprived Israel of reasonable access to most of Asia and Africa and seemed to portend a trampling of Israeli rights.

The American response was to advocate caution. Johnson wrote Eshkol on May 17 explaining that while he understood Israel's worries, he expected Israel to exhaust all peaceful avenues for resolving the crisis. "I know that you and your people are having your patience tried to the limit," Johnson said, before continuing: "I would like to emphasize in the strongest terms the need to avoid any action on your side which would add further to the violence and tension in the area." Johnson warned against rash measures, especially ones that took Washington by surprise. "I urge the closest consultation between you and your principal friends. I am sure you will understand that I cannot accept responsibilities on behalf of the United States for situations which arise as the result of actions on which we are not consulted." A short while later Johnson wrote Eshkol again, calling on the prime minister to demonstrate "steady nerves" and pledging the United States to resist aggression in the Middle East. In this second letter Johnson deleted from the State Department's draft a phrase that declared that the American commitment to regional security "definitely includes Israel." He wanted the Israelis to feel confident of American support, but not too confident.[20]

Johnson also wrote to Nasser. He reiterated America's friendly feelings and his own toward the aspirations of Egypt and of the Arab people. He promised Nasser that the United States would listen sympathetically to Egypt's legitimate grievances. He acknowledged difficulties between the United States and Egypt in the past. These were not the present issue, though. "Right now, of course, your task and mine is not to look back but to rescue the Middle East—and the whole human community—from a war I believe no one wants." He offered American support for new efforts to solve the problems between countries of the Middle East, and he volunteered to send the vice president to the region in such an effort—"if we come through these days without hostilities."[21]

Finally, Johnson contacted Soviet premier Aleksei Kosygin. The administration rightly suspected the Soviets of pushing the Syrians toward violence. Beyond making trouble for Israel and through Israel for the United States, Moscow evidently thought Syrian pressure on Nasser would encourage unity between the principal radical regimes of the region and would weaken the conservative, pro-Western governments. Johnson warned Kosygin that the stakes of conflict between Arabs and Israelis might be higher than the Kremlin had calculated. "The increasing harassment of Israel by elements based in Syria, with attendant reactions within Israel and within the Arab world, has brought the area close to major violence," Johnson told the Soviet leader. "Your and our ties to nations of the area could bring us into difficulties which I am confident neither of us seeks. It would appear a time for each of us to use our influence to the full in the cause of moderation, including our influence over action by the United Nations."[22]

Johnson complemented these private cautions with a public statement of American intentions. On May 23 the president went on television and radio to reaffirm the position established by Eisenhower in 1957 that the Aqaba Gulf was an international waterway open to free navigation by ships of all countries. He declared the Egyptian blockade illegal and gravely threatening to world peace. He said he was distressed at the withdrawal of UNEF. He regretted the violence in the region. He deplored the buildup of military forces in the area. Significantly, however, he stopped short of declaring explicit support for Israel, and he left responsibility for a resolution of the crisis to the international community.[23]

VI

During the last week of May, Johnson followed a delicate strategy. As in the close calls between Greece and Turkey over Cyprus, he had to convince each side of the danger that tough talk and bellicose posturing would get out of control. Threats of war and preparations for war all too often precipitate war. Yet at the same time Johnson had to avoid unduly frightening the potential belligerents. Once either party became convinced of war's inevitability, that party would feel an overwhelming temptation to get in the first blow. This was particularly true of the Israelis, who enjoyed almost no room for retreat if the war started badly. To them Johnson had to offer reassurance they would not be left to fight alone— but only if they did not start an avoidable war.

Staying on the tightrope required knowing what each side intended, as distinct from what each side said. American analysts attempted to discern the former, with indifferent success. The Israelis gave the appearance of thinking the Arabs were really about to attack. On May 25 Rostow relayed what he called a "highly disturbing estimate" from Israeli intelligence via the CIA. According to the Israelis, Nasser evidently had, or thought he had, the backing of Moscow in his plan to squeeze Israel, and the Kremlin apparently underestimated Israel's sensitivity to squeezing. Rostow judged the Israelis mistaken, as did the CIA, whose own assessment threw "a great deal of cold water on the Israeli estimate," as Rostow put it to Johnson. The trouble was that the Israelis were far more likely to act on Israeli perceptions than on those of Americans. Rostow remarked, "The two estimates—Israeli and CIA—both show how explosive are: Israeli anxieties; Nasser's hopes of picking up prestige; U.S.S.R. desires for gaining prestige, short of war."[24]

Slightly less alarming was a CIA evaluation of Nasser and his aims. The intelligence agency considered Nasser more cautious than his speeches sounded. Rostow paraphrased the CIA report for Johnson: "It makes Nasser out to be shrewd, but not mad. As it points out, he is gambling on the Israelis going to the U.N. rather than attacking." Rostow suggested that Nasser was exaggerating his recklessness, perhaps hoping to extort some American aid in exchange for good behavior.[25]

Yet Nasser was hardly a free agent. In the White House view, his Arab critics had pushed him into a corner from which war might afford the only exit. Further, Nasser was acting on information of dubious reliability. Rostow outlined the situation. "The U.A.R.'s brinkmanship stems from two causes," Rostow said. "(1) The Syrians are feeding Cairo erroneous reports of Israeli mobilization to strike Syria. Regrettably, some pretty militant public threats from Israel by Eshkol and others have lent credibility to the Syrian reports. (2) Nasser probably feels that his prestige would suffer irreparably if he failed a third time to come to the aid of an Arab nation attacked by Israel. Moderates like Hussein have raked him over the coals for not coming to Jordan's aid in November or to Syria's when Israel shot down six of its MIGs last month."[26]

Johnson had many fewer informal contacts with Nasser than with the Israelis, but he used those he did have to supplement the information he received from the CIA and other official sources. Robert Anderson, who had served as Eisenhower's back-channel envoy to Nasser before becoming Johnson's Panama negotiator, retained ties to Cairo. Anderson spoke with Rostow shortly after UNEF left Sinai. Anderson said Nasser felt deprived of communication with Washington. Nasser was accustomed to high-level emissaries from the White House, and he had not received any lately. Anderson added that Nasser's recent actions were motivated not by increased hostility toward Israel but by the Egyptian president's political troubles. While Syria and Jordan were berating him abroad, food shortages and other sources of popular discontent vexed him at home. Bread riots were a serious possibility. Anderson offered to go talk to Nasser if Johnson wanted him to.[27]

A few days later Johnson received another sounding from Cairo, in the form of a message from a representative of an American company, ALCO Products, with operations in Egypt. This individual passed along an account of a recent conversation with Nasser. Nasser said Egypt had no intention of attacking Israel. His actions were designed to flush out Hussein and other conservatives and show them to be false friends of the Arab masses. He wanted the United States to avoid direct involvement, such as landing troops or repositioning the American Mediterranean fleet. He said he was willing to cooperate with Washington if Washington would cooperate with him.[28]

Communications such as this one involved obvious pitfalls. Johnson couldn't tell whether he was getting Nasser's message undistorted by the prejudices and interests of the intermediary, whom the president in this case did not know. Assuming the message arrived straight from Nasser, Johnson couldn't tell how much it reflected Nasser's true intentions and how much simply what Nasser wanted the Americans to think he thought.

Communications with Israel, while more direct, were scarcely less complicated. Israeli foreign minister Abba Eban arrived in Washington on May 25. He first visited the State Department for a talk with Dean Rusk, who related the session to Johnson. The Israelis, Rusk said, did not anticipate constructive measures by the United Nations. "They have absolutely no faith in the possibility of anything useful coming out of the U.N." The secretary reported that he had pressed Eban hard about the need for Israel to refrain from preemptive moves. He urged Johnson to do likewise. The president might also underline the fact that if war came, the question of who started it would be vitally important.[29]

Johnson prepared carefully for his meeting with Eban. Just prior to the foreign minister's arrival in Washington, Johnson convened his top advisers. Rusk presented the latest intelligence analysis and characterized the current situation as "serious but not yet desperate." To what the president had already heard about Egypt and Israel, Rusk added that the Soviets were still agitating affairs, albeit more vigorously for public consumption now than for private reality. "Privately we find the Russians playing a generally moderate game, but publicly they have taken a harsh view of the facts and have laid responsibility at Israel's door—and by inference at ours." While Egypt and Syria continued to declare that they had the Kremlin's backing, this backing seemed rather less than complete.

Treasury secretary Fowler suggested using the World Bank and the International Monetary Fund to persuade Nasser to back down. Johnson agreed that the administration should examine "all the cards we have to play in this field." The United States, Johnson said, must seek international cooperation even if the prospects seemed dim. "We should play out the U.N. and other multilateral efforts until they are exhausted. I want to play every card in the U.N." He added, though, that he was not holding his breath for a rescue by the United Nations. "I've never relied on it to save me when I'm going down for the third time." Neither would he count on the Atlantic allies. The British had suggested a joint initiative to break the Egyptian blockade. While Johnson endorsed the idea, he would believe it when it happened. "I want to see Wilson and de Gaulle out there with their ships all lined up," he said. "But all of these things have a way of falling apart."

Johnson asked for a military assessment of matters. Earle Wheeler indicated that forcing passage through the Strait of Tiran would present problems for the American navy during the next couple of weeks. Egypt had two submarines in the area, and the nearest available American antisubmarine vessels were currently at Singapore, fourteen days away. Wheeler said that of course the Mediterranean fleet had antisubmarine capabilities, but if Nasser was serious about closing the strait he was not likely to let American antisubmarine craft through the Suez Canal. Wheeler went on to say that a war Israel started for the purpose of lifting the Egyptian blockade could quickly spread. "If the Israelis move, it might not be possible to localize a strike designed simply to open the straits." Wheeler discussed possible use by Israel of "unconventional weapons"—that is, nuclear weapons. The notes of this portion of the meeting remained classified a quarter century later. But the excisers spared Wheeler's firm conclusion: "The Israelis can hold their own."

Robert McNamara expanded on the prospects of a wider war. The defense secretary predicted that the outset of the conflict would witness a struggle for air superiority. Each side would run through its supply of airplanes, or at least through the rockets and other armaments the planes required. Each would then turn to its prime supplier: Israel to the United States, Egypt and Syria to the Soviet Union. Presumably the Israelis would shoot down more MiGs than they would lose planes of their own. If the MiGs went down with pilots aboard, Moscow might feel obliged to send not only new planes but Soviet pilots. Soviet pilots would complicate matters in one or both of two ways. If they showed greater ability than their Arab predecessors, they might kill lots of Israeli planes and pilots, evening the battle and increasing pressure on the United States to respond in kind. If they did no better, they would get themselves killed, requiring Moscow to escalate or risk humiliation.

Johnson asked for opinions regarding the objectives and motives of the Egyptians and the Soviets. He wondered specifically whether Moscow was aggravating the current situation to distract the United States from Vietnam. Wheeler and CIA director Helms guessed not, though the two agreed that the Kremlin would take full advantage of any distraction the Middle East offered. Helms suggested that the Russians liked the level of tension about where it was. Moscow also liked the idea of forcing a more complete association of the United States with Israel. "The Soviets would like to bring off a propaganda victory as in the 1950s with them as the peacemakers and saviors of the Arabs, while we end up fully black-balled in the Arab world as Israel's supporter." Regarding Egypt, Helms estimated that Nasser had achieved his objectives for the moment. He had adopted a hard line against Israel, thereby confirming his credentials as defender of the Arabs against the Zionist interlopers.

Lucius Battle, recently ambassador to Egypt and now assistant secretary of state for the Middle East, took issue with Helms. Battle said that until a week earlier he would have agreed that Nasser chiefly sought a propaganda victory. But Nasser's announcement of the Aqaba blockade indicated greater seriousness than Cairo had shown previously, and greater danger for Israel and the United States. Either Nasser had a more solid and more sweeping Soviet commitment than the administration knew, or he had gone "slightly insane." Battle elaborated: "It is most uncharacteristic for Nasser not to leave a door open behind him, and that is exactly what he appears to have done in this case." The assistant secretary amplified the list of political problems others in the administration had said Nasser faced—food shortages, a fundamentally failing economy, challenges to Egypt's leadership of the Arab movement from both radicals and conservatives, a loss of prestige in the Third World generally—and indicated that Nasser might be trying to recover his position by means of a stunning stroke against Israel.[30]

Two days later, just prior to Johnson's meeting with Eban, the president gathered his top advisers again. The purpose was "to walk around the problem of the Middle East in an open-minded way," as Rostow put it, "to see all the angles, all the elements." No startlingly new insights emerged, which was hardly surprising since by this time, with high-level meetings on the Middle East a daily occurrence, the angles had been pretty well covered. Rostow did raise the crucial question Eban would ask, probably implicitly: "What can you offer better than a preemptive strike?"[31]

Johnson tried to provide an answer in his meeting with the Israeli foreign minister. The president promised Eban that the United States would apply its "best efforts and best influence" to keep the Strait of Tiran open. But he could do nothing until the United Nations secretary general, now investigating the affair, delivered his report. "If we move precipitously," Johnson declared, "it would only result in strengthening Nasser." The president understood that Israel expected nothing worthwhile to come of U Thant's actions. The United States, he said, had no illusions on the subject either. Still, the United Nations process must run its course. Afterward would be the time for other measures. "When it becomes apparent that the U.N. is ineffective, Israel and its friends, including the United States, who are willing to stand up and be counted can give specific indication of what they can do." In particular the president mentioned an international naval force to challenge Nasser's blockade.

Johnson strove to keep Israel from hitting first. "Our best judgment is that no military attack on Israel is imminent," he said. "Moreover, if Israel is attacked, our judgment is that the Israelis would lick them." Time would not work against Israel. Israel could afford to wait until Thant reported to the Security Council. While maintaining full mobilization was not without disadvantages, the alternatives were worse. "We know it is costly economically, but it is less costly than it would be if Israel acted precipitously and if the onus for initiation of hostilities rested on Israel rather than on Nasser."

Eban asked, "I would not be wrong if I told the Prime Minister that your disposition is to make every possible effort to assure that the strait and the gulf will remain open to free and innocent passage?" Johnson replied that this was correct.

The president reiterated, "We are Israel's friend. The straits must be kept open." Yet an answer to the present predicament required time. "We cannot bring about a solution the day before yesterday." He told Eban it was "inconceivable" that Israel would decide for war while peaceful efforts to end the blockade were still under way. Very deliberately and repeating himself for emphasis, Johnson declared, "Israel will not be alone unless it decides to go alone."[32]

VII

Johnson thought he had handled Eban well. At dinner that night with old friends from Texas, the president recounted the conversation. Speaking of Eban and the Israeli ambassador, who had accompanied the foreign minister to the White House, Johnson said, "They came loaded for bear, but so was I. I let them talk for the first hour and I just listened, and then I finished it up the last fifteen minutes. Secretary McNamara said he just wanted to throw his cap up in the air, and George Christian said it was the best meeting of the kind he had ever sat in on."[33]

Eugene Rostow remembered the meeting differently. The undersecretary of state for political affairs later remarked that Johnson came away from his session with Eban discouraged. Johnson, Rostow recalled, believed that the Israelis were going to attack. Eban had not said they would, but neither had he said they would not. Johnson's instincts, according to Rostow, said they would.[34]

Arthur Goldberg agreed with the discouraged Johnson, less on the basis of instinct than on what he was hearing at the United Nations. Eban traveled from Washington to New York, where he met with Goldberg, Johnson's United Nations representative. In his memoirs Eban described his feelings as he approached the meeting. "Nothing in my talks in Washington had made Israel's tasks lighter or her dilemma less sharp," Eban wrote. The foreign minister's feelings showed in his conversations with Goldberg, and as soon as Eban left, flying back to Israel for a crucial cabinet meeting, Goldberg contacted the White House to convey his conviction that Eshkol's government would decide within hours to strike.[35]

While awaiting the outcome of the Israeli meeting, the Johnson administration continued efforts to calm the international atmosphere. Malcolm Toon of the State Department's Soviet bureau spoke with the chargé d'affaires of Moscow's embassy, Yuri Chernyakov. Toon said the United States was engaged in a "maximum effort" to restrain the governments at odds in the current dispute. He made a point of adding,

"Including Israel." Toon remarked that the United States was encouraged by a recent statement from Moscow calling for peaceful resolution of the Middle Eastern troubles. He said he hoped Moscow's actions mirrored its words.[36]

The meeting of the Israeli government came and went, and nothing happened. "It looks as though they have decided not to go to war at this time," Walt Rostow commented to the president on May 28.[37]

Nasser too indicated a desire to avoid immediate hostilities. At a news conference on the same day—a transcript of which American officials examined carefully—the Egyptian leader placed the present crisis in longer perspective. The problem involving Israel, he said, was not a matter of the Strait of Tiran. The problem was the "aggression which took place and continues to take place against one of the homelands of the Arab nation in Palestine." Nasser did not say so explicitly, but because this was the basic problem that had set Arabs and Israelis at odds for nearly twenty years, its solution presumably did not have to come in the next few days or weeks.

All the same, Nasser was not about to fold his hand. At this same news conference he was asked if he was taking account of possible American intervention in the Arab-Israeli dispute. "I do not take the United States into account," he replied, "because if I take the United States, the Sixth Fleet, the Seventh Fleet and the U.S. generals into account, I shall never be able to do anything or to move." Regardless of what the Americans did, Egypt would not retreat. "If the United States intervenes, we must defend ourselves and defend our rights."[38]

The American embassy in Cairo interpreted these remarks to mean that though Nasser would not back away from a military showdown with Israel, neither would he instigate one. For the time being, a belligerent posture appeared to suit his purposes better than actual belligerence. Nasser added to this impression on May 29 by inviting the American government to send a special envoy to discuss the situation with him. He mentioned Robert Anderson by name.[39]

Johnson quickly dispatched Anderson to Egypt. Simultaneously he weighed a proposal from the State and Defense Departments for opening the Strait of Tiran. Rusk and McNamara offered a three-stage plan for ensuring free passage. Stage one involved more of what the administration was doing already: working through the United Nations. While this failed, as it was and almost certainly would, the administration should prepare to move to stage two: a declaration by the major maritime nations of support for the principle of unfettered navigation of the strait and the Gulf of Aqaba. Such a declaration would convey conviction only to the extent the administration accompanied it with arrangements for stage three: a confrontation against the Egyptians by a multinational fleet. The more nations that contributed vessels to the fleet the better. Rusk and McNamara thought the United States could depend on Britain, while Canada and the Netherlands were maybes. Beyond these three, volunteers would come harder. The two secretaries cautioned that the military risks of blockade-breaking were "not negligible." If shooting started, it might not easily stop. Nor were the political risks negligible. Many Americans would question the wisdom of putting American forces in danger in such a matter. Consequently the administration should purchase insurance. "We believe that a joint congressional resolution would be politically necessary before U.S. military forces are used in any way." Despite expressions of strong support for Israel in the legislature, the administration must take care. Vietnam had lawmakers skittish about attaching American prestige to foreign causes, however worthy. Rusk and McNamara granted that "many congressional doves may be in the process of conversion to hawks"—doves on Vietnam becoming hawks on the Middle East, they meant—but they added, "The problem of 'Tonkin Gulfitis' remains serious." (Though Congress had passed the administration's Gulf of Tonkin resolution by an almost unanimous vote in 1964, more than a few legislators were having second thoughts about this broad delegation of war-making authority to the president.)[40]

Johnson agreed with McNamara and Rusk on the politics of intervention in the Middle East. The president intended to ask Congress for authority to use American ships in the Middle East if the situation came to that. Meanwhile he began checking on congressional views. Representatives Emanuel Celler and Thomas Morgan, after polling their colleagues, described "a clear majority" in the House of Representatives behind Israel. "They feel Israel is being pushed around by Nasser," Celler and Morgan said. "They feel we shall, in the end, have to do something to open the blockade at Aqaba—multilaterally or otherwise." Celler offered Johnson a strong statement backing the administration. He said he had gotten more than one hundred representatives to sign it, "without even trying." Johnson asked Celler to keep the statement handy but not to make it public just yet.[41]

Several factors inclined Johnson to delay. First, he wanted to let support for vigorous measures continue to build. This required allowing U Thant to keep trying until a negotiated settlement became undeniably impossible. Second, even if Congress approved American participation in a multilateral effort to raise the blockade, the requisite fleet could not be formed before the middle of June. Third, breaking the blockade would be difficult and dangerous. Joint Chiefs chairman Wheeler asserted that the United States "must be prepared to conduct strikes against the UAR [Egypt] ranging from discriminating air and naval strikes against selected military targets to full-scale air strikes against all UAR military targets." Wheeler also noted the possibility of encounters with Soviet forces. This was not the sort of thing Johnson wanted to get into if he could avoid it. Fourth, Johnson wished to hear what Nasser had to say to Robert Anderson.[42]

Anderson met with Nasser on June 2. The meeting came against the background of Egyptian displeasure with the United States during the previous few years, augmented by the administration's apparent siding with Israel in the current dispute. Just prior to Anderson's arrival in Cairo, retired ambassador and Middle East veteran Charles Yost had a long talk with the Egyptian foreign minister, Mahmoud Riad. As Yost related the conversation, Riad spoke with "intense and uncharacteristic emotion and bitterness." Yost continued, "He said he had given up hope of the United States ever dealing impartially with Arab-Israeli issues and had concluded that political pressures inside the United States would always make it impossible for the United States Government to support measures in or out of the United Nations which Israel opposes." Riad complained at Israel's violations of armistice agreements and United Nations resolutions. He cited specifically the occupation by Israeli forces of demilitarized zones along Israel's borders with Syria and Jordan. He averred that Egypt as Egypt had no quarrel with Israel. But Egyptians as Arabs could not overlook the fate of the more than one million refugees displaced by Israel's creation. "This can never be forgotten by Arabs," he said. He declared that Israeli attacks on Jordan and Syria had compelled his government to respond. Though the Egyptian military was demanding that the response take military form, his government sought to avoid war. War would cause "great destruction on both sides." All the same, Egypt would not retreat.[43]

Nasser developed these same themes to Anderson. The Egyptian president said he did not desire war. Egypt would not initiate hostilities, although Syria or radical elements among the refugees might. Egypt would wait until Israel moved. Yet Egypt would not allow itself to be caught unprepared, as it had been in 1956. He had ordered troop levels and readiness increased in the Sinai in order to prevent surprise. He said Egypt had developed "elaborate plans" for "instant retaliation" in the event of Israeli attack, and he was confident Egypt's military forces could hold their own if war came.

Nasser explained to Anderson Egypt's position regarding the Strait of Tiran. For eight years after 1948 the strait had been closed to Israel. The channel was less than three miles wide and therefore did not qualify as an international waterway. The strait had been opened only by the "illegal act" of Britain, France, and Israel as part of the 1956 war. In closing the strait, Nasser said, he simply wished to return to the status quo ante bellum. Besides, because the armistice agreements of 1949 and 1956 had never given way to a peace treaty, Egypt remained legally at war with Israel. And as long as Israel insisted on acting like an enemy, it must expect treatment as an enemy.

Anderson asked what Egypt required in order to make peace with Israel. Nasser replied at once: a solution to the Palestinian problem. Anderson, who as Eisenhower's envoy had discussed the same issue with

Nasser in 1956, queried whether permission to return for a limited number of the Palestinians and monetary compensation for the rest would suffice. Nasser said it would not. Nearly all the Palestinians insisted on going home. They would continue to do so even if offered compensation.

Nasser said he wanted friendly relations with the United States. He said he was in no sense a communist despite Egypt's ties to the Soviet Union. He criticized American policy for being unduly influenced by the large Jewish vote in the United States.

After the meeting ended, Anderson cabled his impressions to Johnson. On the crucial question of Nasser's willingness to go to war, Anderson wrote, "He kept reassuring me that he was not going to start a war but that he was not responsible for all groups and that he would intervene in any actual conflict begun." As to whether Nasser might modify his current position, Anderson commented, "For the time being I think he will remain firm." Anderson had stopped in Lebanon on the way to Egypt. In Beirut he had looked up acquaintances from other countries of the region. He had discovered, significantly, that even Saudis, Kuwaitis, Lebanese, and Iraqis who opposed Nasser on most issues were now rallying to his support. Nasser knew this, of course, and the Johnson administration must bear it in mind in formulating American policy. With the backing of nearly all the Arabs, Nasser would probably resist attempts to force passage to the Gulf of Aqaba. "I believe he would regard any effort to open the Straits of Tiran as hostile."[44]

Anderson's message reinforced the Johnson administration's belief that Egypt would not initiate an armed conflict, but it afforded little hope beyond that. Nasser's words suggested he was unwilling to try to control the Syrians and the Palestinians, either of whom might happily provoke a war. Egypt would then join the fray, with much the same result as if Nasser started it.

Even so, Johnson worried more about Israel than about the Arabs. On Eban's visit to Washington, the Israeli foreign minister had indicated less confidence in Israel's ability to defeat the Arabs than American officials thought the balance of forces warranted. The Joint Chiefs predicted an Israeli victory within five to seven days. If Israel struck first, the briefer prediction would hold and Israel would suffer fewer casualties. If Egypt or Syria got in the initial blow, the war would last a few days longer and would exact from Israel a higher price. But by no means was the essential security of Israel at risk.[45]

Yet the Israeli government thought so, or at least chose to give the impression it did. Perhaps the Israelis were simply building a case for teaching the Arabs a lesson. In any event, Johnson felt obliged to repeat his commitment to Israel's safety, in hope that this would ease the pressure for preemption. On June 3 he wrote Eshkol congratulating the prime minister and his associates on their "resolution and calm in a situation of grave tension." Johnson affirmed two basic principles of American policy pertinent to the current crisis: support for the territorial integrity and political independence of all countries in the Middle East, and support for freedom of the seas. He added explicitly that the United States judged the Gulf of Aqaba an international waterway.

In the same letter, Johnson once more urged Eshkol to refrain from hasty action. The United States was seeking international cooperation in fashioning measures to lift the blockade, the president declared. American representatives at the United Nations and in foreign capitals were working around the clock to gain this cooperation. Their efforts required time to yield results. Israel must provide that time. Without it, the United States could not give Israel the guarantees Israel wanted and the United States wanted to give. "Our leadership is unanimous that the United States should not move in isolation," Johnson said.[46]

VIII

Johnson had succeeded in persuading Turkey to stand down when Ankara was on the verge of war with Greece, but he failed with Israel. His failure was partly his fault, partly that of Congress, and partly the Israeli government's.

By contrast to his withering warning to Inonu of the consequences of starting a war, Johnson's message to Eshkol amounted to little more than a polite request. Inonu had no doubt that failure to comply with Washington's wishes would bring down America's wrath. Eshkol had plenty.

Eshkol's doubt reflected not simply Johnson's equivocation. It also reflected the Israeli prime minister's recognition that Congress would not readily allow the president to punish Israel for excessive enthusiasm in its own defense, even had Johnson been so inclined. Johnson could threaten Turkey, confident that few in Congress would object. (Threatening Greece entailed more problems, given the influence of Greek-Americans.) Johnson could not threaten Israel without placing himself in opposition to the many defenders of Israel on Capitol Hill. Eisenhower in 1956 had found it much harder to yank the leash on little Israel than on big Britain and France, due to the protectiveness of the pro-Israel bloc in Congress. A Democratic president would find the yanking harder still.

Yet even if Johnson had thundered and Congress cooperated, the Israelis might have gone to war. Turkey could cancel its war plans in 1964, knowing that the most it hazarded was the welfare of the Turkish Cypriots. Turkey would survive, perhaps with prestige diminished by failure to defend the Turkish community on Cyprus, perhaps with population increased by refugees from that community. For Israel the stakes were higher. Maybe events would prove American confidence in Israel's fighting ability well placed. Maybe not. With no room to retreat, Israel could not rely on guesses. Besides, the savings in lives and money resulting from a first strike might seem small to the Pentagon, which bought body bags by the thousand and spent dollars by the tens of billions. But the savings were not small to a small country with a small population. Lastly, Nasser and his ilk needed a bloody nose. Until they learned not to trifle with Israel, Israelis believed, Israel and the Middle East would never know peace.

On the morning of June 5 Israel attacked Egypt. The Israeli air force struck by surprise, destroying more than three hundred Egyptian planes in the first three hours of the war and losing fewer than twenty aircraft of its own. Shortly thereafter the Israelis flew against Jordan, eliminating that country's air force in minutes. Syria received the same treatment early in the afternoon. Israel's victory in the air essentially guaranteed victory on the ground. Israeli armor supported by Israeli jets invaded the Sinai, severing Egyptian lines and advancing rapidly toward the Suez Canal. Israeli forces occupied the West Bank and seized the Old City of Jerusalem.

Johnson learned of the outbreak of fighting at 4:30 A.M., Washington time, on June 5. He immediately wanted to know who had started it. Walt Rostow, on the other end of the telephone line, couldn't say for certain. The Israeli defense ministry was asserting that Egypt had moved first, and American officials in the area could not confirm or deny. Abba Eban repeated the cover story in a call to the State Department. The administration refused to buy, knowing that Nasser was not foolish enough to tempt fate so egregiously. Johnson's spokesman George Christian told reporters the White House was investigating the matter. Dean Rusk said, "The facts are still very obscure."[47]

Within hours the Israeli story fell apart. The Israelis failed to produce evidence of an Egyptian incursion, while the wrecks of Egyptian planes caught on the ground testified convincingly against it. When Eshkol sent a message to Johnson on the afternoon of June 5 the prime minister did not—quite—say Israel had responded to an Egyptian attack. Yet he did claim Israel had acted out of self-defense. "After weeks in which our peril has grown day by day, we are now engaged in repelling the aggression which Nasser has been building up against us." Eshkol recited the list of Egyptian and Arab provocations against Israel, from the various guerrilla raids to ejection of UNEF to the blockade of Aqaba and Nasser's massing of troops in Sinai. "All of this amounts to an extraordinary catalogue of aggression, abhorred and condemned by world opinion and in your great country and amongst all peace-loving nations." Reminding Johnson of the six million Jews killed by the Nazis, Eshkol thanked the president for America's support of Israel in the past and said he looked forward to America's support in the future. While he indicated that Israeli forces could handle the Arabs, he had a favor to ask the president. "I hope that everything will be done by the United

States to prevent the Soviet Union from exploiting and enlarging the conflict."[48]

Such was precisely Johnson's intention. The president appreciated the diplomatic difficulties the Israelis' preemptive attack created for the United States, but he also realized that their swift success at arms had averted a far more difficult scenario, one in which Israel appeared about to *lose* the war. Walt Rostow recalled the administration's attitude several months after the fact. "President Johnson has never believed that this war was anything else than a mistake by the Israelis," Rostow remembered. "A brilliant quick victory he never regarded as an occasion for elation or satisfaction. He so told the Israeli representatives on a number of occasions. However, at the time, I should say that, war having been initiated against our advice, there was a certain relief that things were going well for the Israelis." The progress of the war confirmed American predictions of a Israeli win and eliminated the possibility the Arabs would drive the Israelis into the sea, Rostow said. "That would have been a most painful moment and, of course, with the Soviet presence in the Middle East, a moment of great general danger."[49]

If the Israelis had appeared about to lose, the Johnson administration would have been sorely tempted to go to their rescue. Both Americans at large and Congress supported what Israel stood for. American prestige, more than ever following the commencement of American weapons deliveries under Kennedy and Johnson, rode into battle with Israel. To acquiesce in Israel's destruction would have been unthinkable.

Moscow confronted an analogous if slightly less excruciating choice. Neither Egypt nor Syria faced national destruction at the hands of the Israelis. Nor was there an Arab bloc in the Soviet Politburo. Even so, the humiliation of Egypt and Syria translated into an embarrassment for the Soviet Union. Johnson was not alone in worrying about credibility. Indeed, Moscow had matters worse in this regard than Washington; in that while there existed but one focus of the "free world," the socialist movement had two centers. The Chinese even more than the Americans would delight at the Soviets' loss of face.

During the first hours of fighting, Johnson's attention concentrated on Moscow. As soon as he got out of bed on the morning of June 5 he sent a message to Kosygin to express America's desire to see the conflict end as quickly as possible. He urged the Soviet Union to join in efforts to that objective.

The Soviet party chairman replied a short while later. Kosygin concurred in Johnson's judgment that protracted hostilities would raise grave dangers. The Soviet Union would work for a truce, Kosygin said. He hoped the United States would use its influence with Israel to do likewise.[50]

Johnson approved of the idea of a truce, but at first the truce terms Moscow sought differed from those the president deemed suitable. The

Soviet delegate on the United Nations Security Council proposed a measure calling not only for the shooting to cease but for invading forces—meaning the Israelis—to withdraw behind the 1956 armistice lines. The Israelis, still smashingly successful in the field, saw no reason to comply. The Israelis remembered the Suez War, following which they had succumbed to international pressure to give up territory won in fighting. They determined this time to establish and retain buffer zones around their borders. Johnson refused to override the Israelis, and he instructed Arthur Goldberg to work for a cease-fire in place. As a tactical matter, Goldberg did not flatly oppose the proposal for a cease-fire and withdrawal. Israel, after all, had started the war and was occupying foreign territory. Instead Goldberg, with Johnson's approval, added requirements that Egypt lift its blockade of Aqaba and demilitarize Sinai. In practice these requirements constituted a veto.

In the initial phase of the war, some American officials believed that the Israeli successes might open new opportunities for solving the Arab-Israeli problem once and for all. Walt Rostow headed the optimists, on this question as on Vietnam. "Our first thought is that the key to ending the war is how well the Israelis do—or don't do—on the ground," Rostow said on the afternoon of June 5. "Up to a point this is correct; but it is not wholly correct because what the Israelis are after is not some abstract military victory but a settlement which, if possible, ensures that this will not happen again in another ten years." The administration should adopt a similar approach. "Our behind-the-scenes work with the Russians and others should consist not merely in negotiating a cease-fire, because a cease-fire will not answer the fundamental questions in the minds of the Israelis until they have acquired so much real estate and destroyed so many Egyptian planes and tanks that they are absolutely sure of their bargaining position." The administration must aim for a comprehensive settlement of the Arab-Israeli dispute, including arrangements for dealing with the Palestinian refugees and an agreement with the Soviets to dampen the arms race in the region. It was an ambitious project. But Rostow thought circumstances favored boldness. "So long as the war is roughly moving in Israel's favor," he concluded, "I believe we can shorten it by getting at the substance of a settlement at the earliest possible time."[51]

A comprehensive settlement, however, required more arm-twisting of Israel than Johnson was willing to apply. The stubbornness of the Israelis and the touchiness of their American partisans showed plainly later on June 5. A State Department spokesman, asked to characterize American policy toward the present conflict, said the United States was "neutral in thought, word, and deed." From the reaction that followed one might have supposed the administration had announced arms deliveries to Egypt—except that the Egyptians complained too. The statement triggered instant outrage among Israel's American backers, who expected far more than neutrality from Washington in what they proclaimed a just war for Israel's existence. Israel may have fired first—although at this stage no one on Israel's side was admitting anything—but Egypt and Syria had provoked the conflict. In Israel's hour of trial, Washington seemed to be reneging on its oft-given promises of support. Blame for the war and pressure to relinquish territory might follow.[52]

Two White House staffers, Ben Wattenberg and Larry Levinson, to whom Johnson assigned the task of measuring the uproar, described the sentiment they encountered. Wattenberg a few days before had suggested to Johnson that a resolute stand in the Middle East might assist in achieving what the administration had been trying to accomplish for two years: bringing Jewish liberals into the fold on Vietnam. "The Middle East crisis can help turn around the 'other other war'—the domestic disaffection about Vietnam," Wattenberg had written. Now Wattenberg joined Levinson in worrying that the careless language by the State Department was sabotaging the administration's efforts. In a memo to the president, Wattenberg and Levinson explained, "The major concern among Jewish leaders is this: that Israel, apparently having won the war, may be forced to lose the peace—again (as in 1956)." Jewish leaders sought reassurance that Johnson would not pull an Eisenhower and require an Israeli rollback without ironclad guarantees of future Arab good behavior. Wattenberg and Levinson urged Johnson to provide the reassurance. Making the same point Wattenberg had made earlier, they added, "It would seem that the Mid-East crisis can turn around a lot of anti-Vietnam, anti-Johnson feeling, particularly if you use it as an opportunity to your advantage."[53]

Obviously Johnson could not declare American unneutrality, but he moved at once to allay the fears of Israel's friends, and of Israel, whose Eshkol sent another message asking understanding of Israel's difficult situation. Via his many contacts with the Jewish community, Johnson spread word that his devotion to Israel had not diminished. Israel could count on Lyndon Johnson, as it always had. The president directed Dean Rusk to announce—from the White House rather than the State Department—a correction to the neutrality statement. Rusk told a news conference that neutrality, while narrowly accurate as a description of American nonbelligerency, did not cover the American attitude. "Neutrality does not imply indifference," Rusk explained. Without specifying Israel by name, Rusk said that the policy of the United States remained unchanged. The American government and people were as committed as ever to the search for a lasting and stable peace in the Middle East, which implied, as Washington had often declared—so often that Rusk did not need to at this ticklish hour—Arab recognition of Israel's right to exist. "There is the position at law that we are not a belligerent," he summarized. "There is the position of deep concern, which we have as a nation and as a member of the United Nations, in peace in that area."[54]

The Arabs never accused the United States of neutrality. Initially they didn't even accuse Washington of nonbelligerency. At the outset of the war the Egyptian government charged that planes from American aircraft carriers had taken part in the raids on Egyptian airfields. Cairo found it difficult to believe—or admit, at any rate—that the Israelis by themselves could have delivered such a blow. But after Johnson requested Kosygin to point out to Nasser what Soviet intelligence knew—that American warplanes were nowhere in the vicinity—the Egyptian government shifted its ground for complaint. It alleged that American support for Israel before and during the fighting rendered the United States a de facto belligerent. On June 6 Cairo broke diplomatic relations with Washington. Syria and Iraq soon followed suit. An anti-American stampede by various other Muslim countries might have followed but for some quick work by the State Department through the Shah of Iran, reminding wavering regimes what they owed to the United States and what they might receive in the future.[55]

Soon the Arabs found another target for criticism. On June 6 the Soviet Union altered its position on the issue of a cease-fire. Reasoning that Egypt would only lose more ground the longer the war lasted, Moscow voted in favor of a Security Council resolution recommending a cease-fire in place. Restoring the status quo, if it ever became possible, would have to wait.

At a White House meeting the next day, Rusk recapitulated the events of the first forty-eight hours of the war. Nasser, the secretary said, had grossly misjudged both the military balance between the Arabs and Israel and the degree to which the Soviets would back him. As a result he had suffered a "stunning loss." There now existed widespread disillusionment among the Arabs with the Egyptian president. Soviet stock in the region had also fallen on account of Moscow's failure to follow through on earlier professions of support. Israel meanwhile was riding high. The Israelis' demands would be "substantial."

Richard Helms focused on the Russian reaction. The CIA director accounted the damage to Soviet prestige almost as great as that to Nasser's. Moscow, Helms said, had badly miscalculated what it was letting itself in for with Nasser and the Syrians. Its error was even greater than the error Khrushchev had made during the Cuban missile crisis.

Llewellyn Thompson, the ambassador to the Soviet Union, in Washington for consultation, thought the Kremlin would be relatively easy to handle despite its present discomfiture. Unlike Khrushchev, the current Soviet leadership did not enjoy gambling or confrontational diplomacy. Barring a direct Israeli threat on Cairo, the Soviets probably would stay out of the war.

Johnson wasn't so sure. The Soviets would have a hard time walking away from their investment in Egypt and Syria, the president said. The United States must keep a close eye on them.

Rusk thought the Israelis would present a bigger problem than the Soviets. The Israeli successes, which had saved the administration from one set of problems, created another. The United States was tied in the Arab mind and in the opinion of most of the world to Israel. The Arabs identified the United States as an aggressor, as the recent severing of relations indicated. The only way to salvage the situation was to keep Israel's demands within reason. This would require the greatest care. Overt and official pressure on Israel would probably fail, even if political conditions in the United States had allowed it. Instead the administration must work from the inside, relying on its many direct and indirect connections to the Israeli government. Administration officials must "make ourselves attorneys for Israel," Rusk said.

Johnson agreed regarding the delicacy of the task. The administration should try to create "as few heroes and as few heels as we can," he asserted. Yet matters could be far worse. "We are in as good a position as we could be given the complexities of the situation." At least Israel was not losing. Significant troubles remained, though. "By the time we get through with all the festering problems we are going to wish the war had not happened."[56]

A new and flabbergastingly unanticipated problem emerged several hours after this meeting. Israeli warplanes and torpedo boats attacked the American intelligence ship *Liberty* off the Egyptian coast. The casualties numbered over two hundred, with thirty-four dead or dying. The ship barely escaped sinking. The attack was almost certainly not a case of mistaken identity, since the vessel was clearly marked and visibility was unlimited. Israeli reconnaissance planes repeatedly flew close overhead prior to the assault.

The most likely explanation for the attack is that the Israelis didn't like the idea of Americans eavesdropping on Israeli communications, a job the *Liberty* was outfitted to do. The war against Jordan had ended on June 7, when Amman accepted the United Nations cease-fire resolution. Egypt was on the ropes and would quit on the day of the *Liberty* attack. Yet the Israelis, predictably full of themselves, had one more goal: the capture of Syria's Golan Heights. The invasion of Syria would commence within hours. If the Americans found out about it in advance, they might object and try to prevent the accomplishment of what the Israeli defense ministry considered a vital task. To prevent any such complication someone in the Israeli chain of command—a CIA report identified defense minister Moshe Dayan—ordered the *Liberty* destroyed.[57]

The Israeli government shrewdly guessed that Washington would not investigate the incident too closely, at least not until too late to do anything about it. The Israelis declared the attack an error. Abba Eban sent Johnson an apology: "I am deeply mortified and grieved by the tragic accident involving the lives and safety of Americans in Middle Eastern waters." Israeli ambassador Avraham Harman similarly told the president of his "heartfelt sorrow at the tragic accident to the U.S.S. *Liberty* for which my countrymen were responsible."[58]

American officials doubted this story as much as they doubted that Egypt had started the war. Clark Clifford, a solid supporter of Israel from before the creation of the Zionist state, told Johnson, "It is inconceivable that it was an accident." Clifford called for an investigation that would set forth the facts and demand punishment of those Israelis responsible. Johnson steamed, "I had a firm commitment from Eshkol and he blew it." The president added, "That old coot isn't going to pay any attention to any imperialist pressures."[59]

Johnson ordered American planes to the area of the attack to find out what they could. To avoid alarming either the Soviets or the Egyptians, he sent Kosygin a message explaining that this deployment had the "sole purpose" of looking into the *Liberty* incident. The United States had no intention of intervening in the fighting. The president told Kosygin he would "deeply appreciate" the chairman's cooperation in passing the message to Nasser.[60]

When the American planes contributed little new knowledge about the *Liberty* affair, Johnson remained angry but decided not to reprimand Israel. The middle of a war seemed an imprudent time for a falling out. The president agreed with Rusk's earlier comment that the only hope for restraining the Israelis—short of a politically inconceivable application of major sanctions—was to remain on friendly terms with them. Consequently he chose to accept the Israeli government's apologies, and he ordered the incident kept quiet.[61]

The June 9 Israeli invasion of Syria initiated the final phase of the war and produced a final set of problems for the Johnson administration. At the time of the invasion Arthur Goldberg was explaining to the United Nations "the extreme urgency of bringing the fighting to an end." Israel's invasion of Syria did not reflect favorably on the United States. Either America lacked the will to stop the Israelis, in which case its professions of evenhandedness were a sham, or it lacked the ability, in which case it was not much of a superpower.[62]

More worrisome was the Soviet response. The Kremlin, provoked beyond endurance by the humiliation of its allies—a humiliation the Chinese were making much of—decided it needed to do something about this most recent outrage. On news of the Israeli invasion the Soviets broke diplomatic relations with Israel. Shortly afterward Kosygin called Johnson to declare that the situation in the Middle East had reached a "very crucial moment." The chairman warned of a "grave catastrophe" and announced that unless the Israelis halted operations immediately the Soviet Union would take "necessary actions, including military."[63]

This move caught the administration by surprise. Just a day earlier the State Department had sent a circular to all American diplomatic and consular posts summarizing the administration's understanding of the situation in the Middle East. On the matter of Soviet actions and intentions the circular explained, "Our characterization of Soviet behavior to this point is that the Soviets became increasingly concerned as the fighting progressed to get it stopped in recognition of the fact that any effort on their part to retrieve the Arab military situation would have to be massive and hence would carry unacceptable risk of confrontation with us." On June 8 the CIA declared flatly, "There is no danger of Soviet military intervention in the Middle East."[64]

Following Kosygin's threat, Johnson responded in two ways. He ordered the Mediterranean fleet, hovering off the Syrian coast, to move closer to shore. What the fleet would do when it got there, he hadn't decided. He hoped he wouldn't have to. The point was to convince the Russians that two could play the brinkmanship game. At the same time, Johnson told Kosygin the United States was working on getting Israel to accept a cease-fire. An end to the fighting, he said as convincingly as he could, was imminent.[65]

Fortunately for the United States, for the Soviet Union, for Israel, and for Syria, Johnson was right. The Israelis decided they had gained all the ground they needed. Early in the afternoon (New York and Washington time) on June 10 Israel and Syria signed a truce accord. Fighting continued for some hours afterward, but by June 11 all was still.

IX

With the end of the war, Johnson's Middle East problems moved off the critical list to the merely serious. Until very recently some administration officials had retained hope that the war's jolting might have shaken loose a solution to the Arab-Israeli conflict. Just before the Israeli invasion of Syria, Walt Rostow remarked that the basic question was "whether the settlement of this war shall be on the basis of armistice arrangements, which leave the Arabs in the posture of hostilities towards Israel, keeping alive the Israel issue in Arab political life as a unifying force, and affording the Soviet Union a handle on the Arab world; or whether a settlement emerges in which Israel is accepted as a Middle Eastern state with rights of passage through the Suez Canal, etc." The administration's objective, Rostow said, should be to push for "as stable and definitive a peace as is possible." This would require concessions from Israel on territory taken, a shift in the political balance in the Arab countries from radical leaders to moderates, a Middle East arms control agreement, and "the emergence of a spirit of regional pride and self-reliance to supplant the sense of defeat and humiliation engendered in the Arab world in the wake of the failure of Nasser, his strategy, and his ideological rhetoric." It would also require—this most fundamentally—"a broad and imaginative movement" by Israel on the question of the Palestinians.[66]

The problem was that none of Rostow's requisites appeared likely to obtain, less than ever after the Israeli seizure of the Golan Heights. Israel showed scant inclination to give up territory seized in battle or to exercise imagination regarding the Palestinians. The Soviets, having suffered a severe political defeat, would probably not have much interest in collaboration with the Americans to limit arms sales to the region. A moderation of Arab politics would have to await healing of the wounds of the war.

Regarding Israel's uncompromising mood, Johnson aide Harry McPherson delivered a personal report from the Israeli front. McPherson had arrived in Israel hours before the June 5 attack on Egypt. Israeli officials inadvertently—or arrogantly—blew their own cover story the morning the war began. McPherson was talking outdoors with the chief of Israel's military intelligence when air raid sirens sounded. McPherson began looking for shelter. The intelligence chief checked his watch, then calmly continued his briefing. The reaction puzzled McPherson until he figured out what was happening. The intelligence officer, knowing what time the surprise attack had been scheduled, knew that the sighted plane could not be an enemy, because Egypt now lacked an air force. It must be an Israeli plane and hence no cause for alarm. He was right.[67]

As McPherson explained to Johnson, the Israelis were flushed with victory. "The spirit of the army, and indeed of all the people, has to be experienced to be believed," McPherson stated. "The temper of the country, from high officials to people in the street, is not belligerent, but it is determined, and egos are a bit inflated—understandably. Israel has done a colossal job." The military wanted to keep all the territory seized. Everyone wanted to keep Sharm el Sheikh and the Old City of Jerusalem. "Regaining the Old City is an event of unimaginable significance to the Israelis. Even the non-religious intellectuals feel this way." McPherson sensed a bit of room for give on Sinai and the West Bank, among the politicians if not the generals. At some point Israel might consent to hand back a demilitarized Sinai. Annexing the West Bank, with its large population of Arabs, presented problems the government had not figured out how to deal with. But government and people were united in opposition to a return to the prewar status quo. "There are constant references and comparisons to 1956. The Israelis do not intend to repeat the same scenario—to withdraw within their boundaries with only paper guarantees that fall apart at the touch of Arab hands." The United States might as well forget about persuading the Israelis to relinquish territory they did not freely choose to give up. "We would have to push them back by military force, in my opinion, to accomplish a repeat of 1956," McPherson predicted. "The cut-off of aid would not do it."[68]

Events of the week after the war effected little change in this view. Walworth Barbour, Johnson's ambassasor to Israel, reported a conversation with Abba Eban in which the foreign minister explained that his government was still sorting out its options. "There must of necessity be a lack of precision in Israel's thinking as to detailed policies," Eban said. The dramatic changes of the previous several days had raised "opportunities which were inconceivable before and for which Israel is unprepared." On June 15, after additional discussions with Israeli officials, Barbour commented, "They genuinely believe that a completely new situation has been created which offers an opportunity to move forward to their goals of durable peace and security such as never existed for Israel to date."[69]

Johnson declined to try to convince the Israelis that this new situation included the same old problems that had led to three Arab-Israeli wars. The Palestinian question, to cite the most intractable, was farther from solution than ever following the Israeli seizure of the Gaza Strip and the West Bank. Johnson was not the man to take on the Israelis and their American supporters in the moment of triumph.

But he had to say something. On June 19 the president delivered an innocuous speech outlining five predictable principles for peace: the right of all countries to physical security; justice, of a nature unspecified, for refugees; the right of innocent passage through international waters; restraint on arms sales to countries of the Middle East; and respect for the political independence and territorial integrity of all countries. Johnson disavowed any primary role for the American government in efforts to achieve a settlement. The United States would provide good offices, but the burden would rest elsewhere. "There is no escape from this fact: The main responsibility for the peace of the region depends upon its own peoples and its own leaders." To underline this point, Johnson added, "What will be truly decisive in the Middle East will be what is said and what is done by those who live in the Middle East."[70]

XI

A sidebar to the story of the 1967 Middle East war was Johnson's only meeting with a leader of the Soviet Union. Just after the shooting in Syria ended, the Kremlin announced that Kosygin would travel to the United Nations to address the international body. The Johnson administration anticipated a diplomatic salvage operation. Walt Rostow told the president, "I believe he is coming because he believes only a peacemaker stance can really retrieve something of the Soviet position from the Mid-East debacle—a little like the Test Ban Treaty after the Cuba missile crisis."[71]

Johnson's advisers urged the president to show himself to be no less devoted to peace than the Soviet chairman. Rusk reminded the president of his frequently reiterated pledge to go anywhere in the search for peace. The president had traveled across the globe for peace. Now with Kosygin coming to the United States, the president could hardly avoid a meeting. For Kosygin to leave the country without a conversation with the president would be an "enormous political loss to you." Rostow similarly suggested that a meeting with Kosygin would have significant political benefits. "At home it will cover your flank to the left and among the columnists," Rostow said. "If you don't do it, they will blame every difficulty that follows on the lack of a meeting." Alluding to Eisenhower's 1952 pledge to go to Korea, Rostow added, "The Republicans will run on: 'I will go to Moscow.'"[72]

No one in the administration expected much substantive to come from a meeting with Kosygin. The State Department's intelligence bureau thought that if the Soviet leader agreed to meet Johnson it would be mostly to size up the president. The Soviet chairman's basic purpose would be "reconnaissance," a department analysis asserted. Averell Harriman, the dean of American dealers with the Kremlin, described the Russian approach to diplomacy. "With Russians," Harriman said, "it takes three meetings to make a deal: the first, courteous; the second, rough; the third, the deal is made." Johnson would not have time for the full treatment, and therefore the likelihood of deals was small.[73]

The matter of who should invite whom produced some diplomatic to-and-froing. The Soviets opened the dialogue at a low level, with the press attaché of the Soviet embassy floating a hint of interest to Carl Rowan of the United States Information Agency. The administration reciprocated carefully at a correspondingly low level. A suggestion by John Roche of the White House staff captured the administration's caution: "Let Kosygin come to us, if he can. (He can get the telephone number from Information.)"[74]

Eventually the mutual desirability of a meeting was established. Arranging a venue consumed several more days. Johnson invited Kosygin to Washington. Kosygin didn't like Washington, since with the Arabs, Chinese, and various opinionated others already castigating Moscow for capitulationism, it would never do for the leader of the Soviet Union to be seen in the headquarters of imperialism. Johnson offered Camp David. Kosygin replied that even the general area of Washington was out of the question. Kosygin countered: What was wrong with New York? The United Nations was neutral territory. Why not there? Johnson nixed New York. The president was having trouble with antiwar protesters. New York contained lots of them. To their ranks would probably be added demonstrators on any number of additional topics, from greater-Israelists to bomb-banners. The appropriate ambience for a summit meeting would be lacking. Johnson suggested an air force base in New Jersey. It was isolated, secure, easy of access. Kosygin, envisioning what the Chinese would make of photos of the Soviet chairman and the American president shaking hands in front of planes from the same air force that was currently ravaging North Vietnam, rejected the idea at once.

Finally the two sides concurred on Glassboro, New Jersey. The town was about halfway between New York and Washington and was the home of Glassboro State College. Both the town and the college were quiet, unlike many campuses and college towns at the time.

The summit filled two days in the fourth week of June. The pressing issue of the moment was the recent Middle East war, for which Johnson and Kosygin each blamed the other, at least in part. When Johnson suggested that the superpowers should act as "older brothers" to the countries in their respective spheres and "provide proper guidance" to these lesser powers, Kosygin said the Soviet Union had done precisely that with regard to the Arabs. The Arabs were an "explosive group of people," but Moscow had successfully discouraged them from attacking the Israelis. The United States had not been so successful in restraining the Israelis. Johnson responded that the Arabs had triggered the crisis by closing the Strait of Tiran. Kosygin declared that the Soviet Union, unlike the Western countries, had no pecuniary motives in dealing with the Middle East, since the Soviet economy was self-sufficient in oil. He said that peace would come to that troubled region only if the Israelis were made to withdraw from the territories they had captured. Otherwise the region would be afflicted by a long and bitter war.

Johnson realized he was in a weak position having to defend Israel, and he turned the conversation to arms control. In particular he advocated limiting the development of antiballistic missile systems. Kosygin replied by challenging the most frequently forwarded argument for banning missile defenses—namely that they would merely spur additional construction of offensive missiles, which were cheaper to build—and saying that the American emphasis on offensive missiles represented "a commercial approach to a moral problem." This approach was invalid. He stated that the Soviet Union did not want war, but would defend itself if war were thrust upon it.

On Vietnam, Johnson explained the American view that South Vietnam was being attacked by North Vietnam, and that the United States had every right to accept Saigon's request for protection. Kosygin replied that "the Vietnam problem must be solved by the Vietnamese people themselves." He said that if Johnson halted the American bombing, peace talks could begin at once. Johnson perked up at this statement, asking if the Soviet leader really thought so. Kosygin said he did. Kosygin went on to say that the Vietnam issue had spoiled relations between the United States and the Soviet Union and had allowed China "a chance to raise its head with consequent great danger for the peace of the entire world." Johnson agreed that China constituted a danger to both superpowers.[75]

The sessions were nearly devoid of lighter moments, but not entirely. When Robert McNamara questioned one point of Soviet logic, Kosygin quipped, "I guess you'll understand when you become a Marxist." McNamara, formerly chief executive of Ford Motor Company, replied, "Perhaps I will—if you'll ever become a capitalist."[76]

As administration officials had guessed, the meetings produced little of substance. Johnson admitted as much in his report to the American people afterward. "Meetings like this do not themselves make peace in the world," the president said. Yet he went on to claim that some good had emerged from his sessions with Kosygin. "It does help a lot to sit down and look at a man right in the eye and try to reason with him, particularly if he is trying to reason with you. We may have differences and difficulties ahead, but I think they will be lessened, and not increased, by our new knowledge of each other."[77]

Eight | **Vietnam in Context**

WHEN THE FRAMERS of the American constitution placed the United States on a biennial election calendar, they did not envision the federal government deciding complicated issues of war and peace on a regular and continuing basis. Nor, prescribing indirect selection of senators and an upper house initially just a few members shy of half as large as the House of Representatives, did they realize that by the 1960s more than 87 percent of American legislators would have to face their constituents every second year.

Other features the framers wrote into the constitution, especially as they evolved later, might have mitigated the sensitivity of the American foreign policy apparatus to popular passions. The most important was the role assigned to the president as commander in chief of the armed forces. During the early part of the twentieth century, presidents sent American troops all over Central America and the Caribbean without asking Congress for declarations of war. From 1950 to 1953 Presidents Truman and Eisenhower waged a major war in Korea on their own authority.

During Lyndon Johnson's presidency, though, popular sentiment as a check on foreign policy regained ground. While Johnson cited the examples of Truman, Eisenhower, and Kennedy to justify many of his actions, in his attention to the constraints public opinion placed on foreign policy he most resembled Franklin Roosevelt. Like Roosevelt, Johnson refused to allow foreign affairs to divert him, Congress, or the American people from enacting his domestic reform program. In Roosevelt's case, the president's reluctance to move faster than public opinion had contributed to the blow the United States suffered at Pearl Harbor. In Johnson's case, it led to the debacle of Vietnam.

II

Johnson inherited America's Vietnam problem under the most inauspicious of circumstances. Following the 1954 Geneva conference the Eisenhower administration had neither fished nor cut bait: rather than sign the Geneva accords, as chief American negotiator Walter Bedell Smith urged, or repudiate them, as many on the Republican right demanded, Eisenhower attempted a third route. He and John Foster Dulles declined to put pen to a pact countenancing communist control over the northern portion of Vietnam, yet they indicated they would consider as a serious matter the violation of the accords by any party. Any party, that is, except the government of Ngo Dinh Diem, which with Washington's blessing refused to allow the 1956 country-reunifying elections the accords specified. Washington itself crossed the line, in spirit if not in letter, only a few months after the Geneva conference when the American government organized the Manila Pact—SEATO to most—for the defense of South Vietnam, despite the Geneva accords' forbidding either half of the country to enter military alliances. (Because South Vietnam was not a formal member of SEATO, the Geneva accords were technically unviolated.) At the same time, Eisenhower sent large quantities of economic and military aid to Diem's regime, and the CIA initiated a covert war against Diem's enemies.

By the time Kennedy became president, the political and moral infrastructure for large-scale American military involvement was securely in place. In 1954 Eisenhower could have walked away from Vietnam, blaming the communist victory in the north on the French and taking no responsibility for what became of the rest of the country. Conservatives would have complained, but most Americans would have been happy at missing a second Korea. Kennedy could not walk away. Eisenhower had made Vietnam a proving ground for American credibility, and Kennedy, having attacked Eisenhower for laxity in the face of the communist threat, could do no less than Eisenhower had done to keep South Vietnam in the American camp.

Even less could Johnson do less. A Kennedy inclined to exit Vietnam would have had to worry about Republicans and conservative Democrats. A Johnson seeking an out would have had Kennedy's ghost and its liberal partisans to contend with as well. Eventually liberals would abandon the war, but only after it turned bad. Kennedy's contribution to the disaster consisted in making it a liberal cause as well as a conservative one.

Had Johnson been looking for reasons to extricate America from Vietnam, he needn't have looked far. The assassination of Diem three weeks before Kennedy's assassination had only made America's Vietnam problems worse. The dissident generals who toppled Diem, with the Kennedy administration's encouragement, claimed to be frustrated by the corruption of the Diem regime, but in the aftermath of the coup it appeared that their real complaint was that not enough of the corruption had flowed their way. Johnson sent Robert McNamara to Vietnam in December 1963 to survey matters on the ground. "The situation is very disturbing," McNamara reported on his return." The new government of General Duong Van Minh was "indecisive and drifting." Minh and his associates had not made the changes necessary to stem the Vietcong insurgency. "Current trends, unless reversed in the next 2–3 months, will lead to neutralization at best and more likely to a Communist-controlled state."[1]

Neutralization seemed not a bad idea to some observers of Vietnamese affairs. The great powers had devised a neutralization scheme for Laos in 1962, and recently the always helpful Charles de Gaulle had suggested something similar for Vietnam. To Johnson's way of thinking, de Gaulle's advocacy of neutralization immediately rendered it suspect, especially since the French president declined to specify precisely what neutralization would mean for Vietnam. Did de Gaulle intend the withdrawal of all foreign forces from Vietnam? The abstention of both Vietnams from alliances? The end of foreign aid to North and South Vietnam? Would North Vietnam be counted as foreign relative to South Vietnam? Who would guarantee compliance? The superpowers? The United Nations? France? De Gaulle, desiring not to torpedo his plan before talking even started, left details obscure.

De Gaulle's scheme gained support in the United States from a quarter Johnson could not ignore. In December 1963 the president received a letter from Mike Mansfield urging the administration to accept de Gaulle's advice. Mansfield, a longtime student of East Asia before becoming Senate majority leader, judged South Vietnam a losing proposition. The Diem regime, Mansfield pointed out, had failed to stem Vietcong activities despite large and increasing amounts of American aid. Diem's successors seemed likely to have little more luck. Absent improbable changes in South Vietnam's government, the United States would find itself drawn further and further into the conflict. How did the president propose to justify such an extension of the American commitment? "What national interests in Asia would steel the American people for the massive costs of an ever-deepening involvement of that kind?" Mansfield called instead for a negotiated settlement along the lines suggested by de Gaulle. "France is the key country," he declared.[2]

The Montana senator repeated his advice during the early weeks of 1964. The Minh government, since taking power in Saigon at the beginning of November, had shown itself less amenable to suggestions from Washington regarding how to beat down the insurgency than American officials had expected. To make matters worse, Minh had begun hinting at a compromise settlement with the Vietcong. This was hardly what the Kennedy administration had had in mind in approving Diem's overthrow, and it sat no better with Johnson. With most of his advisers, Johnson considered compromise a way station to defeat and completion of the communist takeover of Vietnam. For this reason neither the Pentagon nor the president saw fit to discourage a second group of officers, led by Nguyen Khanh, from toppling Minh at the end of January. The CIA, which got word of the coup some time before the event, reported the good news: "It is safe to say that Khanh's group will be essentially pro-American, anti-communist and anti-neutralist in general orientation."[3]

Mansfield thought the Khanh takeover additional reason for backing away from Vietnam. "It is far from certain that this recent military coup will be the last," Mansfield warned Johnson. "On the contrary, it is likely to be only the second in a series, as military leaders, released from all civilian restraint, jockey for control of the power which resides in United States aid." The result would be continued erosion of popular support for the South Vietnamese government, already distressingly low. "This process of coup upon coup may be expected to become increasingly divorced from any real concern with the needs of the Vietnamese people. If the people do not go over actively to the Viet Cong, they will at best care very little about resisting them, let alone crusading against them. Indeed, the bulk of the Vietnamese people, as well as the lower ranks of the armed forces, may already be in this frame of mind."

Mansfield knew that some administration officials were advocating a much larger role for the American military in Vietnam. This was precisely the wrong approach, he told Johnson. "A deeper military plunge is not a real alternative. Apart from the absence of sufficient national interest to justify it to our own people, there is no reason to assume it will settle the question. More likely than not it will simply enlarge the morass in which we are already on the verge of indefinite entrapment." Again Mansfield urged Johnson to listen to the French, who had learned firsthand the futility of fighting in Vietnam. Mansfield went on to note that Johnson had said that Americans did not want another China. "Neither do we want another Korea." The president and the United States must make the fateful choice soon. "We are close to the point of no return."[4]

Johnson deemed Mansfield's warnings sufficiently important to have the administration's top people prepare rebuttals. Dean Rusk challenged Mansfield's support of neutralization. The National Liberation

Front—to use the Vietcong's proper name—advocated neutralization simply as a device to cut the South Vietnamese government off from American aid, Rusk told Johnson. "What the communists mean by 'neutralization' of South Vietnam is a regime which would have no support from the West and would be an easy prey to a communist takeover."[5]

Robert McNamara agreed that neutralization was a phony issue. If accepted for South Vietnam, the defense secretary said, it would produce a government "that would in short order become Communist-dominated." He rejected Mansfield's claim that the conflict in Vietnam was unwinnable. "The security situation is serious," McNamara conceded to Johnson. "But we can still win, even on present ground rules." Moreover, to back off in Vietnam would call into question American resoluteness worldwide. "South Vietnam is both a test of U.S. firmness and specifically a test of U.S. capacity to deal with 'wars of national liberation.'"[6]

McGeorge Bundy perceived an entire constellation of evils in the direction of neutralization. To waffle in South Vietnam, Bundy said, would produce "a rapid collapse" of anticommunist forces in that country and the completion of the communist conquest. Other dominoes would follow. Laos would quickly go down, while Thailand would be forced to accommodate Hanoi and Beijing. Malaysia, already under siege by Indonesia, would be additionally beset. The Philippines and Japan would distance themselves from the United States. South Korea and Taiwan would be unnerved and would demand reassurance of American intentions, which would not come cheap.[7]

Johnson concurred regarding the unwisdom of neutralization, but even had he thought otherwise he was in no position to strike out on a new path in Vietnam. At the beginning of 1964 the president had to deal with the riots in the Panama Canal Zone and the subsequent break of U.S.–Panamanian relations, with the Guantanamo tiff with Castro, with the impending war between Greece and Turkey over Cyprus, and with Sukarno's anti-American antics in Indonesia. He did not need the kind of trouble over Vietnam a proposal to neutralize would produce.

Consequently, while the president pressed the Khanh government to get tougher on the insurgents, he essentially held to a steady course. "I am glad to know that we see eye to eye on the necessity of stepping up the pace of military operations against the Viet Cong," Johnson wrote to Khanh before it was entirely evident that such was the case. He added, "We shall continue to be available to help you carry the war to the enemy and to increase the confidence of the Vietnamese people in their government."[8]

Robert McNamara encouraged the president to help Khanh carry the war to the enemy. The defense secretary began by reviewing the rationale for administration policy in Vietnam. He listed four alternatives, the first three of which he predictably proceeded to demolish. The United

States might withdraw from Vietnam. The consequences of this course were clear. "Vietnam will collapse, and the ripple effect will be felt throughout Southeast Asia, endangering the independent governments of Thailand and Malaysia, and extending as far as India on the west, Indonesia on the south, and the Philippines on the east." The United States might agree to neutralization. This would produce the same results, if slightly more slowly. "We all know the communists' attitude that what's mine is mine and what's yours is negotiable." The United States might invade North Vietnam. "If we do, our men may well be bogged down in a long war against numerically superior North Vietnamese and Chinese Communist forces." Finally, the United States might stick with current strategy, augmented and improved. "We can continue our present policy of providing training and logistical support for the South Vietnam forces. This policy has not failed. We propose to continue it."

McNamara went on to recommend a package of measures designed to raise the effectiveness of South Vietnam's counterinsurgency efforts. Just back from Saigon, he specified twelve areas in which Khanh, with American help, could increase popular backing for the government and diminish the appeal of the rebels. McNamara's plan specified no major new departures, consisting principally of additional kinds and amounts of American economic and military aid. The plan would strengthen the security situation in South Vietnam, but of equal importance would be the fact that it would signal American steadfastness. The essential object, McNamara told Johnson, was "to make it clear that we are prepared to furnish assistance and support to South Vietnam as long as it takes to bring the insurgency under control."[9]

Johnson quizzed McNamara about his plan at a White House meeting on March 17. The president asked for an estimate of the time required for the plan to produce beneficial results. McNamara responded that good things would begin to happen in a short while if the program was implemented vigorously. "Khanh can stem the tide in South Vietnam, and within four to six months improve the situation there."

Johnson solicited the professional military judgment from Maxwell Taylor, at this point chairman of the Joint Chiefs. Taylor said the McNamara plan was necessary but perhaps not sufficient. The general suggested that to make a serious dent in the insurgency, South Vietnam or the United States would have to hit directly at North Vietnam.

Johnson refused to countenance such a change in strategy. The president had no desire to widen the war any more than absolutely necessary to prevent Saigon's fall. He described the administration's current middle-of-the-road approach as "the only reasonable alternative." McNamara's plan would do the job, at least for now. "It will have the maximum effectiveness with the minimum loss," Johnson declared.[10]

After the Pentagon and the other departments and agencies worked out the price tag for McNamara's plan, Johnson asked Congress for the money. In requesting the $125 million for the South Vietnamese, the president asserted, "Duty requires, and the American people demand, that we give them the fullest measure of support."[11]

III

At least one group of Americans certainly was demanding the fullest measure of support for South Vietnam. As the Republicans prepared to nominate Barry Goldwater for president, most of them insisted that the United States take whatever steps were necessary to preserve freedom in Southeast Asia. Johnson understood the dangers Democrats ran of being branded soft on communism, and while he hoped not to escalate the war in Vietnam, he recognized the need to avoid losing ground to the Republicans on the issue.

At the moment, however, Johnson had other political worries. The spring and summer of 1964 found the president in one of the toughest legislative battles of his career. The most important civil rights bill in a century passed the House of Representatives without great difficulty, but Republicans and southern Democrats were threatening a filibuster of the measure in the Senate. Though Johnson appreciated the problems voting for civil rights would cause valued friends like Richard Russell, he determined to get this bill. He recalled his thoughts afterward, perhaps a little melodramatically: "A President cannot ask the Congress to take a risk he will not take himself. He must be the combat general in the front lines, constantly exposing his flanks. I gave this fight everything I had in prestige, power, and commitment."[12]

An additional complication arose when George Wallace announced his candidacy for president on a platform of opposition to federal meddling in civil rights matters. In several Democratic primaries the Alabama governor ran surprisingly strongly. Wallace's showing did not appear at all likely to deny Johnson the Democratic nomination, but it indicated the depth of the opposition to what Johnson was trying to accomplish.

Johnson redoubled his efforts, and at the beginning of June he won Everett Dirksen over to the administration's side. When the minority leader called for cloture, thereby preventing a filibuster, the battle effectively ended. On July 2 Johnson signed the historic 1964 Civil Rights Act into law.

The fight for civil rights legislation combined with the Goldwater nomination, which arrived on schedule two weeks later, to push Johnson toward a stronger position on Vietnam. The president had no intention of keeping pace with Goldwater, stridency for stridency, on the anticommunist question. Johnson aimed instead to stick to the middle of the road. But Goldwater's nomination pulled the road, middle and all, to the right. At the same time, Johnson wanted to ease the pressure the civil rights fight had placed on old Democratic friends like Russell and recent Republican allies like Dirksen. Neither Russell nor Dirksen was as hawkish as Goldwater, yet neither did either wish to see Saigon slip away.

Events in Vietnam meanwhile conspired to render deeper American involvement more likely. Michael Forrestal, McGeorge Bundy's assistant for Southeast Asia, wrote Johnson that the situation was not improving as hoped. "I don't think that there is in the near future a danger of military collapse," Forrestal explained. "But there is an increasing danger of political accident or upheaval unless there is some dramatic change in the atmosphere in which the struggle is being waged." Forrestal went on, "What I think is needed fairly soon (i.e., within the next month or six weeks) is action by the United States in some part of Southeast Asia which gets across forcefully to the Vietnamese a sense that we believe the Communist insurgency can be contained and that we will do whatever is required to insure this."[13]

Bundy agreed. The national security adviser urged Johnson to approve use of "selected and carefully graduated military force against North Vietnam." Bundy justified his recommendation on three grounds: first, that the United States couldn't afford to see South Vietnam go communist; second, that on present trends Saigon would do so in the not distant future; and third, that a firm public decision to use whatever force circumstances required offered "the best present chance of avoiding the actual use of such force."[14]

American military leaders wanted more force too—much more. Earle Wheeler, the new chairman of the Joint Chiefs, sent Johnson a request for approval of an air offensive against North Vietnam by unmarked American planes and for support of South Vietnamese troops in cross-border operations in Laos. Wheeler discounted the likelihood of a major communist reaction. The North Vietnamese, he predicted, would do something to register disapproval, but their fear of even greater American action would keep them in line. On this point CIA director John McCone concurred.[15]

Between the rightward influence of political developments in the United States and the downward trend in Vietnam, a presidential decision for greater American involvement in the war was nearly assured. A sequence of events in the Gulf of Tonkin provided the precipitant. On August 1 an American destroyer assigned to provide covert electronic support to South Vietnamese raids on the North Vietnamese coast encountered North Vietnamese fire. The attack on the American ship took place in international waters, and the Johnson administration chose to interpret it as a challenge. When a second American vessel, sent to reinforce the American presence in the gulf, reported fire from enemy PT boats on August 4, the administration swung quickly into action.

Shortly after noon the president convened a session of the National

Security Council. McNamara explained the latest incident. Some confusion clouded the affair, he conceded, but evidence indicated a substantial engagement involving several North Vietnamese PT boats. The defense secretary said, "Two of the PT boats were reportedly sunk and three to six were fired upon. So far we have no casualties."[16]

There was good reason for the absence of casualties. The reported attack probably never took place. The American commander on the scene, noting the bad weather and darkness surrounding whatever had happened, began questioning his own reports almost as soon as he sent them, and subsequent confirmation efforts failed. In follow-up messages to Washington he urged a complete reevaluation of all evidence before retaliating against North Vietnam.[17]

But Washington, primed for action, was disinclined to reevaluate. At a second August 4 meeting, Dean Rusk declared, "An immediate and direct reaction by us is necessary. The unprovoked attack on the high seas is an act of war for all practical purposes." Johnson asked, "Are we going to react to their shooting at our ships over forty miles from their shores?" The president wondered aloud at the reasoning behind the North Vietnamese attack. "Do they want a war by attacking our ships in the middle of the Gulf of Tonkin?" McNamara thought not, but the secretary of defense guessed that Hanoi wanted to show the Americans that North Vietnam was not to be tangled with. Carl Rowan of the United States Information Agency, the administration's point man in making America's case to a skeptical international audience, inquired about the firmness of the evidence of an attack. "Do we know for a fact that the provocation took place? Can we nail down exactly what happened?" Rowan added a warning, "We must be prepared to be accused of fabricating the incident." McNamara responded, "We will know definitely in the morning."[18]

In fact no one knew anything definitely in the morning. The fog obscuring the incident only thickened. But the military situation in Southeast Asia and the political situation in the United States were such that had the Tonkin incident not occurred the administration might well have invented it—as the administration apparently did, albeit half by accident. Whether or not the North Vietnamese had committed aggression in this instance, Johnson and other administration officials believed that they certainly were responsible for aggression against South Vietnam in plenty of other instances. Hanoi needed a whacking, both for the communists' edification and for the edification of others watching.

The others watching included not only foreign governments but senators and representatives, the press, and voters who might be tempted to go with Goldwater in November. Johnson wished to demonstrate that while he did not welcome a wider war, his patience had limits. A series of questions scribbled by McGeorge Bundy at the noon meeting on August 4 indicated what was on the minds of top administration officials. Bundy jotted, "Goldwater?—Congress?—Press?—Diplomatic?"[19]

Johnson approved a two-pronged response. Immediately he ordered air strikes against the bases of the North Vietnamese PT boats. The strikes succeeded in destroying several vessels and disrupting fuel supplies. They succeeded equally in generating political plaudits for the president. James Greenfield, monitoring the domestic reaction for the State Department, accurately described "overwhelming support" from the press, Congress, labor and veterans' organizations, and the like for the reprisals.[20]

This favorable public reaction encouraged the president to launch the second prong of his Tonkin counteroffensive: a drive for congressional authorization for broader use of American military force in Vietnam. As usual Johnson tested the water carefully before venturing in. The president invited the leaders of Congress to the White House for a briefing on the situation in Southeast Asia and to solicit their opinions on how he ought to react. Because he did not reveal the doubts administration officials were having about the attack, the outcome of the gathering was largely foreordained. But it was nonetheless important. Speaker McCormack declared of the incident, "There is no question but it is an act of war—an attack on American vessels." Senator Bourke Hickenlooper thought the communists were testing America's resolve. The Iowa Republican likened the affair in the Tonkin Gulf to the Cuban missile crisis of two years before. Richard Russell worried that the president might not be devoting sufficient military equipment and personnel to counter the challenge. "Are you sure you have enough stuff to do this job?" the Georgia Democrat asked. "We don't want to do it half way." Johnson had read the group a draft resolution allowing the president to undertake "limited" reprisals against North Vietnam. Everett Dirksen didn't like the qualifier. "I would put our references to the word 'limited' in the deep freeze," the minority leader advised. "It connotes we would be like sitting ducks." William Fulbright registered approval of the president's resolution. "I will support it," Fulbright declared.[21]

Buoyed by this backing, Johnson formally submitted his resolution to Congress. The submitted version lacked any reference to "limited" American action, instead authorizing the president to take "all necessary measures to repel any armed attack against the forces of the United States and to prevent further aggression." The legislators nearly fell over themselves in rallying around the flag and the president. Like Johnson, the great majority of them had elections coming up. Of the 535 senators and representatives only two dissented. Significantly or not, the two dissenters—Senators Wayne Morse of Oregon and Ernest Gruening of Alaska—would not face the voters until 1968.

IV

The Tonkin resolution served Johnson's purposes perfectly, for the moment. A more powerful vote of confidence by Congress would have been difficult to conceive. Goldwater might criticize the president's handling of the war, but now his complaints would bounce harmlessly off. Beyond the initial reprisals, Johnson had no intention of using his new authority, for the moment. Just as in the case of Cyprus, again frothing at this time, Johnson would make every effort to postpone major violence until after the November election. Voters who are worried at the possibility that they or their sons or husbands might go off to war usually don't make happy voters. Unhappy voters usually don't vote for incumbents. By the middle of August it was evident Johnson did not really have to fear *losing* to Goldwater, but Johnson had big plans for January 1965 and after, and big plans required a big victory.

Unfortunately the Vietcong did not follow the script. The insurgency continued to sap Saigon's security, and absent American escalation the Khanh regime didn't appear likely to survive long. Khanh didn't help matters by using the Tonkin flap as cover to crack down on dissidents, and by doing so in what Maxwell Taylor, now ambassador to South Vietnam, called an "inept manner." The dissidents, in turn, used the crackdown to justify further antigovernment demonstrations. The demonstrations eventually led to severe riots, forcing Khanh to resign and touching off another round of squabbling among South Vietnam's clique of power brokers.[22]

Johnson's generals again called for a larger American role to suppress the insurgency, but the president's civilian advisers doubted that Saigon's fragile government could stand the strain of escalation. At a White House meeting in the second week of September, Taylor cautioned against early escalation. The communists would retaliate with increased pressure, Taylor predicted, rendering a solution to the South Vietnamese government's problems less likely than it was already. CIA director McCone spoke similarly. McCone said that American attacks on North Vietnam would force Hanoi to respond with greater efforts in South Vietnam and might trigger "major increases" in Chinese participation in the war.

Johnson asked whether there was anything the United States could do to quell the feuding among South Vietnamese officials. Taylor saw little hope. The South Vietnamese, the ambassador said, had no sense of public duty. They viewed government service chiefly as a means of individual enrichment.

Johnson, obviously frustrated, wondered to Taylor whether South Vietnam deserved the effort the United States was devoting to its defense. Taylor thought so, despite Saigon's deficiencies. America, Taylor said, could not walk away from South Vietnam. Whatever one thought of the gang running the country, an American withdrawal would have a grave impact on American interests in Southeast Asia and elsewhere. Earle Wheeler agreed with Taylor. "If we should lose in South Vietnam, we would lose Southeast Asia," the Joint Chiefs chairman said. "Country after country on the periphery would give way and look toward Communist China as the rising power of the area." Dean Rusk concurred also, as did John McCone.

Rusk had recommended to Johnson a program designed to demonstrate American determination to defend South Vietnam without unduly provoking Hanoi. Under this plan the United States navy would continue patrols in the Tonkin Gulf, South Vietnamese vessels would resume suspended coastal operations against North Vietnam, and the South Vietnamese army would conduct limited forays against Vietcong positions and supply routes in Laos. At the same time the United States would prepare for sharper responses against North Vietnam. The preparation, Rusk explained, would enable American forces to move "at five minutes' notice."

Johnson polled the group to see if anyone dissented from Rusk's recommendation. No one did, and the president approved.[23]

Yet through the autumn of 1964 the situation in South Vietnam grew no better. The jostlers for place in Saigon continued throwing elbows and anything else they could think of to push their way to the front. The CIA described increased infiltration across the 17th parallel. Communist forces, the agency said, were "steadily building" toward the time when the communists would drop their guerrilla tactics for a conventional assault on at least the northern portion of South Vietnam.[24]

Johnson, who since 1963 had been uneasy about the American role in the overthrow of Diem, waxed audibly nostalgic for the stability of former days. "We could have kept Diem," the president said at a Tuesday lunch meeting at the beginning of December. "Should we get another one?"[25]

Had another Diem been available Johnson would have grabbed him quick. But none was. Business continued as usual in Saigon, with civilians and generals grabbing for power. As 1965 began, George Ball commented, "The regime has the smell of death."[26]

The regime's opponents smelled it too, and during January 1965 large protests erupted against both the South Vietnamese government and the United States. Rioters attacked American offices in South Vietnam and demanded the ouster of Ambassador Taylor. Johnson, more frustrated than ever, found the ingratitude of the South Vietnamese infuriating. After watching clips of the demonstrations he exploded, "Taylor go home?! It makes your blood boil!"[27]

V

If Johnson had wanted out of Vietnam, the beginning of 1965 afforded his best opportunity. Now elected overwhelmingly in his own right, he no longer labored under the burden of having to finish what Kennedy—himself following Eisenhower and Truman—had begun. To be sure, there remained the troublesome issue of official American commitments to the security of South Vietnam. But given the persistent unwillingness or inability of the South Vietnamese to live up to their end of the bargain and create a viable state, Johnson might reasonably have declared the contract broken. Assuming that South Vietnam soon fell to the communists, his critics would have charged him with losing more ground to the reds. Yet the charge needn't have been fatal. Eisenhower lost Cuba to Castro and retired as popular as ever. Johnson, the owner of a larger mandate than Eisenhower had ever enjoyed, could have faced down his critics. The obvious analogy to Truman's experience after the communist victory in China, while suggestive, was not entirely apt. Truman's narrow escape from Dewey in 1948 had rendered him susceptible to the slightest rocking of the boat. Johnson in 1965 could have weathered a much larger storm without swamping.

There were two problems with this scenario. The first was that Johnson, about to launch his all-out effort to build the Great Society, needed every bit of congressional and public support he could wheedle or cajole. The fight for the Civil Rights Act of 1964 had shown the necessity of cultivating Republicans and southern Democrats, groups that did not take kindly to weakness in Vietnam. Johnson's present plans for reform, which amounted to a piggy-backing of the New Deal onto Reconstruction and envisioned more profound changes in American life than any single president had ever attempted, would strain even his huge majority.

The second problem with the withdrawal scenario was that it contradicted everything Johnson knew and believed about America's obligations to the world. The president fully accepted the policy of containment and the philosophy on which it rested. He perceived communism as a plot to conquer the world, piecemeal or whole depending on circumstances. He remembered Munich and saw little evidence that aggression, whether communist or fascist, had changed much since 1938. Global security would begin to crumble if the United States did not stand firm in South Vietnam.

At a more personal level, Johnson was a politician who followed the politician's golden rule: Once a person gives a promise, the promise must be kept. The rule applied especially to the president of the United States, who was not simply a politician but the leader of the Free World. Johnson had promised to stand firm against aggression in Southeast

Asia. Whether the promise had been wise or foolish now mattered less than the fact that it had been given. Honor dictated honoring it. The shortcomings of the South Vietnamese government were beside the point. As Johnson remarked just after his 1965 inauguration, "Stable government or no stable government, we'll do what we ought to do."[28]

What Johnson came to think the administration ought to do was to escalate American involvement in the war. Since the reprisal raids of the late summer of 1964 the Pentagon had been planning a more thorough air offensive against North Vietnam. Johnson had put off the bombers as long as he believed he could, but by the early part of the new year it seemed now or never. Bundy described a policy of "sustained reprisal" as "the one possible means of turning around a desperate situation which has been heading toward a disastrous U.S. defeat." Johnson wished he could disagree, but couldn't.[29]

A February 1965 Vietcong attack on an American barracks and helicopter base at Pleiku provided the same kind of catalyst for action the Tonkin Gulf affair had provided the previous summer. On February 6, shortly after news of the attack reached Washington, Johnson called a special meeting of the National Security Council. At issue was the manner of the American response. George Ball, usually a dissenter regarding Vietnam, at this meeting joined the majority. "We are all in accord that action must be taken," the undersecretary of state declared. Suggesting how the administration should explain a decision to move against North Vietnam, Ball said, "We must make clear that the North Vietnamese and the Vietcong are the same. We retaliate against North Vietnam because Hanoi directs the Vietcong, supplies arms and infiltrates men." Earle Wheeler outlined an immediate response consisting of a series of raids involving 132 American planes and 22 South Vietnamese aircraft. Johnson approved attacks on four targets.[30]

During the next few days the president pondered, then accepted, the Pentagon's proposal for a continuous bombing campaign. Bundy, who supported the proposal, summarized the prevailing view in the administration. "The prospect in Vietnam is grim," Bundy wrote Johnson on February 7. "The energy and persistence of the Viet Cong are astonishing. They can appear anywhere—and at almost any time. They have accepted extraordinary losses and they come back for more." The weaknesses of the South Vietnamese government did not require reiteration. Saigon's weaknesses had become America's weaknesses, and there was little the administration could do about most of them. But not all. "There is one grave weakness in our posture in Vietnam which is within our own power to fix—and this is a widespread belief that we do not have the will and force and patience and determination to take the necessary action and stay the course." A bombing offensive against North Vietnam would demonstrate America's will. This said, Bundy did not argue that bombing North Vietnam would save South Vietnam in the short or medium term. "At its very best the struggle in Vietnam will be long." American officials and the American people would have to reconcile themselves to this fact. "There is no shortcut to success in South Vietnam."[31]

Johnson agreed that the administration must demonstrate America's resolve. "We face a choice of going forward or running," the president remarked on February 8, just before a meeting with congressional leaders. "All of us agree on this." In a nod to Ball, whose dissent was again showing and now took the form of dilatory tactics, he added, "There remains some difference as to how fast we should go forward."[32]

To the legislators, Johnson explained that he believed that the attack at Pleiku justified the reprisals he had ordered. "We had to respond," the president said. "If we had failed to respond we would have conveyed to Hanoi, Peking, and Moscow our lack of interest in the South Vietnamese government. In addition the South Vietnamese would have thought we had abandoned them." While Johnson admitted the deficiencies of the regime in Saigon, he refused to accept these as reasons for failing to fulfill America's commitments. "There is a bad governmental situation in Vietnam, but it is our hope that current U.S. actions may pull together the various forces in Saigon and thus establish a stable government."

Johnson did not think the communists would react recklessly to American air attacks, although he did not rule out the possibility. He said the administration would deal with whatever arose. In part this was why he had called the legislative leaders to the White House. "If the response to our action is larger than we expect, we will then of course make a request for a larger amount of U.S. military assistance and will need additional personnel."

Johnson did not state explicitly that the raids already ordered were the start of a sustained air war against North Vietnam. Had his conscience tugged, he might have taken refuge in the fact that he had not yet irrevocably committed himself. He did tell the legislators he did "not intend to limit our actions to retaliating against Vietcong attacks."[33]

VI

Two significant characteristics marked the process that led to the decision for regular bombing of North Vietnam, which decision Johnson in fact finalized a short while later. The first was the strained nature of the reasoning producing the decision. Previously the administration had decided against escalation on grounds that it might provoke counterescalation by North Vietnam and endanger the shaky government in Saigon. During the first part of 1965 the South Vietnamese government was not noticeably more secure than before. But now, as Johnson explained to the legislative leaders, the administration hoped the bombing of North

Vietnam would "pull together" the competing factions in South Vietnam and allow the creation of a "stable government." Why it should do so at this time, when it had not been expected to six months earlier, the president did not explain.

The second characteristic was the secrecy in which Johnson shrouded the decision-making process. At nearly every meeting during the period when the air campaign was under consideration, the president made a major issue of the need to avoid leaks. Having co-opted Congress by the Tonkin resolution, he had no desire to reopen debate on the war. He did not wish the February raids to be seen as a turning point, at least not for the present. At a meeting with Johnson on February 10, Bundy suggested the strategy the president adopted. "At an appropriate time we could publicly announce that we had turned a corner and changed our policy," Bundy said. But that time had not yet come. "No mention should be made now of such a decision," Bundy concluded.[34]

A single theme united the logic chopping and the fact fudging. Not surprisingly, in light of Johnson's overall approach to foreign policy, it related to the political situation in the United States. On January 17 Johnson unveiled eighty-eight new programs to fight poverty in America. On January 25 he sent Congress a budget specifying the largest expansion of social welfare services in thirty years. On February 6—the day of the Pleiku attack—the White House announced Johnson's intention to seek sweeping new voting rights legislation. A few weeks later Johnson went before Congress personally to deliver the most moving and effective speech of his career, a speech in which he called on the American people to "overcome the crippling legacy of bigotry and injustice" that blighted the land and belied the American promise. "And we *shall* overcome!" he vowed.[35]

Johnson's domestic concerns did not determine his policy toward Vietnam. As noted, and as Johnson repeatedly restated in public and private, he wholeheartedly embraced the containment philosophy on which American involvement in the war rested. He would have sent American bombers against North Vietnam regardless of his domestic agenda if he thought bombing necessary to prevent a communist victory. "We'll do what we ought to do," he had said, and he meant it.

But circumstances in the United States did influence the timing, public presentation, and announced justification of Johnson's decisions regarding Vietnam. There was no way in the world he would have commenced regular bombing before the November elections, regardless of what the Joint Chiefs advised. When he granted approval in February, he chose to ease the country into this new phase of the war gently so that a debate over where the president was going in Vietnam would not distract Congress from where he wanted to go on civil rights and social welfare. As to the logic of bombing North Vietnam to stabilize the government of South Vietnam, Johnson needed a rationale for doing in

February 1965 what he had refused to do in August 1964, and this one was available.

Circumstances outside both the United States and Southeast Asia also influenced Johnson's response to the declining situation in Vietnam. Walt Rostow, still heading the State Department's thinkers, tied communist activities in Vietnam to a broader leftist challenge to the West. Rostow wrote:

> I am struck by the following, in addition to the continued operations of Hanoi in Laos and South Vietnam:
>
> a. The truly ominous pincers movement being operated in Africa via Algeria, Ghana, Mali and Brazzaville, on the one hand, and through Egypt, the Sudan and East Africa, on the other. It is evidently designed to permit nationalist-extremists to take over the heart of Africa. . . .
>
> b. Che Guevara in New York feels comfortable in openly stating that Cuba is aiding revolutionary movements in Latin America, OAS resolutions or not.
>
> c. Sukarno is content openly to acknowledge a confrontation with Malaysia based on the illegal movement of arms and men across frontiers.

Rostow, with most other administration officials, was sufficiently perceptive to recognize that the communist monolith, if it had ever existed, no longer did. But the Moscow-Beijing rift made the radical challenge to the United States and world peace only more threatening. As the administration would discover in the Middle East during the 1967 war, Chinese goading could force the Soviets into positions the Kremlin would not have taken on its own, positions that drove the superpowers uncomfortably close to major war. In Rostow's memo the policy planning director commented that the confluence of revolutionary situations in Africa, Latin America, and Asia had raised the stakes in the competition between China and the Soviet Union. "The leaders in Moscow are more and more pressed to get out in front, lest they be outflanked by Peiping."[36]

The outbreak of the Dominican insurrection in April 1965 seemed still more evidence of the danger confronting the democratic countries. In Johnson's mind, the Dominican Republic and Vietnam were two battlegrounds in the same global struggle. Dominican communists looked to Castro for guidance, while communists in Vietnam looked to Ho Chi Minh, but the totalitarian aims of the two groups were effectively identical. Johnson reacted as strongly as he did to events in Santo Domingo not simply out of fear of a radical takeover there. He believed, doubtless rightly, that Hanoi, Beijing, and Moscow were watching to see what the American president would do in the face of this newest challenge to an American-supported regime.

If the situation in Santo Domingo confirmed Johnson's determination to stand fast in Vietnam, the Dominican crisis also encouraged him to exercise caution in escalating further in Vietnam. Having just sent twenty-two thousand troops into the Dominican Republic, Johnson was in no hurry to dispatch more troops to Vietnam.

But increasingly the troops appeared necessary. American officials had not expected the air offensive against North Vietnam to win the war, and it didn't. Although Johnson loosened his initially tight rein on targeting and authorized ordnance, the bombing did not diminish infiltration from North Vietnam. It may have demonstrated American will, but like German bombing of Britain during the Second World War, it redoubled the resolve of the bombed to fight to victory.

Johnson's generals predictably asked for more muscle. The first request was innocuous enough: two battalions of marines to guard an air base at Danang. Johnson approved this deployment in March without requiring inordinate convincing. But the generals wanted much more. William Westmoreland, the commander of American forces in Vietnam, argued that the war would have to be won on the ground. The Joint Chiefs of Staff agreed. Westmoreland requested two army divisions to augment forces already in the country. The chiefs went Westmoreland one division better, calling for three.[37]

Johnson at first resisted these larger requests, against the counsel of most of his advisers. For a time the president's resistance received the support of Maxwell Taylor, whose military background and firsthand experience in South Vietnam allowed him to credibly dispute the counsel of Westmoreland and the Joint Chiefs. Taylor worried that the introduction of American combat troops would turn the conflict into an American war, rather than leaving it a war for South Vietnam to fight with American help. "For both military and political reasons we should all be most reluctant to tie down Army/Marine units in this country," the general-ambassador wrote from Saigon. The president should give the order "only after the presentation of the most convincing evidence" of the necessity for the troops.[38]

Taylor softened his objections at an April meeting with Westmoreland, Wheeler, McNamara, and Bundy in Honolulu. Westmoreland and Wheeler continued to press for a major infusion of American combat forces, and McNamara and Bundy seconded their argument. Taylor agreed to a compromise that would increase the number of American troops from 33,000 to 82,000. McNamara passed this recommendation to the president.[39]

The Honolulu program fell considerably short of what Westmoreland and the Joint Chiefs had been advocating, but Johnson accepted even the increases it proposed reluctantly and gradually. Rather than give his approval all at once, the president stretched his decision out over most of a month.[40]

To support the increase, Johnson asked Congress for a supplemental appropriations bill for South Vietnam. He couched his request in terms that made it nearly impossible for the legislature to refuse. He reminded the senators and representatives of their almost unanimous approval of the Tonkin resolution, which formed the basis for his actions in Vietnam. The money he was now requesting would allow American troops, carrying out the wishes of Congress, to defend themselves against attack. "To deny and delay this means to deny and to delay the fullest support of the American people and the American Congress to those brave men who are risking their lives for freedom in Vietnam."[41]

Johnson got his money, by a margin nearly as overwhelming as in the Tonkin case. Only a total of ten senators and representatives voted nay.

Equally important, the president got what amounted to another vote of confidence in his Vietnam policy. This might come in handy soon. In asking for the appropriation, Johnson suggested that the present request would not be the last. "If our need expands I will turn again to the Congress. For we will do whatever must be done to insure the safety of South Vietnam from aggression. This is the firm and irrevocable commitment of our people and Nation."[42]

VII

As with the bombing campaign, the Honolulu program failed to stabilize the situation in South Vietnam. Despite the unannounced commencement of offensive operations by American forces in the country, South Vietnam slid closer to the edge, leading Johnson's Pentagon advisers to advocate still more troops. At the beginning of the summer Westmoreland called for raising the ceiling on troops to 115,000. A short while later the Joint Chiefs recommended pushing to 180,000. McNamara, following a July visit to Saigon, proposed going as high as 200,000 in the near term, with more to follow in 1966. McNamara additionally proposed calling up the reserves and expanding America's regular armed forces, with the goal of making available 600,000 troops for use in Vietnam.[43]

Johnson would not even consider the McNamara plan without a complete airing of views. During the fourth week in July the president held lengthy meetings with his usual advisers, with outside consultants, and with members of Congress. The discussions began on the morning of July 21. Johnson indicated at once that he would have to be convinced that McNamara's proposals were the right way to go. "What are the alternatives?" he demanded. "We must make no snap judgments. We must consider carefully all our options." Johnson encouraged dissenters to speak up and heard them out at length. Referring to the paper describing McNamara's plan, the president asked, "Is anyone of the opinion we should not do what the memo says? If so I'd like to hear from them."

George Ball accepted the invitation. "I can foresee a perilous voyage ahead—very dangerous," Ball declared. "I have great apprehensions that we can win under these conditions." Johnson responded, "But is there another course in the national interest that is better than the McNamara course? We know it's dangerous and perilous. But can it be avoided?" Ball answered, "There is no course that will allow us to cut our losses. If we get bogged down, our cost might be substantially greater. The pressures to create a larger war would be irresistible." Johnson rejoined, "What other road can I go?" Ball said the administration should accept its losses in South Vietnam. "Let their government fall apart. Negotiate." Ball conceded that such a course would probably lead to a takeover by the communists. "This is disagreeable, I know." Johnson wanted Ball to elaborate further. "Can we make a case for this and discuss it fully?" the president asked. Ball replied, "We have discussed it. I have had my day in court." Johnson wanted Ball to have another day, and he directed the undersecretary to ready his brief for a second meeting after a break for lunch. "We should look at all other courses."

Johnson queried McNamara as to whether the administration really needed to call up the reserves. Couldn't America's allies put in more troops? "Have we wrung every soldier out of every country we can?" the president demanded. "Who else can help? Are we the sole defenders of freedom in the world?" Johnson asked the defense secretary what the Vietcong would do if the United States upped its troop levels as much as the defense secretary desired. McNamara said he could not say for sure. "We don't know what Vietcong tactics will be when confronted by 175,000 Americans." Johnson aimed a question at Earle Wheeler. "What makes you think if we put in 100,000 men Ho Chi Minh won't put in another 100,000?" The Joint Chiefs chairman agreed that Ho might act just as the president indicated. But such a response would be the communists' undoing. "This means greater bodies of men, which will allow us to cream them."[44]

At the beginning of the afternoon session Johnson gave Ball the floor again. "We can't win," Ball declared flatly. "The most we can hope for is a messy conclusion." The undersecretary perceived a "great danger" of Chinese intervention if the president followed McNamara's recommendation. The United States would suffer a damaging erosion of prestige in the eyes of the world if the war dragged on for years more. And Ball believed the war would, even under the best of circumstances. "I have serious doubt that an army of Westerners can fight Orientals in an Asian jungle and succeed." Ball urged Johnson to resign himself to the inevitable. "Every great captain in history is not afraid to make a tactical withdrawal if conditions are unfavorable to him." Ball suggested telling the government in Saigon in the bluntest terms to shape up or the United States would get out. The South Vietnamese, he predicted, would reject reform in favor of neutralization, which the United States could acquiesce in. If future conditions so dictated, the United States could make a military stand in Thailand, which was far more stable and defensible than South Vietnam. Washington could also concentrate on bolstering South Korea, Taiwan, and Japan.

Johnson asked, "Wouldn't all these countries say Uncle Sam is a paper tiger? Wouldn't we lose credibility, breaking the word of three presidents if we set it up as you proposed? It would seem an irreparable blow. But I gather you don't think so." Ball replied, "The worse blow would be that the mightiest power in the world is unable to defeat guerrillas." Johnson came back: "You are not basically troubled by what the world would say about us pulling out?" Ball answered, "If we were actively helping a country with a stable, viable government, it would be a vastly different story." Ball urged the president to accept short-term losses to avoid the long-term bleeding of an impossible war.

At this point Johnson's other advisers jumped in. Bundy predicted, "The world, the country, and the Vietnamese would have alarming reactions if we got out." The national security adviser could not accept a withdrawal. "It goes in the face of all we have said and done." Dean Rusk asserted that America's good word was on the line. "If the communist world finds out we will not pursue our commitments to the end, I don't know where they will stay their hand." Henry Cabot Lodge, returning as ambassador to South Vietnam after an election-year sabbatical, raised the appeasement issue. Lodge reminded those at the meeting of the consequences of "our indolence at Munich," and he asserted, "There is a greater threat of World War III if we don't go in."[45]

The first day of meetings ended with Johnson perhaps leaning toward accepting McNamara's proposal, or at least much of it, but still requiring to be convinced that no better alternative existed. As the president's encouragement of Ball demonstrated, Johnson wanted to be sure he knew the disadvantages of what he might be getting the country into. If Johnson was a prisoner of circumstances regarding Vietnam—he certainly felt he was, in many respects rightly—he was not a prisoner of carelessness.

Johnson remained skeptical in a meeting with the Pentagon's top guns on July 22. The president wondered why he should believe the military's prognostications about the effect of major troop deployments. "Have the results of bombing actions been as fruitful and productive as we anticipated?" he asked. General John McConnell, the air force chief of staff, admitted they had not. Johnson pressed, "Doesn't it really mean if we follow Westmoreland's requirements we are in a new war? This is going off the diving board." Robert McNamara granted that his program amounted to "a major change" in American policy. "We have relied on South Vietnam to carry the brunt. Now we would be responsible for a satisfactory military outcome."

General Harold Johnson, the army chief of staff, described three options. The least desirable was to pull out, he said. The next bad was to continue on the present course. The best was to go in and get the job done. The president responded doubtfully, "I don't know how we are going to get that job done." Sending more men hardly seemed a guarantee of success. "There are millions of Chinese. I think they are going to put their stack in." The president wondered about a showdown with the communists. "Is this the best place to do this? We don't have the allies we had in Korea." Wouldn't the escalation the Pentagon called for require the Chinese and maybe the Russians to reply in kind? "If we come in with hundreds of thousands of men and billions of dollars, won't this cause them to come in?" General Johnson answered, "I don't think they will," which provoked the president to recall Korea to the gathered minds: "MacArthur didn't think they would come in either." General Johnson dismissed the analogy, saying Vietnam was not comparable to Korea, although he did not spell out why it wasn't. The president insisted on knowing what the generals planned to do if they got it wrong and China did enter the war. After a long silence, General Johnson offered, "If so, we have another ball game." The president said, "I have to take into account they will."

The president wondered where escalation would lead. "Are we starting something that in two or three years we can't finish?" He questioned whether Congress and the American people would support the idea of sending several hundred thousand men and many billions of dollars to a small country ten thousand miles away. American presidents had made a commitment to South Vietnam, but that was before they knew what the commitment would entail. "If you make a commitment to jump off a building and you find out how high it is, you may withdraw that commitment."

After arguing against Americanizing the war, Johnson seemed to decide in favor. "There are three alternatives," he said. "1. Sit and lose slowly. 2. Get out. 3. Put in what needs to go in." Yet he wanted to pursue peace even as he pushed war. "It's like a prize fight. Our right is our military power, but on our left must be our peace proposals. Every time you move troops forward, you move diplomats forward. I want this done."[46]

Three days later Johnson gave the opposition another chance, inviting Clark Clifford to Camp David. Like Ball, Clifford opposed the war on prudential grounds. Clifford's political instincts told him the war was a loser for the Johnson administration—or any administration, for that matter. The United States simply could not win a war in Southeast Asia. "If we send in 100,000 more, the North Vietnamese will meet us. If the North Vietnamese run out of men, the Chinese will send in volunteers." In some abstract military sense the United States might possess the capacity to prevail, but in the real world it did not. The American people simply would not tolerate the sacrifices victory would require. Nor should they. "If we lose 50,000 plus, it will ruin us. Five years, billions of dollars, 50,000 men—it is not for us." Clifford warned Johnson of dire consequences down the road of Americanization: "I can't see anything but catastrophe for my country."[47]

After hearing the arguments of Clifford and Ball, Johnson did not underestimate the hazards of escalation. As his comments to his generals indicated, he couldn't see victory in the near or medium term. But he decided in favor of escalation—though not in favor of the whole Mc-Namara program—because it appeared to him the least bad of the options he faced. Johnson could not escape his recollections of the late 1930s. He believed that only American steadfastness in the face of aggression had prevented a third world war since 1945, and he believed that America must continue to demonstrate that steadfastness. During the summer of 1965 Johnson could look around the globe and see peace under threat almost everywhere. The Dominican Republic still bubbled, its revolution frustrated for the moment but the impulses that produced it hardly dead. India and Pakistan had recently skirmished in the Rann of Kutch, evidently preparing for another blowup. Indonesia's confrontation with Malaysia threatened to expand into a regionwide war, and even if it didn't, Sukarno seemed to be delivering Indonesia into the hands of the communists. In the Middle East the Yemen war persisted, affording Nasser opportunity to undermine the conservatives in Saudi Arabia and Jordan. The emergence of the Baathists in Syria promised to ratchet up the Arab-Israeli conflict. Turkish jets were not strafing Cypriot villages at the moment, but the Cyprus problem remained as volatile as ever.

Johnson was not so foolish as to think the United States held the key to solving all the world's problems. As his Great Society acknowledged, the United States had a huge job simply fixing its own ills. Nor did he believe all the world's troubles were the result of a communist conspiracy. Of the current disputes, those in South Asia, the Middle East, and the eastern Mediterranean clearly had nothing significant to do with Marxism-Leninism, either indigenous or imported from the Soviet Union or China. Yet with most of his generation he believed that the United States offered the best hope for preserving international stability and world peace. Perhaps the task was too great. Perhaps the world was destined for destruction. Even so, someone had to try to halt the aggression. If not Americans, who? If not in Vietnam, where? Johnson summarized his thoughts in one sentence of a televised address on July 28: "We did not choose to be the guardians at the gate, but there is no one else." While Johnson was a bit off on the not choosing, he hit the mark on there being no one else.[48]

Consequently Johnson, with reluctance and foreboding, agreed to a major increase in the number of American troops in Vietnam. He approved the immediate dispatch of 50,000 additional troops, with 50,000 more to follow by the end of the year.

As in the earlier decision to commence regular bombing of North

Vietnam, the content of Johnson's decision to Americanize the war followed principally from his interpretation of world affairs, while the context he provided for the decision reflected his perceptions of American domestic politics. Initially he announced only the dispatch of the first 50,000 additional troops, and he declined to call up the reserves or to go to Congress for more money. In a meeting on July 27 Johnson defended his soft-pedaling on grounds that making a national emergency out of Vietnam would force the Chinese and the Soviets to respond similarly. "I don't want to be dramatic and increase tension," he said.[49]

What Johnson really didn't want to do was sidetrack his domestic legislation. The Great Society was racing forward at full steam. Johnson knew he couldn't maintain the pace for long, and he desired to get as much out of Congress as he could while the reforming spell lasted. In April he had signed into law a program injecting the federal government in an unprecedented way into elementary and secondary education. In July his pen put Medicare on the books, creating another huge area of federal responsibility. In August the Voting Rights Act would send federal officials into the South to guarantee access to the polls to millions of disfranchised African-American citizens. Dozens of other pieces of legislation, singly less significant but cumulatively sweeping, were in various stages of consideration. Johnson did not wish to slow things down. Alarming Americans excessively about Vietnam would do just that.

Congress appeared willing enough to accept Johnson's handling of the Vietnam issue. At a meeting of legislative leaders just before the president announced his troop decision, Democratic senator Russell Long of Louisiana urged Johnson to do whatever he had to do to keep South Vietnam from falling into the hands of the communists. "If we back out, they'd move somewhere else," Long said. "Are we ready to concede all Asia to the communists?" Long avowed that he was not. Upon Vietnam, he said, hung the future of the entire continent. "If a nation with 14 million can make Uncle Sam run, what will China think?" Speaker McCormack said, "I don't think we have any alternatives." The road into Vietnam held hazards, McCormack admitted. But the path of appeasement held more. "The lesson of Hitler and Mussolini is clear. I can see five years from now a chain of events far more dangerous to our country." Everett Dirksen backed Johnson's decision to send additional troops, although the Senate minority leader thought the president ought to request supplementary funding. Only Mike Mansfield objected to the basic thrust of Johnson's decision. Mansfield said the United States had pledged to *assist* in the defense of South Vietnam, not carry the major portion of the burden. By its actions, the South Vietnamese government had forfeited any claim on the United States. "We owe this government nothing," Mansfield declared.[50]

VII

Johnson's July decision to Americanize the war was the most important action he took regarding Vietnam. What came afterward followed naturally, almost inevitably, from this decision. In defending his decision to his closest advisers, the president encapsulated his entire approach to the conflict in Southeast Asia: "We have chosen to do what is necessary to meet the present situation, but not to be unnecessarily provocative to either the Russians or the Communist Chinese."[51]

In the summer of 1965 this strategy appeared reasonable. None could fault Johnson's caution in trying to forestall direct Soviet or Chinese involvement in the fighting, and South Vietnam's position did not seem irretrievable. "The present military situation is serious but not desperate," Maxwell Taylor asserted at the beginning of August.

Yet Taylor at the same time sounded a warning and underscored the weakness of Johnson's gradualist approach. "No one knows how much Viet Cong resilience is still left," Taylor said. The troops Johnson was dispatching to South Vietnam would prevent the communists from winning—at the communists' current level of fighting. "By the end of 1965, the North Vietnamese offensive will be bloodied and defeated without having achieved major gains," Taylor predicted. But once the communists adjusted to the American escalation, they might well up the stakes themselves. "Hanoi may then decide to change its policy."[52]

While waiting for the communists to react to his troop decision, Johnson had to deal with responses at home. The Americanization of the war sparked the first sustained antiwar protests in the United States. Shortly after Johnson's troop announcement, some one thousand demonstrators took a page from the manual of the civil rights movement and held a sit-in outside the White House. The protests mounted during the succeeding weeks, reaching a crescendo, for the moment, in a November 1965 march of twenty thousand in Washington. William Fulbright, having recently attacked the president on the Dominican issue, added his voice to the protest. The Arkansas senator denounced the escalation of the war and called on the president to stop the bombing.

The criticism had an effect on Johnson, though he didn't like to admit it. Robert McNamara, after August visits to the foreign affairs committees of the Senate and House, described the sentiment there regarding the administration's war policies. "In both the Senate and the House Committees," McNamara told Johnson, "there is broad support, but it is thin. There is a feeling of uneasiness and frustration." McNamara noted that the indications of uneasiness and frustration came from the right as well as from the left. Suggestions that the president was going too far in Vietnam were countered by complaints that he was not doing enough.[53]

Johnson, as earlier, sought the middle of the road. At the end of September the administration announced the second installment of the

July escalation package. This move further evidenced the president's resolve and presumably pleased the right. Two months later Johnson followed up the dispatch of new troops with a halt to the bombing, a move calculated to placate the left. In addition, the bombing halt would remind the world that what the United States wanted in Vietnam was not war but peace.

Johnson stopped the bombing reluctantly. He had ordered a brief bombing halt the previous May, and that suspension had produced no positive response from the communists. But it had elicited loud objections from American supporters of the war who complained that if this was really a war, the United States ought to fight it vigorously. Johnson soon gave in to the complaints and ordered the bombing resumed. "My judgment is that the public has never wanted us to stop the bombing," he commented. "We have stopped in deference to Mansfield and Fulbright, but we don't want to do it too long or else we'll lose our base of support."[54]

When the president in December again grounded the bombers, the military once more opposed the idea, predicting that the communists would simply use the respite to restock and regroup. The president's civilian advisers recognized this risk, but for the most part argued that the possible diplomatic and political benefits outweighed the military costs. "We think this is the best single way of keeping it clear that Johnson is for peace while Ho is for war," McGeorge Bundy explained.[55]

Abe Fortas and, somewhat surprisingly, Clark Clifford were skeptical. The two veteran negotiators suggested that since North Vietnam had shown no desire to talk peace, the president would be wasting his time trying to entice them. Fortas feared that if the bombing halt failed to produce moves toward peace—and he thought it would indeed fail—then the administration would come under pressure to take more drastic action than heretofore. Clifford contended that a bombing halt would look gimmicky after the president had to renew the bombing, as he surely would.[56]

Johnson sided with the halters. Although a suspension of bombing would not satisfy the hard-core antiwarriors, who demanded that the president withdraw American forces from Vietnam and leave the South Vietnamese to their fate, it would place the weight of responsibility for the war's continuation on the communists. Or so Johnson hoped.

To further demonstrate his pacific intentions, the president accompanied the bombing halt with a well-publicized diplomatic offensive. He sent envoys to European, Asian, African, and Latin American capitals carrying a message of America's desire for peace. He specially contacted Soviet chairman Kosygin to convey the same message. He emphasized his great desire for peace in his January 1966 State of the Union address.

Neither the pause nor the professions moved Hanoi, which refused to retreat from its demand that the American imperialists exit Vietnam at once, and by the fourth week of January the president was thinking seriously about resuming the bombing. Objections from American generals were increasing. With every day that North Vietnam resupplied the communists in South Vietnam unhindered, the generals argued, American soldiers came under more serious danger. The bombing must recommence forthwith. Johnson felt the force of this argument, but he wanted to check congressional opinion before making a decision.

On January 25 he again invited the congressional leaders to the White House. Mike Mansfield and William Fulbright, as Johnson had expected, remained fundamentally opposed to the war. Mansfield said the president should call for a cease-fire and prepare to withdraw American forces after arranging terms of amnesty for the South Vietnamese involved in the fighting. Fulbright dismissed the administration's antiappeasement analogies. The situation in Vietnam, Fulbright asserted, had little in common with the situation in Europe in the 1930s. The United States was fighting a colonial war, not defending a country against fascist aggression. He asked Johnson what victory in Vietnam would look like. "If we win, what do we do? Do we stay there forever?" He warned that additional escalation might easily lead to a third world war.

As Johnson had also expected, Mansfield and Fulbright found themselves in a small minority at this meeting. Everett Dirksen declared that to withdraw from Vietnam would be a "disaster." Dirksen told Johnson, "If we are not winning now, let's do what is necessary to win. I don't believe you have any other choice. I believe the country will support you." Richard Russell arrived at the same conclusion by a somewhat different route. "This is the most frustrating experience of my life," the Georgia senator said. "I didn't want to get in there, but we *are* there." The president had an obligation to protect American soldiers by going after the communists. "I'd rather kill them than have American boys die." Russell found American concern over civilian casualties in Vietnam a bit surprising. "We killed civilians in World War II and nobody opposed it." Some people had suggested a partial resumption of bombing. Russell warned Johnson to have nothing to do with such advice. "For God sakes, don't start the bombing half way. Let them know they are in a war." He concluded, "Please, Mr. President, don't get one foot back in it. Go all the way."

Leverett Saltonstall, the ranking Republican on the Senate Armed Services Committee, agreed with Russell, as did Carl Albert, the majority leader in the House of Representatives. Albert registered his confidence in the administration, telling Johnson, "You and your advisers have my trust." Leslie Arends, the ranking Republican on the House Armed Services Committee, encouraged the president to restart the bombing lest the American military position suffer further. Frances Bolton, the top Republican on the House Foreign Affairs Committee, thought bombing would send a needed signal that the United States planned to see its commitment in Vietnam through to the end. "Don't let them think we won't fight," Bolton said.[57]

For all the encouragement at this meeting, Johnson recognized that support for the war was not what it had been. On January 28 the president predicted to his advisers that resumption of the bombing might lead to "deep divisions in our government." Opponents remained far from a majority, but he guessed that if something like the Gulf of Tonkin resolution were placed before the legislature now, some forty senators and representatives would vote against it. Although this was not a great many out of 535 legislators, it was still considerably more than the two naysayers of August 1964.[58]

If any single sentiment captured American feelings about the war early in 1966, it was frustration. As the remarks of Richard Russell and others at the January 25 meeting indicated, Americans felt frustrated over the inconclusive character of the fighting. At this same session representative Mendel Rivers put the mood best. "Win or get out," the South Carolina Democrat said.

Eventually Johnson would admit that the issue came down to this stark choice, but for the time being he still thought he could finesse things. He refused to court the dangers associated with a unrestrained effort to win the war. For fear of provoking the Chinese or the Soviets or both, he rejected all ideas of invading North Vietnam. For fear of killing his domestic reforms, he rejected proposals to place the country on a war footing. Yet though he wouldn't win, neither would he get out. Getting out would demoralize America's allies and hearten America's enemies, and it would subject his administration to devastating criticism at home.

At the end of January 1966 Johnson judged the costs of getting out higher than the costs of staying in. And he judged that staying in required resumption of the bombing. His decision to resume reflected a felt obligation to American soldiers in the field. "I don't want to fail the men out there," he said. It also reflected fear of a conservative backlash against his fruitless peace efforts. Such a backlash could easily make the antiwar movement seem a minor annoyance. "I certify that the Fulbrights and the Morses will be under the table and the hardliners will take over, unless we take initiatives," Johnson asserted. The initiative he selected on January 31 was renewed bombing.[59]

VIII

As long as he possibly could, Johnson continued to avoid making the choice Mendel Rivers described. As before, he sought the middle course between the opponents of the war, who advocated withdrawal, and his military advisers, who recommended, if not all-out war, at least considerably more war than they were fighting. "What do you want most?" the president had asked General Johnson during discussions of bombing resumption. "A surge of additional troops," the army chief of staff had replied. "We need to double the number now and then triple the number later. We should call up the reserves and go to mobilization to get the needed U.S. manpower. This involves declaring a national emergency." During the spring of 1966 the generals agitated for permission to bomb petroleum facilities in the Hanoi-Haiphong area, previously off limits for fear of forcing the hand of China or the Soviet Union, the latter perhaps by accidentally hitting a Soviet ship. In June 1966 General Wheeler suggested going still further and mining Haiphong harbor. "Do you think this will involve the Chinese Communists and the Soviets?" the president asked. "No, sir," Wheeler replied. "Are you more sure than MacArthur was?" came Johnson's counter. "This is different," Wheeler explained. "We had ground forces moving to the Yalu."[60]

Johnson gave the military part of what it wanted, but not all. He authorized additional increases in American troop levels, to almost 400,000 by the end of 1966, and to nearly half a million during the following year. He let the bombers hit the petroleum facilities and much else besides. But he refused to declare a national emergency and to go to full mobilization, and he rejected the mining of Haiphong.

The actions the president did allow more than sufficed to spur the growing antiwar movement to unexampled efforts of opposition. By 1966 many liberals of the Kennedy persuasion were repenting of their earlier backing for the war. For the most part their disapproval did not signal a rejection of the antiappeasement policies that had formed the basis of America's approach to the world since 1947. Instead it reflected a judgment that Vietnam was not the place to make the Free World's stand. American material, moral, and psychological resources were not infinite, and those resources could find better use in other areas and in support of other causes.

The liberals in the antiwar movement—represented in Congress by Mansfield and Fulbright, in the media by the *New York Times* and the *Washington Post,* in the intellectual community by the likes of Arthur Schlesinger, Jr., and John Kenneth Galbraith—gave critical credibility to a movement until recently dominated by New Leftists and other Cold War rejectionists. The protests of the combined phalanx took a variety of forms. Poet Robert Lowell publicly boycotted a White House arts festival, at which literary and war critic Dwight McDonald circulated an anti-Johnson petition. Fulbright held televised hearings allowing opponents of the war equal time to challenge administration officials. Journalist Harrison Salisbury visited Hanoi and reported heavy bomb damage to schools, hospitals, and homes, contradicting administration claims that American bombardiers carefully avoided civilian installations and activities. More than six thousand academics and professionals took out a three-page ad in the Sunday *New York Times* to register disapproval of Johnson's policies. War resisters burned draft cards and effigies of the president. Marchers and sitters-in clogged campuses. The biggest single protest to date occurred in October 1967 when approximately one hundred thousand activists, pacifists, and curious onlookers crowded the mall in Washington, with a large contingent crossing the Potomac to invest the Pentagon.[61]

Johnson took some solace in the knowledge that it was a president's duty at times to make unpopular decisions. Walt Rostow bucked up his boss by likening Johnson's position to that of Abraham Lincoln during the Civil War—another necessary but controversial conflict. Criticism came with executive responsibility.[62]

On numerous occasions Johnson attempted to answer the opposition. In April 1965, at Johns Hopkins University in Baltimore, the president proposed a billion-dollar development program for Southeast Asia, deliverable as soon as the communists agreed to accept the independent existence of South Vietnam. In the same speech he declared America's readiness to engage in unconditional discussions with representatives of the Vietcong and North Vietnamese. In January 1966, amid the bombing halt of that period, he promised, "We will meet at any conference table, we will discuss any proposals—four points or fourteen or forty—and we will consider the views of any group. We will work for a ceasefire now or once discussions have begun." In January 1967 he confessed to a small group of legislative leaders that he wished he had "never heard of Vietnam" and that the United States had never gotten involved in that country. "But we are there," he continued, and the country had to make the best of the situation. In September 1967, in a speech at San Antonio he offered to stop the aerial campaign in return for Hanoi's pledge to negotiate.[63]

Johnson's efforts to defend his policies did little good, and by the end of 1967 American public support for the war had dropped drastically. In March 1965 Gallup had reported 66 percent approval for the president's handling of the war. An August 1967 poll showed only 39 percent of the American public approving of the president's performance.[64]

The criticism of the war crescendoed in February 1968, in the wake of the most extensive Vietcong-North Vietnamese offensive of the war. During the Tet holiday, the communists launched attacks on most of South Vietnam's provincial capitals, nearly all its major cities, and over one hundred smaller villages and hamlets. They assaulted the presidential palace and the airport in Saigon, and for several hours occupied the American embassy compound in the heart of that city.

After initial setbacks, American and South Vietnamese forces dug in, then successfully counterattacked. Fierce fighting consumed most of February, producing hundreds of deaths among American and South Vietnamese troops, thousands among communist forces, and thousands among noncombatants—with some four thousand persons executed by communist units in Hue—and throwing tens of thousands of new refugees into the maelstrom of wartime homelessness.

Tactically, the Tet offensive proved a severe defeat for the communists, but strategically it marked the turning point of the war. Johnson caught a glimpse of the transformation on the eleventh day of the offensive. After a meeting with his top military advisers, the president remarked, "The chiefs see a basic change in the strategy of the war. They say the enemy has escalated from guerrilla tactics to more conventional warfare." This first glimpse was correct but also misleading, for it suggested that the communists were now playing to America's strength. America's advantages in firepower and maneuverability would soon begin to tell, and, indeed, it was exactly this telling that allowed the American and South Vietnamese troops to beat back and then punish the Tet attackers.[65]

But more important, even in defeat the breadth and strength of the communist assault knocked the political wind out of the American effort in Vietnam. (It also threatened to knock out the economic wind, straining the dollar internationally and provoking a run on gold.) Johnson did not have to read the polls to know the American popular mood. "There has been a dramatic shift in public opinion on the war," the president said. Clark Clifford explained the administration's public-opinion problem succinctly: "The major concern of the people is that they do not see victory ahead." Dean Rusk put the matter more strongly: "The element of hope has been taken away by the Tet offensive. People don't think there is likely to be an end."[66]

Johnson's generals, however, did not appear discouraged. William Westmoreland advocated more of the same: two hundred thousand troops more. With the enemy shifting to conventional warfare, Westmoreland argued, now was the time to press hard for victory. Earle Wheeler forwarded Westmoreland's request to Johnson with recommendation for approval.

IX

The last thing Johnson desired to hear at this stage was a request for more of the same. At a meeting of the Tuesday lunch some months before the Tet offensive, the president had directed General Johnson to "search for some imaginative ideas to put pressure to bring this war to a conclusion." He told the army chief of staff not merely to recommend nuclear weapons or more men. The president said he could think of those ideas himself. The demand was a measure of Johnson's frustration, for on sober reconsideration he must have realized he was asking the impossible. He had ruled out an invasion of North Vietnam to prevent provoking the Chinese and the Russians. For the same reason he refused to widen the war into Laos and Cambodia to interdict the Ho Chi Minh

trail or to destroy communist sanctuaries there. The one genuinely original idea the Pentagon proposed—an electronic fence along the 17th
parallel—proved even more unworkable than most of America's high-
tech approaches to the conflict, and was abandoned.[67]

Forced to confront the fact that his policies had led him and the
United States into a dead end, Johnson reassessed the whole business.
For help he turned to his council of Wise Men. Until now the group had
almost unanimously endorsed maintaining the present course. Asked in
November 1967 whether the United States should pull out of Vietnam,
Dean Acheson asserted, "Absolutely not." Douglas Dillon declared,
"Definitely not." Henry Cabot Lodge answered, "Unthinkable." Mc-
George Bundy replied, "As impossible as it is undesirable." Abe Fortas,
also present, said, "The public would be outraged if we got out." Robert
Murphy, who during the Eisenhower years had been in on CIA plots
against troublesome Third World governments, suggested that the administration expand current operations to include assassination of
North Vietnamese leaders. As Murphy put it, the administration's "303
committee"—the device for keeping the president's hands, if not so
easily his conscience, clean—should "study the elimination of the men
responsible in the North."[68]

The CIA was already booked up trying to eliminate the leadership of
the Vietcong in South Vietnam, and there is no available evidence that
Murphy's recommendation went any farther than this meeting. Perhaps
Johnson recognized a fundamental drawback behind assassinating heads
of foreign governments: that the successors encounter the same kind of
pressures Johnson had felt after Kennedy's death, to continue the policies of the martyred hero. Johnson wanted to talk peace, and he knew
that the only person who could talk back was Ho Chi Minh.

In any event, the Tet offensive threw thoughts of expanded operations
out nearly all administration windows. The Joint Chiefs wanted to escalate further, but they could not even convince their boss at the Pentagon.
Robert McNamara, increasingly discouraged at the failure of the policies
he had championed, had drifted gradually in the direction of George
Ball. In the autumn of 1967 McNamara wrote Johnson that continuation
on the present path would be "dangerous, costly in lives, and unsatisfactory to the American people." Johnson, at this pre-Tet point, was not
ready for such dismal realism, and when the World Bank, with White
House encouragement, offered McNamara its directorship, Johnson
gladly accepted the offer.[69]

Johnson replaced McNamara with Clark Clifford. A thoroughgoing
realist in political matters, Clifford had stuck by Johnson after the president had overruled his opposition to the Americanization of the war.
Clifford had tried to moderate the more extravagant recommendations
of administration hawks, but by no means was he a dove. While he lacked
the organizational flair and experience McNamara had brought to the

Defense Department, he possessed political sensitivities second only to Johnson's. At the beginning of 1968, in fact, Clifford's senses were sharper than Johnson's, since the president's had been numbed by four difficult years of responsibility. To some degree Johnson's appointment of Clifford to replace McNamara indicated an understanding on the president's part that the war had become more a domestic political problem than a foreign military one.

Clifford lost no time after the Tet offensive in telling Johnson that the political problem of the war had grown insoluble. At a White House meeting early in March, Clifford said the Tet offensive had come as "a shock." No one had thought the communists could mount such an operation. No one expected them to be able to repeat the performance soon, but obviously they had a lot more fight left in them than American analysts had recognized. The American public wondered whether the administration knew where it was going. Clifford urged Johnson to reject Westmoreland's request for additional troops. If the general got these two hundred thousand, he would soon be back asking for another two hundred thousand. Americans would not stand for a war of attrition. Vietnam was a "sinkhole," Clifford said. "We put in more—they match it. We put in more—they match it." Absent a major change in policy, only grief lay ahead. "I see more and more fighting with more and more casualties on the U.S. side, and no end in sight to the action."[70]

When the Wise Men met a short while later, they agreed with Clifford. McGeorge Bundy spoke for the majority of the group when he told the president that there had been "a very significant shift in our position" since the previous autumn. "When we last met we saw reasons for hope. We hoped then that there would be slow but steady progress." The Tet offensive had shaken these hopes, if not shattered them entirely. "A great many people—even very determined and loyal people—have begun to think that Vietnam really is a bottomless pit. We have had no solid and visible successes there." Nor did any seem likely soon.[71]

Dean Acheson put the matter more directly. "We can no longer do the job we set out to do in the time we have left," the former secretary of state said. Acheson did not go so far as to say the war was unwinnable, but he was certain the American public, after the blow of Tet, would not allow the administration the time victory required. From the moment in 1947 when Acheson had helped persuade Truman to give a fighting speech in support of aid to Greece and Turkey, Acheson had understood the necessity of public backing for foreign policy. That backing was evaporating rapidly. "We must begin to disengage," he concluded.[72]

Most but not all of those at this meeting agreed with the Bundy-Acheson view. George Ball, Arthur Dean, Cyrus Vance, and Douglas Dillon assented, for reasons similar to those cited by Bundy and Acheson. Robert Murphy, Maxwell Taylor, and Omar Bradley thought the disengagers were allowing themselves to be stampeded by the unwarranted pessimism that had seized the American political system. Events had proven the early reports of military disaster wrong. To grant the communists politically what they had failed to achieve militarily would be disastrous. Now was the time to move forward to victory.

Johnson, after considerable ego-wrestling, accepted the essence of the disengagers' case. He rejected Westmoreland's troop request and decided, on Rusk's recommendation, to cut back significantly on the bombing of North Vietnam. In a March 31 televised address he explained that only enemy staging areas just north of the 17th parallel would remain subject to American attack. In the same speech he reiterated his desire to commence negotiations to stop the fighting. To this end he appointed Averell Harriman his personal negotiator. Most significantly, he declared that he would not be a candidate for the presidency in the 1968 election.

The special significance of the final announcement consisted in the fact that it took Johnson out of the war. Nixing more troops was no great new departure, since the president consistently had given the generals less than they had wanted. Besides, the situation in South Vietnam had continued to stabilize in the month since Wheeler had endorsed Westmoreland's request, making additional American soldiers less necessary than they had seemed at the time. Further, with American prodding, the South Vietnamese government of President Nguyen Van Thieu had unveiled plans to substantially enlarge the South Vietnamese army, thus laying the basis for what would be called "Vietnamization."

Neither was the bombing halt revolutionary. Since February 1965 Johnson had grounded the bombers or reduced their range of targets more than a dozen times. The longest complete halt was the one running from December 1965 through January 1966. Other groundings lasted from twenty-four hours to several days. Four times the president had declared Hanoi off-limits to American attacks. In any event, the bombing halt was revocable at presidential discretion.

Johnson's announcement of his desire to negotiate an end to the war was equally unnovel. Repeatedly the president had asserted his administration's willingness to engage in unconditional discussions with the communists. Although the diplomatic grapevine had recently carried hints that Hanoi might be more interested than previously in talking, none of the hints conveyed any intimation Hanoi was modifying its demand that the United States get out of Vietnam and leave the country to the Vietnamese. Put otherwise, the communists might talk, but they would not deal.

By contrast, Johnson's declaration of lame-duckhood was both novel and for all practical purposes irrevocable, and it opened the door to substantial change in American policy. While the basic American objective in Vietnam remained as before—an independent, noncommunist South Vietnam—Johnson's departure hinted that Washington might be adjusting its sights downward. Johnson constantly compared Vietnam to Korea, and he couldn't have helped remembering how in the Korean case a new chief executive—Eisenhower—had been able to accept a settlement that would have elicited calls for Truman's impeachment. In 1969 a new president, Democrat or Republican, would enter the White House unburdened by much of the political baggage of Johnson's policies.

This is not to say Johnson was jettisoning his policies. He may even have felt his withdrawal would give those policies the greatest chance of success. Vain as he was, he understood that much of the animus against the war focused on him personally. At the least, protesters who chanted, "Hey, hey, LBJ, how many kids did you kill today?" would have to rewrite their rhymes after January 1969.

Yet whether or not he judged his policies still feasible, Johnson appreciated that a change was necessary. His withdrawal from the political arena would facilitate that change.

Nine

Hegemony's End

JOHNSON'S ABDICATION marked both a defeat for the president personally and a defeat for the policy of global containment that had formed the basis of America's approach to international affairs for more than two decades. In March 1947 Harry Truman had committed the United States to stopping the spread of communist influence throughout the world. In March 1968 Lyndon Johnson conceded, albeit implicitly, that the job was more than America could handle.

The implications of Johnson's announcement, as they related to Vietnam, took time to surface. While Hanoi surprised many by accepting Johnson's offer to negotiate, the talks that began in May 1968 in Paris went nowhere slowly. With most of North Vietnam off-limits, American planes now pounded enemy forces in South Vietnam more ferociously than ever. Shortly before the November election, Johnson turned off the rest of the bombing of North Vietnam, but after January 1969 Richard Nixon turned it back on and added Cambodia and Laos to the list of targets. Nixon continued the Vietnamization begun during Johnson's last months, yet the process was not complete until January 1973. The final failure foreshadowed by Johnson's March 1968 speech occurred only in April 1975, when communist forces captured Saigon and terminated the war.

In other areas, the end of the period of America's ability to shape the world according to American design came sooner. Serious slippage showed in Europe during Johnson's presidency. De Gaulle's various declarations of French independence from the United States demonstrated that Washington would have to deal with the Europeans on a new basis of equality. Although de Gaulle's assertions of French sovereignty got much of the restiveness out of the Gallic system, France continued to demonstrate its independence of the United States, and rights American leaders had taken as a matter of course until the mid-1960s required Paris's permission afterward. Permission could be withheld, as Ronald Reagan discovered in 1986 when he unsuccessfully sought Paris's consent for American jets to fly through French air space en route to an attack on Libya.

Johnson's offset troubles with West Germany suggested that in certain areas of relations, notably involving economic matters, Washington could not take even equality for granted. Throughout the 1970s and 1980s the issue of "burden sharing" persisted, and the problem grew larger as the German economy grew stronger. Germany's economic independence begot other forms of independence. Though Richard Nixon did not like Willy Brandt's *Ostpolitik*, he recognized there was little he could do to block it.

In the eastern Mediterranean, the erosion of the Cold War paradigm continued after Johnson left office. In 1964 Johnson had just managed to force the Turks to stand down by threatening to withdraw American protection against the Soviet Union. Ten years later Washington held no such terror for Ankara, and amid Washington's post-Watergate confusion the Turkish army invaded and partitioned Cyprus. Makarios, after having mused with George Ball regarding assassination prospects, in fact died in bed in 1977, but not before surviving at least four assassination attempts by various enemies and a 1974 overthrow by Greek army veterans in the Cypriot national guard—which ouster triggered the Turkish invasion.

In the Middle East, American influence diminished drastically during the 1970s and early 1980s. The October War of 1973 produced an Arab embargo of oil to the United States, which showed the oil producers the advantages of common action and contributed to a rapid quadrupling of petroleum prices that sent the American economy spinning. The Iranian revolution of 1979, directed as much against the United States as against America's client the Shah of Iran, banished American influence from the largest of the Persian Gulf countries and led to the spectacle of the world's greatest power being held hostage by a gang of religious zealots.

While the India-Pakistan war of 1965 revealed Johnson's inability to prevent two governments on the American dole from going to war against each other, a thorough alienation of India (without doing the Pakistanis any measurable good) required the skills of Richard Nixon and Henry Kissinger. In 1971 India and Pakistan went to war again. The conflict's catalyst this time was not Kashmir but a revolt in East Pakistan against the Pakistan government. When refugees from the fighting threatened to bankrupt never-too-liquid New Delhi, Indira Gandhi sent troops into Dacca. Nixon, immersed in secret plans to normalize relations with China and requiring the covert good offices of Pakistan president Agha Yahya Khan in the normalization, chose to overlook widespread atrocities committed by Pakistan's soldiers against Pakistan's own citizens. Nixon added intimidation to insensitivity when he ordered the aircraft carrier *Enterprise* into the waters off India in a show of force designed to rescue Pakistan. The action did not prevent the birth of Bangladesh, ripped from Pakistan, but it outraged the Indians in a way Johnson's mere neutrality in 1965 had not even approached.

In Latin America, Johnson's invasion of the Dominican Republic checked, without stopping, the spread of support for radical alternatives to a status quo tied to Washington. Though the Dominican Republic itself calmed down, the revolution Castro had pioneered jumped in the 1970s to the Central American mainland, where it toppled the pro-American regime of Anastasio Somoza in Nicaragua and generated a civil war in El Salvador that lasted through the 1980s into the 1990s.

Suharto's consolidation of power in Indonesia continued past Johnson's exit from office, into the 1970s, then the 1980s, then the 1990s. American aid also continued, although it gave the United States hardly more leverage over events in Indonesia than American officials had exercised during Sukarno's last years. Like one of those Manhattan-sized icebergs that periodically break off the Antarctic ice sheet and roam the southern seas, Indonesia moved at a pace and with an inertia that largely defied external prodding.

II

In 1947 Truman made a bad bargain with history. Eisenhower and Kennedy made the bargain worse. But before the mid-1960s the bargain's badness—the overcommitment of American resources inherent in the universalist rhetoric of the Truman doctrine, in the multiplying military pacts of the 1950s, and in the ambitious pledge to pay any price and bear any burden in the defense of liberty—remained largely hidden.

By Johnson's turn, however, the ill consequences of the open-ended policy of supporting an anticommunist status quo planetwide were growing painfully apparent. The world of the 1960s was not the world of the early postwar period. What America could conceive of in the flush of its victory over fascism, of its atomic monopoly, and of its enormous economic lead over its closest competitors, the America of Johnson's day could not deliver. To Johnson fell the job of overseeing the end of the era of American hegemony—of directing the reentry of the United

States into an international order unskewed by the conditions that had peculiarly favored America during the twenty years immediately after the war.

Johnson handled the job sometimes capably, sometimes not. Considering the dimensions of the task, it is surprising he did as well as he did. Johnson's years in the White House, especially the first two when he was just getting the feel of the place, were dauntingly eventful. From the winter of 1963–1964 through the autumn of 1965 the president had to deal with wars, near-wars, coups, and assorted other crises–each of which had potentially serious implications for the United States–in Panama, Cyprus, Brazil, Vietnam, the Dominican Republic, India and Pakistan, and Indonesia. Lesser contretemps tangled American relations with the Congo, when Johnson in the fall of 1964 sent American planes to ferry Belgian paratroops into action to free Western hostages captured by rebel forces fighting against the American-backed government; and with Rhodesia (and through Rhodesia with Britain), following the white-ruled colony's November 1965 unilateral declaration of independence from London.

The pace of events slowed slightly after 1965, yet their import did not diminish. De Gaulle's eviction of the United States from France, the offset difficulties with West Germany and Britain, and the continued escalation in Vietnam filled 1966. The spring of 1967 brought war in the Middle East, followed by another close Cyprus shave in November. The first crisis of 1968 involved North Korea's seizure of the American intelligence ship *Pueblo*, but Washington's worries over the fate of the eighty-two crew members (who came home in December after exasperating but largely unremarkable negotiations) were quickly swallowed up in the alarm surrounding the Tet offensive in Vietnam. The Soviet invasion of Czechoslovakia in the summer of 1968 marked the finish of Johnson's string of crises—and, along with Russia's rumbling at the close of the Middle East war of 1967, marked the closest approach to confrontation between the superpowers during Johnson's tenure.

Johnson managed to dodge most of these bullets. It would be difficult to fault the president's handling of de Gaulle. Whether from his instinctive understanding of a political mind like de Gaulle's, or from the advice of Elysée-watchers like Charles Bohlen, Johnson perceived the futility of confronting the general publicly. He disagreed with de Gaulle on many issues of world affairs, but he recognized that public debate would only aggravate tensions between the two countries. Franco-American relations would survive de Gaulle.

Johnson likewise did well with the Germans and the British. Delicately balancing Bonn's and London's demands for offset readjustment against congressional complaints that America was spending too much in Europe, Johnson succeeded in keeping American and British troops on the Rhine and beyond while trimming American expenses. By doing so he kept Germany untempted to provide for its own defense in its own, perhaps unsettling, way.

Forestalling war between Greece and Turkey called for less subtlety and more table-thumping than did the talks with West Germany and Britain, but Johnson adjusted. Though a lasting answer to the Cyprus question would have been nice, Johnson joined a large crowd—namely everyone at least since Homer—in failing to find it.

South Asia went to war despite Johnson's efforts, but the India-Pakistan conflict was essentially beyond America's control. In other respects the president's policies have to be counted on the success side of the ledger. If his evenhandedness during the 1965 war won America few friends in the subcontinent, neither did it create lasting enemies. His skeptical treatment of foreign aid for India, while thoroughly unappreciated by the Indian public and most of India's supporters in the United States, produced the dual benefits of greater backing in India for reform and a broader base for Indian aid in Congress.

A favorable shift in the strategic balance in island Southeast Asia rewarded Johnson's patience regarding Indonesia. Credit for the shift—and blame for the deaths of the hundreds of thousands killed in Suharto's purge of the PKI—belonged to the Indonesians, not to Americans. But sometimes wisdom consists in leaving things alone and playing for the breaks, which Johnson did regarding Indonesia.

Johnson could not have left the Middle East alone even if he had wanted to, which he didn't. His approval of arms sales to Israel reflected his own convictions as well as those of Congress and the American public. He might have done more to try to restrain the Israelis after the beginning of the crisis of May 1967, but by then the situation was substantially out of his hands. Most Israelis feared they were facing a national—and perhaps personal—death threat. Against this fear American calls for moderation availed little. Moscow may or may not have been genuinely contemplating intervention on the last day of the fighting. If it was, Johnson's warning words and gestures were an appropriate response. If not, they were an overreaction but one that did no harm.

Johnson's overreaction to events in the Dominican Republic did more harm, though not a great deal more. The president's dispatch of twenty-two thousand troops to subdue a revolt against an illegitimate military government, on grounds that the rebels included a few communists, confirmed the conviction common in Latin America that Washington would always oppose hemispheric reform. But in light of the long history of intervention by the United States, few Latin Americans required additional evidence to maintain this conviction, and after the troops withdrew and the Dominicans fairly elected a new government, the complaints lost much of their bite. Meanwhile, the talks with Panama consequent to Johnson's agreement to renegotiate the canal treaties showed that the *norteamericanos* were not entirely unreasonable.

The only major area of Johnson's policy that falls clearly into the negative category is Vietnam. Vietnam's negatives, of course, were sufficient for a whole generation of presidents. In Johnson's case the trouble followed principally from a failure to balance two requisites of world leadership: determination and judgment. Johnson had too much determination and not enough judgment. Put more precisely, he failed to make an accurate judgment of the point at which determination became counterproductive. Johnson could do nothing about the wisdom or unwisdom of the American promise to defend South Vietnam against communism. The promise was on the books when he entered the White House. It remained for Johnson to decide how far and how long the promise committed the United States.

What he discovered in doing so was what early critics of the Truman doctrine had charged about Truman's founding statement of America's Cold War policies, namely, that ideologically based commitments have almost no limits short of the exhaustion of the committer. Further, any president who suggested the existence of limits would face potentially punishing political attack. He would be pilloried and perhaps run from office, as Truman was pilloried for arguing that the United States could not save China from communism and run from office after concluding that the United States could not liberate North Korea. Johnson did not desire to risk Truman's fate—which made ironic the fact that he ultimately shared it.

III

Johnson's refusal to recognize limits in Vietnam, at least prior to March 1968, contrasts intriguingly with his acceptance of limits on American power in other areas of international affairs. His unilateral truce with de Gaulle, his compromises with Bonn and London, his low profile in Indonesia, his refusal to take sides in the India-Pakistan war, his agreement to negotiate with Panama—all reflected an understanding that the United States had to accommodate itself to changing times and changing circumstances. What was so noticeably lacking in Johnson's approach to Vietnam constituted the basis for his handling of much of the rest of the world.

A possible explanation for the discrepancy is that Vietnam preoccupied Johnson, leaving little presidential attention or presidential energy for the other areas of American policy. These other areas consequently became the province of the full-time foreign affairs advisers, who understood the transformations reshaping the world better than Johnson did. They worked out policy among themselves, then presented an agreed-upon set of recommendations to the president, who snatched time from Vietnam to grant approval.

This explanation has some merit. Vietnam definitely did preoccupy

Johnson, convincing him that energetic initiatives in most other areas of foreign affairs would be imprudent if not impossible. And certain areas of policy Johnson did indeed leave largely to his advisers. He knew next to nothing about Indonesia, and he knew how little he knew. When Dean Rusk, McGeorge Bundy, Ellsworth Bunker, and George Ball said to hunker down and wait out Sukarno, Johnson was willing to accept their counsel. Johnson knew Germans better than Indonesians, but when John McCloy asserted that Bonn had gone as far as it could go on the offset issue, the president decided to cut a deal.

Yet there are two weaknesses of this preoccupied-Johnson, prudent-advisers argument. The first is that Johnson was not uniformly restrained in non-Vietnam policy. His handling of the troubles in the Dominican Republic, just as the war in Vietnam was heating up, showed a president willing and able to take strong action where he felt it was required. Sometimes he rejected the advice of the professionals. Regarding aid to India, he ignored repeated pleas by all his top people to release grain, refusing to do so until long past the time they had predicted would produce dire results for India and thereby for America. He threw out the recommendation of Ball and Dean Acheson to let Turkey invade and partition Cyprus. Sometimes he stepped in to settle disputes among various factions of the professionals. He sided with Bohlen against Rusk and Ball and Harriman in declining to take on de Gaulle. He utilized McCloy to resolve the difference between the State Department and the Pentagon over how many troops America could afford to bring home from Europe.

The second weakness of the argument that Johnson's successes in areas besides Vietnam resulted from his leaving such areas to his foreign policy experts is that it was precisely the experts who designed America's Vietnam policy. In no case did Johnson take the initiative regarding Vietnam and change policy over the objections of a majority of his advisers. In most cases he acted as a drag on policy, requiring his advisers to reexamine their arguments and demonstrate that he had no real alternative to what they recommended.

Another, more persuasive, explanation for the discrepancy between Johnson's Vietnam disaster-making and his general success in other areas of foreign policy starts from the president's fundamental approach to foreign affairs. As noted throughout the preceding chapters, and as he repeatedly demonstrated during his five years in office, Johnson conducted American diplomacy as a branch of American politics. This is not to say he subordinated American interests abroad to his own political interests at home, at least not more than most presidents. But it *is* to say he viewed foreign policy within the context of his total program as president. Johnson believed that his greatest gifts lay in the domestic realm, and that in the domestic realm he would make his largest contribution to America's welfare. Accordingly he sought the line of least political resistance in foreign affairs, aiming to keep Congress and the American people concentrated on creating the Great Society.

The line of least resistance led to the policies Johnson pursued in several areas. In some cases diplomatic considerations pointed to these same policies—which demonstrates that there is not always a strain between what is politically convenient and what is diplomatically judicious in American foreign relations. Though a confrontation with de Gaulle might have blipped the president's approval rating momentarily upward, it would have encouraged the neoisolationists who wanted to shake the dust of Europe from the soles of American combat boots. This might have sparked a repeat of the "great debate" of 1951, in addition to damaging the Europeans' faith in the United States. A brawl over America's European commitment was the last thing Johnson wanted. Regarding the offset issue, while Johnson personally would have chosen to keep the full measure of American troops in Germany, he gave enough ground to the retrenchers in Congress and got enough money from Bonn to make everyone sufficiently satisfied to shut up.

Opinion in the United States offered relatively little guidance on the Cyprus question. Greek-Americans naturally supported Greece, without eliciting any meaningful countervailing influence from the far smaller number of Turkish-Americans. But there weren't all that many Greek-Americans, and in any event their interests lay on the side of keeping Turkish marines in their barracks, as did Johnson's.

In Latin America the line of least resistance led not to placation but to confrontation. The difference between Latin America and Europe in American policy followed from two factors: first, the greater proprietary sense most Americans felt toward events in their own hemisphere, and second, the staggering disparity between American power and that of the other countries of the hemisphere. After the Cuban revolution, American politics demanded that leftists not succeed elsewhere in Latin America. Johnson recognized the demand and intended to meet it. In Panama, negotiations sufficed. In Brazil, the Brazilian army did the job. In the Dominican Republic, the outcome appeared less certain, and Johnson, possessing the capacity to stamp out uncertainty, did just that. William Fulbright and other liberals didn't approve, but Johnson could console himself that the carpers constituted an inconsequential faction. They did at that point, at any rate.

In South Asia, the line of least resistance led initially to neutrality between India and Pakistan in the 1965 war, and subsequently to caution in dispensing American aid. Johnson cannily placed himself in the position of finally being pushed by Congress to accept what he could have squeezed out of the legislature only at substantial cost earlier.

Regarding Indonesia, he procrastinated as long as he could before choosing to cut back aid lest Congress cut it off completely. His advisers warned him of the dangers of provoking Sukarno, but he decided that what Sukarno did was basically beyond his control, while what Congress did was not.

The line of least resistance in the Middle East ran to Israel, as always since 1948. Pressure from the pro-Israel lobby encouraged Johnson to approve arms deliveries to Israel. His attempts to arrange third-party suppliers suggest that he might not have approved unpressured, despite his own personal concern for Israel's safety. Pro-Israel pressure made it impossible for him to apply strong sanctions to prevent a preemptive Israeli attack in June 1967. In this instance Johnson probably did not need the pressure to act as he did, since he sympathized with Israel's predicament. The extent of his sympathy showed in his decision to over-look the attack on the *Liberty*.

On each of these issues, American congressional and public opinion, such as it was from case to case, did not seriously conflict with what events proved to be American national interests. The line of least resistance did not lead Johnson far astray. This was partly good luck. More fundamentally it was a matter of the American political system broadly figuring out what was good for the United States.

Unfortunately for Johnson, for the United States, and for nearly all other parties involved in the war in Indochina, the least-resistance approach broke down on application to Vietnam. At every turning point in the war Johnson took pains to hew to the middle of the road. Keeping the hawks on his right and the doves on his left, he gave his military advisers less than they wanted but enough to prevent Saigon's fall. Initially the distance between hawks and doves was relatively slight, as the vote on the Gulf of Tonkin resolution suggested. At this stage the middle of the road was comfortable and safe. Escalation in 1965 and after, however, pushed the hawks and doves apart, sparking political combat in the United States. Johnson stayed in the middle of the road, but the middle of the road increasingly resembled the middle of a battlefield. As indicated by the 1968 New Hampshire Democratic primary, which Johnson came close to losing, after the launching of the Tet offensive the middle of the road was essentially untenable, with hawks and doves damning Johnson with equal energy. For a middler like Johnson, calling it quits was almost his only option.

Why did Johnson's least-political-resistance approach fail in Vietnam, even as it succeeded in several other areas? The answer to this question lies with the other theme of Johnson's foreign policy: his preference for the status quo. In every area of international affairs Johnson attempted to preserve the status quo from fundamental change. He rejected de Gaulle's plan for a remodeled Europe, in favor of the Atlantic framework that had safeguarded the continent for fifteen years. He resisted efforts to pull American troops out of Germany. He demanded that Greece and Turkey leave each other and Cyprus alone. He refused to risk letting radicals take control of the Dominican Republic. He helped maintain the balance in South Asia by halting American weapons aid before either India or Pakistan gained a decisive edge in their war. He did what little he could to encourage the Indonesian army to prevent Sukarno from taking Indonesia communist, and he restored aid after Suharto stopped the slide. He strove mightily to keep South Vietnam independent and noncommunist.

In all areas but the last Johnson largely achieved his goals. His success followed not from any personal diplomatic virtuosity but from the fact in these areas enough remained of the status quo to make a status-quo policy feasible. In nearly every area the dike was straining, but fingers here and corks there could keep the flood back for a few more years. In Vietnam, by contrast, leakage had so undermined the structure as to render it unsalvageable. That sector of the status quo simply had to be abandoned.

But doing so required altering attitudes pounded into the American people since the beginning of the Cold War. Endlessly Americans had been told that yielding ground to communism would lead to enslavement or world war or both. The United States, assisted by the other democracies if possible, alone if necessary, must stand firm. Because communists figured in nearly every movement for fundamental change around the globe, and because communists were thought to be incomparably clever in perverting even moderate reform movements to their godless and tyrannical purposes, containing communism often came to mean opposing basic change.

The message of containment, so simple and morally reassuring, was remarkably effective in mobilizing American public and congressional opinion for the protracted effort waging the Cold War required. And as long as American interests lay in the direction of defending the status quo, the opinion thus mobilized afforded a reasonable guide to American policy. A president like Johnson could follow the path of least resistance and succeed reasonably well in securing American interests.

Trouble set in, however, when the status quo in a particular region—such as Indochina—required reinforcing that cost more than the region was worth to the United States. At this point securing American interests necessitated changing course. But changing course demanded a long, hard pull on the rudder of public opinion, to overcome the momentum that had developed during the two decades of the Cold War. A president committed to the line of least resistance would avoid such pulling. Johnson avoided the pulling, and produced the colossal wreck of Vietnam.

Johnson's several successes and his signal failure reflect on more than the performance of one president. They reflect also on the capacity of a democracy to devise effective foreign policies. For better, in most cases, and for worse, in the case of Vietnam, the policies Johnson pursued were very much America's policies. To have asked Johnson to do better would have been to ask the American political system to do better.

It might have, but only by producing a president with the insight to see where America was going wrong and the courage to say so. While democratic politics does not necessarily select against insight, it offers few prizes for courage. On rare occasions voters insist on being told the truth about their situation and prospects, but usually they prefer self-congratulation. No one likes prophets of doom. People listen to doom-sayers only when there is no other option. By then too often, as with Vietnam, the doom is nearly unavoidable.

The difficulty is probably intractable, given the human inclination to prefer good news to bad. No one ever said democracy was efficient, and it approaches its least efficient in effecting basic changes in foreign policy. Yet an awareness of the problem may help in minimizing evil consequences. If Lyndon Johnson's experience yields some lesson of lasting value, this might be it.

Notes

1. Great Expectations

1. Dean Acheson, *Present at the Creation* (New York, 1969), 255.

2. Sam Houston Johnson, *My Brother Lyndon* (New York, 1969), 117.

3. Rusk interview, 7/28/69, Johnson Library (Austin, Texas) oral history collection. Unless otherwise stated, all interviews cited are from this collection.

4. Bundy to Johnson, 6/24/64, box 2, Memos to the president file, Johnson Library. Unless otherwise stated, all files and papers cited are from the Johnson Library.

5. George W. Ball, *The Past Has Another Pattern* (New York, 1982), 377–78.

6. Bruce interview, 12/9/71; Gordon interview, 7/10/69.

7. Clark Clifford with Richard Holbrooke, *Counsel to the President* (New York, 1991), 459; Thomas J. Schoenbaum, *Waging Peace and War: Dean Rusk in the Truman, Kennedy, and Johnson Years* (New York, 1988), 285.

8. Rusk interview, 7/28/69; Ball interview, 7/9/71.

9. Thomson interview, 7/22/71.

10. Bundy to Johnson, 10/13/65, box 5, Memos to the president file.

11. Bundy to Rusk, 11/30/65, ibid.

12. Valenti interview, 3/3/71; Rusk interview, 7/28/69.

13. Memo of Johnson meeting with Jack Leacacos, 10/14/67, box 2, Meeting notes file.

14. McCone interview, 6/19/70; Helms interview, 4/4/69.

15. Clifford interview, 7/2/69.

16. Ball interview, 7/9/71.

17. Smith interview, 4/49/69.

18. Bundy to Goodpaster, 8/19/65, box 3, Name file.

19. Memo of meeting, 2/17/65, box 9, Augusta series, Eisenhower post-presidential pa-

pers, Eisenhower Library, Abilene, Kansas; notes of telephone conversation, 7/2/65, box 10, ibid.; H. W. Brands, "Johnson and Eisenhower: The President, the Former President and the War in Vietnam," *Presidential Studies Quarterly* (Summer 1985).

20. Ball interview, 7/9/71; notes of meeting with Bob Thompson, 8/21/67, box 2, Meeting notes file.

21. Notes of meeting with congressional leaders, 7/25/67, box 1, Tom Johnson file.

22. Johnson interview, 8/12/69.

23. Kathleen J. Turner, *Lyndon Johnson's Dual War: Vietnam and the Press* (Chicago, 1985), 160.

24. Clifford interview, 12/15/69.

25. William Bundy interview, 6/2/69; Eugene Rostow interview, 12/2/68.

26. David C. Humphrey, "Tuesday Lunch at the Johnson White House," *Diplomatic History* (Winter 1984); Rusk interviews, 7/28/69 and 9/26/69.

27. William Bundy interview, 6/2/69.

28. Read interview, 6/13/69.

29. Rostow in Humphrey, "Tuesday Lunch."

30. Anderson interview, 7/28/69.

31. Robert Kennedy in Richard N. Goodwin, "The War Within," *New York Times Magazine,* 8/21/88; Eric F. Goldman, *The Tragedy of Lyndon Johnson* (New York, 1969), 525; Helms interview, 4/4/69; McCloy interview, 7/8/69; Rusk interview, 7/28/69.

32. Meeting with Max Frankel, 9/15/67, box 2, Meeting notes file; Bruce interview, 12/9/71; McCloy interview, 7/8/69.

33. Eugene Rostow interview, 12/2/68; Bruce interview, 12/9/71; Ball interview, 7/9/71; Read interview, 6/13/69.

34. Rusk interview, 7/28/69; Helms interview, 4/4/69.

35. William Bundy interview, 5/29/69.

2. Who Lost Cuba?

1. Stuart to Department of State (DOS), 1/10/64, Mann to Rusk, 1/11/64, USCINCSO to JCS, 1/16/64, box 1, NSC history file.

2. FBIS report, 1/10/64; Solis to Rusk, 1/10/64, ibid.

3. Mann to Rusk, 1/12/64, ibid.

4. Meeting notes, 1/10/64, box 18, NSF McGeorge Bundy file; Lyndon Baines Johnson, *The Vantage Point* (New York, 1971), 182–83.

5. Mann to Rusk, 1/11/64; Johnson to Mann, 1/11/64, box 1, NSC history file.

6. Mann to Martin, 1/14/64, ibid.

7. Stuart to DOS, 1/17/64; White House to CINCSO, 1/14/64, ibid.

8. Mann to Martin, 2/8/64, ibid.

9. Mann to Rusk, 2/1/64, ibid.

10. William J. Jorden, *Panama Odyssey* (Austin, 1984), 73–74.

11. *Congressional Record,* 1/14/64, 379–80; 1/31/64, 1472, 1476–77.

12. Mansfield to Johnson, box 1, Memos to the president file.

13. Johnson statement, 1/23/64, Department of State *Bulletin,* 2/3/64.

14. Rusk to Panama City, 1/22/64; Mann to Martin, 1/24/64 and 1/25/64, ibid.

15. *Washington Post* clipping, 2/12/64, ibid.

16. Jorden, *Panama Odyssey,* 80–82.

17. Johnson statement at news conference, 3/21/64, *Public Papers of the Presidents: Lyndon B. Johnson,* 1964 (Washington, 1965).

18. Jorden, *Panama Odyssey,* 87.

19. Johnson, *Vantage Point,* 184.

20. NSC meeting notes, 4/3/64, box 1, NSC meetings file.

21. Roa to UN Security Council, 2/3/64; FBIS report, 2/5/64; AP wire report, 2/6/64; memo of conversation, 2/4/64, box 3, NSC history file.

22. Johnson, *Vantage Point*, 187.

23. Phyllis R. Parker, *Brazil and the Quiet Intervention* (Austin, 1979), 10–11.

24. Johnson to Goulart, 12/19/63, box 9, Country file.

25. Ball to Gordon, 12/19/63, ibid.

26. Gordon to Mann, 3/4/64, ibid.

27. Parker, *Brazil*, 61–63.

28. Gordon to DOS, 3/17/64, box 9, Country file.

29. *New York Times*, 3/19/64. (This meeting's transcript eluded the present author).

30. CIA cable, 3/30/64, box 9, Country file.

31. Gordon to Rusk, Bundy, et al., 3/27/64; Gordon to DOS, 3/31/64, ibid.

32. Rusk to Brasilia, 3/30/64, ibid.

33. Ball telephone conversation with Johnson, 3/31/64, box 1, George Ball papers.

34. Ball et al. to Gordon, 3/31/64, box 9, Country file; Johnson telephone conversation with Ball, 3/31/64, box 1, Ball papers.

35. Parker, *Brazil*, 68–70.

36. Ibid., 75–76; JCS to CINCLANT, 3/31/64, box 10, Country file.

37. NSC meeting notes, 4/2/64, box 1, NSC meetings file.

38. Johnson to Mazzilli, 4/2/64, box 9, Country file.

39. NSC meeting notes, 4/3/64, box 1, NSC meetings file.

40. CIA memo, 4/3/64; CIA intelligence bulletin, 4/6/64, box 10, Country file.

41. Chase to Bundy, 4/2/64; Hughes to Ball, 4/10/64, ibid.

42. John Bartlow Martin, *Overtaken by Events* (New York, 1966), 347; Ball, *Past Has Another Pattern*, 327.

43. Rowland Evans and Robert Novak, *Lyndon B. Johnson: The Exercise of Power* (New York, 1966), 513.

44. Connett to DOS, 4/24/65, 5:04 P.M., and 4/25/65, 4:20 A.M., box 4, NSC history file.

45. Connett to DOS, 4/25/65, 5:00 P.M. and 11:23 P.M., ibid.

46. Connett to DOS, 4/26/65, 10:08 A.M., ibid.

47. CIA cables, 4/24/65 and 4/25/65, ibid.

48. CIA cables, 4/25/65 and 4/26/65 (2), ibid.

49. Bennett to DOS, 4/27/65, 3:58 P.M., ibid.

50. Bennett to DOS, 4/28/65, 12:28 A.M. and 12:59 A.M., ibid.

51. Abraham F. Lowenthal, *The Dominican Intervention* (Cambridge, Mass., 1972), 76–81; Piero Gleijeses, *The Dominican Crisis* (Baltimore, 1978), 195ff.

52. Bennett to DOS, 4/48/65, 1:43 P.M., box 4, NSC history file.

53. Bennett to DOS, 4/28/65, 5:16 P.M., ibid.

54. Ball, *Past Has Another Pattern*, 328.

55. Rusk to Santo Domingo 4/27/65; Bennett to DOS 4/28/65, box 4, NSC history file.

56. Minutes of meeting with congressional leadership 4/28/65, box 1, Meeting notes file.

57. Johnson address, 4/28/65, *Public Papers*.

58. Johnson addresses, 5/2/65 and 5/3/65; Johnson comments to congressional committee members, 5/4/65; Johnson news conference, 6/17/65, *Public Papers*.

59. Rusk to Bennett, 4/29/65, 11:11 P.M., box 5, NSC history file.

60. Johnson address, 4/30/65, *Public Papers*.

61. Notes of meeting, 5/2/65, box 13, Office files of the president.

62. Notes of meeting, 4/30/65, ibid.

63. Johnson address, 5/2/65, *Bulletin*, 5/17/65.

64. CIA memo, 5/7/65, box 49, Country file.

65. Dan Kurzman, *Santo Domingo: Revolt of the Damned* (New York, 1965), 193–97.

66. Jerome Slater, *Intervention and Negotiation: The United States and the Dominican Revolution* (New York, 1970), 71ff.

67. *Congressional Record*, 9/15/65, 23855–61.

3. The Trojan War (cont.)

1. Memo of telephone call, 8/22/61; Komer to Wilkins, 6/7/63, box 16, NSC history file.
2. Johnson to Makarios and Kutchuk in DOS to Nicosia, 12/25/63, ibid.
3. Memo of conversation, 1/25/64, ibid.
4. Mansfield to Johnson, 1/31/64, box 1, Memos to the president file.
5. DOS to Nicosia, 12/28/63, box 16, NSC history file.
6. Memo of conversation, 1/25/64, ibid.
7. Memo of conversation, 1/25/64, ibid.
8. Ball telephone conversation with Johnson, 1/28/64, box 2, Ball papers; Ball to Athens and Ankara, 1/26/64, box 16, NSC history file.
9. Johnson memo, 1/31/64, ibid.
10. Johnson to Makarios in DOS to Nicosia, 2/2/64, ibid.
11. Wilkins to DOS, 2/3/64, ibid.
12. Ball to Johnson in Athens to DOS, 2/9/64, CIA intelligence memo in Nicosia to DOS, 2/11/64, ibid.
13. Ball to Johnson in Athens to DOS, 2/9/64, ibid.
14. Ball, *Past Has Another Pattern*, 340–46.
15. Laurence Stern, *The Wrong Horse* (New York, 1977), 84.
16. Ball to DOS, 2/12/64, box 16, NSC history file; Ball, 345.
17. Ball to Johnson in Nicosia to DOS, 2/13/64 and 2/16/64, box 16, NSC history file.
18. Memo for record, 2/17/64, ibid.
19. Johnson to Fulbright, 5/21/64, ibid.
20. Rusk to Ankara, 6/4/64, ibid.
21. Ball, *Past Has Another Pattern*, 350.
22. Johnson to Inonu, 6/5/64, released by the White House 1/15/66, published in *Middle East Journal*, Summer 1966.
23. Hare to DOS, 6/6/64, box 16, NSC history file.
24. Hare to DOS, 6/6/64, ibid.
25. Memo of conversation, 6/11/64, ibid.
26. "History of the Johnson Administration: Cyprus Problem," ibid.
27. Andreas Papandreou, *Democracy at Gunpoint* (Garden City, N.Y., 1970), 134.
28. NSC meeting notes, 7/28/64, box 1, NSC meetings file; "Summary of Acheson Plans I and II," box 16, NSC history file.
29. Bundy to Johnson, 7/7/64, box 2, Memos to the president file.
30. Durbrow to DOS, 8/8/64, box 16, NSC history file.
31. Belcher to DOS, 8/9/64, ibid.
32. Hare to DOS, 8/9/64, ibid.
33. Johnson to Inonu, Papandreou, and Makarios, 8/9/64, ibid.
34. Ball, *Past Has Another Pattern*, 357.
35. "Summary of Acheson Plans I and II," box 16, NSC history file; notes of meeting, 8/13/64, NSF Bundy files; H. W. Brands, "America Enters the Cyprus Tangle," *Middle Eastern Studies* (July 1987).
36. Ball, *Past Has Another Pattern*, 349; NSC meeting notes, 8/25/64, box 1, NSC meetings file.
37. Acheson to Battle, 12/7/64 in David S. McLellan and David C. Acheson, eds., *Among Friends: Personal Letters of Dean Acheson* (New York, 1980), 284.
38. Lawrence S. Wittner, *American Intervention in Greece, 1943–1949* (New York, 1982), 303.
39. Bundy to Johnson, 8/25/64, box 2, Memos to the president file; Tuesday lunch notes, 9/8/64, box 18, NSF Bundy files; memo for record, 9/8/64, ibid.; Komer and Bundy to Johnson, 8/18/64, box 1, NSC meetings file; Brands, "Cyprus Tangle."
40. Memo for record, 9/8/64, box 18, NSF Bundy files.
41. Ball telephone conversation with Dirksen, 3/17/66, box 3, Ball papers; Komer to Johnson, 3/19/66, box 6, Memos to the president file.

42. Vance interview, 11/3/69.
43. NSC meeting notes, 11/29/67, box 1, NSC meetings file.
44. NSC meeting notes, 12/1/67, box 1, Tom Johnson file.
45. NSC meeting notes, 11/29/67, box 1, NSC meetings file.

4. Suffer the General

1. Howard Jones, *The Course of American Diplomacy* (New York, 1985), 387.
2. Ball interview, 7/9/71.
3. Bruce interview, 12/9/71.
4. Stephen E. Ambrose, *Eisenhower* (New York, 1984), 2:539; Brands, *Cold Warriors: Eisenhower's Generation and American Foreign Policy* (New York, 1988), 89.
5. Ball, *Past Has Another Pattern*, 271.
6. Bohlen memo, 12/13/63, box 169, Country file.
7. Bohlen memo, undated (attached to Bundy to Johnson, 3/11/64), ibid.
8. Dillon memo, undated [December 1965–January 1966], box 6, Memos to the president file.
9. Tyler to Bundy, 3/12/64, ibid.
10. Harriman to Rusk, 3/18/64, ibid.
11. De Segonzac memo, undated, box 1, Memos to the president file.
12. Ball interview, 7/9/71; Edward Weintal and Charles Bartlett, *Facing the Brink* (New York, 1967), 105.
13. Memo of conversation, 5/25/64, box 171, Country file.
14. Johnson to Eisenhower, 4/29/67, ibid.
15. Johnson to Bohlen in Rusk to Bohlen, 3/24/64, box 1, Memos to the president file.
16. CIA special memo, 2/5/64, box 169, Country file.
17. Johnson to Bohlen in Rusk to Bohlen, 2/25/64, ibid.
18. Johnson to Ball, 6/4/64, box 170, Country file.
19. Ball to Johnson, 6/5/64, ibid.
20. Bohlen to DOS, 3/12/65, box 171, Country file.
21. "De Gaulle's Foreign Policy: 1964," 4/20/64, box 169, Country file.
22. Bundy to Johnson, 12/2/64, box 2, Memos to the president file.
23. Ball memo, 12/5/64, ibid.
24. Bundy to Rusk, McNamara, and Ball, 11/25/64, ibid.
25. CIA special report, 11/27/64, box 170, Country file.
26. Ibid.
27. Strategy paper for NATO ministerial meeting, 5/4/64, box 33, International meetings file.
28. Voice of America transcript of Fulbright speech, 11/30/60, box 1, Memos to the president file.
29. Bohlen to DOS, 3/31/65, box 171, Country file.
30. Bohlen to DOS, 6/3/65, ibid.
31. Ball, *Past Has Another Pattern*, 331–33.
32. Bundy memo, 10/18/65, box 5, Memos to the president file; Bundy to Johnson, 1/28/66, box 6, ibid.
33. Smith interview, 4/29/69.
34. Bohlen to DOS, 12/7/65, box 172, Country file.
35. De Gaulle to Johnson, 3/7/66, in Ambassade de France, *French Foreign Policy*, 1966 (New York, [1967]), 24–25.
36. Johnson to de Gaulle, 3/7/66, box 3, Heads of state file.
37. Johnson, *Vantage Point*, 305; notes of meeting with Johnson, 4/5/66, box 1, Meeting notes file.
38. Rusk draft speech, 4/2/66, box 7, Memos to the president file.
39. Ball interview in *Bulletin*, 4/11/66; Johnson speech, 4/4/66, *Public Papers*, Johnson

speech transcript, 6/16/66, box 7, Memos to the president file.

40. Owen memo, 4/30/66, box 7, Memos to the president file.

41. Johnson to Rusk and McNamara, 5/4/66, ibid.

42. Rusk to Johnson, 4/27/66, ibid.

43. Rostow and Bator to Johnson, 5/18/66, ibid.

44. Bohlen to DOS, 6/13/66, box 8, Memos to the president file.

45. Gregory F. Treverton, *The Dollar Drain and American Forces in Europe* (Athens, Ohio, 1978), 32–34; Ball interview, 7/9/71.

46. Johnson to Wilson, 8/26/66, box 50, NSC history file; Johnson to Erhard, 8/25/66, ibid.

47. *Congressional Record,* 8/31/66, 21442–43.

48. Ball to Bonn, 9/3/66, box 50, NSC history file; McNamara to Johnson, 9/19/66, ibid.

49. Bator to Johnson, 9/25/66, ibid.

50. Memo of conversation, 9/26/66, ibid.; Johnson-Erhard communiqué, 9/27/66, ibid.

51. Johnson to Wilson, 10/6/66, ibid.

52. Eugene Rostow interview, 12/2/68.

53. Hillenbrand to DOS, 4/16/66, box 20, Subject file.

54. McCloy to Johnson, 10/21/66, box 50, NSC history file.

55. Johnson to Wilson, 11/15/66, ibid.

56. Notes of meeting, 11/14/67, box 3, Meeting notes file.

57. Bator to Johnson, 2/23/67, box 50, NSC history file; Eugene Rostow to Rusk, undated (February 1967), box 17, Francis Bator papers.

58. McCloy to Johnson, 2/23/67; notes of meeting, 3/9/67; Bator to Johnson, 2/23/67, box 50, NSC history file.

59. Memo for record, 3/2/67, box 18, Bator papers.

60. Johnson to McCloy, 3/1/67, box 50, NSC history file.

61. McCloy to Johnson, 3/22/67, ibid.

62. DOS release, 5/2/67, box 51, NSC history file.

63. Bator summary of congressional positions on Kennedy round talks, undated (June 1967), box 13, Bator papers.

64. Notes of meetings with legislative leaders, 11/19/67 and 11/20/67, box 3, Walt Rostow files.

65. Johnson to Kiesinger and Moro, 3/15/68, box 53, NSC history file.

66. Notes of meeting, 9/9/68, box 2, Tom Johnson file.

67. Notes of NSC meeting, 8/20/68, box 2, NSC meetings file.

68. Johnson, *Vantage Point,* 462.

69. Clifford to Johnson, 8/24/68, box 39, Memos to the president file; notes of NSC meeting, 9/4/68, box 2, NSC meetings file.

70. Johnson, *Vantage Point,* 489–90.

5. When the Twain Met—Head-on

1. Zhou in S. M. Burke, *Mainsprings of Indian and Pakistani Foreign Policies* (Minneapolis, 1974), 91.

2. Kennedy to Nehru, 10/28/62, box 24, NSC history file.

3. Komer to Talbot, 10/24/62, ibid.

4. Komer to Bundy, 11/6/62, ibid.

5. Joint Chiefs to McNamara, 12/23/63, ibid.

6. Bowles to Rusk, 2/20/64, box 1, Memos to the president file.

7. Komer and Bundy to Johnson, 6/12/65, box 3, Memos to the president file.

8. Summary of conversation in DOS to Karachi, 12/2/63, box 24, NSC history file.

9. Johnson to Ayub, 12/9/63, excerpted in Komer to Johnson, 9/9/65, ibid.

10. Komer to Johnson, 2/24/64, ibid.

11. Bundy and Komer to Johnson, 4/24/64, box 1, Memos to the president file.

12. Philip Geyelin, *Lyndon B. Johnson and the World* (New York, 1966), 90.
13. *New York Times*, 4/22/64; memo of conversation, 4/27/64, box 24, NSC history file.
14. Saunders to Komer, 3/31/65, box 24, NSC history file.
15. Rusk to New Delhi and Karachi, 4/14/65, box 3, Memos to the president file.
16. Bowles to Rusk, 4/15/65, ibid.
17. Komer to Johnson, 4/16/65, box 24, NSC history file.
18. Johnson to Ayub, 5/15/65, ibid.
19. Johnson note on Bundy to Johnson, 6/8/65, box 3, Memos to the president file.
20. McNamara to Johnson, 7/6/65, box 3, Memos to the president file.
21. Komer to Johnson, 8/22/65, box 24, NSC history file.
22. Ibid.
23. CIA cable, 8/31/65, box 150, Country file.
24. Komer to Johnson, box 24, NSC history file; Komer to Bundy, 8/31/65, box 4, Memos to the president file.
25. Bowles to DOS, 9/2/65, box 129, Country file.
26. Rusk to New Delhi, 9/2/65, ibid.
27. CIA cable, 9/2/65, box 150, ibid.
28. CIA cable, 9/7/65, box 129, ibid.
29. CIA to White House situation room, 9/10/65 and 9/15/65, ibid.; Komer memo, undated, box 150, ibid.
30. Johnson to Ayub and Johnson to Shastri, 9/4/65, box 24, NSC history file.
31. CIA to White House, 9/15/65, box 129, Country file; CIA cable, 9/16/65, box 130, ibid.
32. CIA cable, 9/13/65, box 150, ibid.
33. DOS to New Delhi, 9/11/65, box 129, ibid.
34. CIA memo, 9/15/65, box 150, ibid.
35. DOS to New Delhi, box 129, ibid.
36. Rusk to New Delhi, ibid.
37. CIA cable, 9/13/65, box 150, ibid.; CIA to White House, 9/17/65, box 150, ibid.
38. Bowles to Bundy, 11/25/65, box 15, Bundy file.
39. Karachi to DOS, 9/12/65, box 150, Country file.
40. Stephen P. Cohen in Lloyd I. Rudolph, ed., *The Regional Imperative* (New Delhi, 1980), 103; Bowles to Bundy, 10/2/65, box 15, Bundy file; William Bundy telephone conversation with Ball, 9/27/65, box 5, Ball papers.
41. Department of State administrative history, vol. I, chap. 4.
42. Bundy to Johnson, 9/12/65, box 15, Bundy file.
43. Komer to Johnson, 9/22/65, box 24, NSC history file.
44. Ibid.
45. Bundy and Komer to Johnson, 10/5/65, box 24, ibid.
46. Memo of conversation, 12/20/65, box 25, ibid.
47. Komer to Bundy, 10/21/65, ibid.
48. Johnson to Kennedy, 5/23/61, box 18, Bundy file.
49. Komer memo, 9/27/65, box 24, NSC history file.
50. Bundy and Komer to Johnson, 10/1/65; Komer to Bundy, 10/10/65; Komer to Johnson, 10/28/65, ibid.
51. Johnson greeting speech, 12/14/65; Johnson toast, 12/14/65, *Public Papers.*
52. Johnson-Ayub communiqué, 12/15/65, ibid.; Johnson telephone conversation with Ball, 12/14/65, box 5, Ball papers; Komer to Johnson, 12/14/65, box 24, NSC history file.
53. Komer to Johnson, 11/15/65; Bowles to DOS, 11/11/65, box 5, Memos to the president file.
54. Notes of meeting, 12/14/65, box 13, President's office file.
55. Johnson to Ayub, 2/11/66, box 24, NSC history file.
56. H. W. Brands, *India and the United States* (Boston, 1990), 120.
57. Saunders to Bundy, 12/2/65, box 25, NSC history file.
58. Freeman to Johnson, 1/4/66, ibid.

59. Memo for record, 2/3/66, ibid.
60. Komer to Johnson, 2/4/66, ibid.
61. Komer to Johnson, 3/21/66, ibid.
62. Komer to Freeman, 3/24/66, ibid.
63. Memo of conversation, 3/28/66, ibid.
64. Johnson message to Congress, 3/30/66, *Public Papers*.
65. Johnson to Wilson, 3/31/66, box 25, NSC history file.
66. Rusk to New Delhi, 4/6/66, ibid.
67. Memo of conversation, 8/12/66, ibid.
68. Rostow to Johnson, 7/19/66, box 9, Memos to the president file; Johnson to Gandhi, 8/31/66, box 10, ibid.
69. Rusk, Freeman, and Gaud to Johnson, 8/22/66; Rostow to Johnson, 9/15/66, box 1, Rostow file.
70. Johnson note on Smith to Johnson, 8/24/66, box 26, NSC history file.
71. Saunders to Rostow, 8/31/66, ibid.
72. Rostow to Johnson, 8/31/66, ibid.; Rostow to Johnson, 10/15/66, box 11, Memos to the president file; Rostow to Johnson, 9/15/66, box 1, Rostow file; Rostow to Johnson, 10/13/66, box 26, NSC history file; Saunders to Rostow, 10/14/66, ibid.; Rostow to Johnson, 11/28/66, box 11, Memos to the president file; Rostow to Johnson, 11/9/66, ibid.; Bowles to Johnson, 11/4/66, box 26, NSC history file.
73. Saunders to Rostow, 9/15/66; Wriggins to Rostow, 9/22/66, box 26, NSC history file.
74. Johnson notes on Rostow to Johnson, 9/26/66, 10/13/66, and 10/15/66, ibid.
75. Saunders and Wriggins to Rostow, 9/26/66, ibid.
76. Rostow to Johnson, 12/16/66, box 12, Memos to the president file.
77. Poage, Miller, and Dole to Freeman, 12/20/66, box 26, NSC history file.
78. Rostow to Johnson, 12/27/66, box 12, Memos to the president file.

6. Bloody Good Luck

1. Ball, *Past Has Another Pattern*, 285–86.
2. Rusk to Johnson, 1/6/64, box 1, NSC meeting file.
3. Forrestal to Bundy, 1/6/64; Komer to Bundy, 2/25/64, box 246, Country file.
4. CIA report, 2/7/64, ibid.
5. NSC meeting notes and record of action, 1/7/64, box 1, NSC meeting file.
6. Howard Palfrey Jones, *Indonesia* (New York, 1971), 205–6.
7. Ibid., 300–303.
8. NSAM 278, 2/3/64, NSAM file.
9. NSC meeting, 4/3/64, box 1, NSC meeting file.
10. Robert J. McMahon, *Colonialism and the Cold War* (Ithaca, N.Y., 1981), 10–11.
11. Rusk to Jakarta, 12/20/63, box 246, Country file.
12. Rusk to Jones, 3/3/64, ibid.
13. Jones to DOS, 3/3/64, ibid.
14. Jones to DOS, 3/6/64, ibid.
15. Jones to DOS, 4/3/64, ibid.
16. Jones to DOS, 3/19/64 and 3/26/64, ibid.
17. Jones to DOS, 2/17/64, ibid.
18. Jones, *Indonesia*, 321.
19. Ibid.; Rusk press conference, 4/3/64, *Bulletin*, 4/20/64; *Congressional Record*, 1964, 6361, 6679–80; Ryan to Johnson, 3/28/64, box 246, Country file.
20. CIA cable, 4/15/64, box 246, Country file.
21. Rusk to Jakarta, 4/21/64; Jones to DOS, 5/19/64; Ball to Jakarta, 7/6/64; CIA cable, 7/25/64, ibid.; Jones, *Indonesia*, 326.
22. Johnson-Rahman communiqué, 7/23/64, *Bulletin*, 8/10/64.
23. Galbraith to DOS, 8/18/64, box 246, Country file.

24. CIA memo, 8/20/64, ibid.

25. Komer to Johnson, 8/19/64, ibid.

26. Thomson to Bundy, 8/25/64, ibid.

27. Rusk to Johnson, 8/30/64, ibid.

28. Bundy to Johnson, 9/2/64, box 2, Memos to the president file; Bundy to Johnson, 8/31/64, box 246, Country file.

29. Bundy to Rusk, 9/3/64, box 246, Country file.

30. CIA cable, 1/13/65, ibid.

31. Rusk to Jakarta, 1/14/65, ibid.

32. CIA memo, 1/26/65, ibid.

33. Johnson to Wilson, 1/25/65, ibid.

34. Komer to Bundy, 2/8/65, ibid.

35. Bunker in Jakarta to DOS, 4/6/65 and 4/14/65, box 247, Country file.

36. Bunker to Johnson, 4/23/65, ibid.

37. Ball to Johnson, undated, ibid.

38. Brands, "The Limits of Manipulation: How the United States Didn't Topple Sukarno," *Journal of American History* (December 1989), 800.

39. Bundy to Johnson, 6/30/65, box 3, Memos to the president file.

40. Jones to DOS, 4/20/65 and 5/1/65, box 247, Country file.

41. Green to DOS, 8/17/65, ibid.

42. Rusk to DOS, 8/9/65, 9/10/65, and 9/13/65, ibid.

43. Jones, *Indonesia*, 373.

44. Green to DOS, 10/1/65, box 247, Country file.

45. Murray to Johnson, 9/30/65; CIA memo, 10/2/65, ibid.; Ball telephone conversation with Rusk, 10/1/65, box 4, Ball papers.

46. Green to DOS, 10/3/65; Indonesia working group reports, 10/4/65 and 10/6/65, box 247, Country file.

47. Green to DOS, 10/5/65, ibid.

48. Ball to Jakarta, 10/6/65, ibid.

49. CIA memo, 10/8/65, ibid.

50. Rusk to Jakarta, 10/13/65, ibid.

51. CIA memo, 10/28/65; Green to DOS, 11/4/65, ibid.

52. CIA memo, 11/12/65, ibid.

53. Green to DOS, 12/1/65, ibid.

54. Brands, "Limits of Manipulation," 803.

55. Rusk to Jakarta, 12/8/65, box 248, Country file.

56. Green to DOS, 12/21/65; Ball to Jakarta, 12/16/65, ibid.; Bundy to Johnson, 12/30/65, box 5, Memos to the president file; CIA memo, 1/3/66, box 248, Country file.

57. Komer to Johnson, 2/10/66, box 6, Memos to the president file; memo of conversation, 2/15/66; CIA memo, 2/5/66, box 248, Country file.

58. Rusk to Jakarta, 2/15/66, box 248, Country file.

59. Komer to Cooper and Thomson, 1/14/66, ibid.

60. Komer to Johnson, 3/12/66, box 6, Memos to the president file; CIA memo, 4/1/66, box 248, Country file.

61. Thomson to Rostow, 4/2/66; Thomson to Moyers, 3/31/66; Ropa to Rostow, 4/18/66, box 248, Country file.

62. Paper by Policy Planning Council, 3/25/66, ibid.

63. DOS circular telegram, 5/24/66, ibid.; DOS summary attached to Rostow to Johnson, 6/8/66, box 8, Memos to the president file.

· 64. NSC meeting, 8/4/66, box 1, NSC meeting file.

65. Rostow to Johnson, 8/31/66; Rusk to Johnson, 9/1/66; Gaud and Freeman to Johnson, 2/11/67; box 248, Country file.

66. CIA memo, October 1966, ibid.

67. Rostow to Johnson, 5/1/67; Green to DOS, 12/21/66, ibid.

68. Rostow to Johnson, with Johnson note, 2/20/67; Johnson to Rostow, 2/22/67, ibid.
69. Katzenbach to Johnson, 2/23/67; Jorden to Johnson, 2/23/67, ibid.
70. Johnson note on Jorden to Johnson, 2/23/67, ibid.
71. McNamara to Johnson, 3/1/67, ibid.
72. Rostow to Johnson, 6/27/67, ibid.
73. Meeting notes, 7/25/67, box 1, Tom Johnson file; NSC meeting notes, 8/9/67, box 73, President's appointment file; NSC meeting notes, 8/9/67, box 2, NSC meetings file.
74. Johnson to Rostow, 11/21/67, box 248, Country file.
75. McNamara to Johnson, 3/1/67, ibid.
76. Helms to Rostow, 5/13/66, ibid.
77. NSC meeting notes, 11/8/67, box 2, NSC meetings file.

7. Six Days in June

1. H. W. Brands, *The Specter of Neutralism: The United States and the Emergence of the Third World, 1947–1960* (New York, 1989), 284.
2. Steven L. Spiegel, *The Other Arab-Israeli Conflict* (Chicago, 1985), 102.
3. Douglas Little, "From Even-Handed to Empty-Handed," in Thomas G. Paterson, ed., *Kennedy's Quest for Victory* (New York, 1989), 170–171.
4. CIA memo, 2/25/64, box 1, Memos to the president file.
5. Feldman to Johnson, 3/14/64, ibid.; Komer to Johnson, 3/4/64, ibid.
6. McPherson interview, 1/16/69.
7. Bundy to Johnson, 6/12/64, box 2, Memos to the president file.
8. Spiegel, *Other Arab-Israeli Conflict*, 131–32.
9. Bundy to Johnson, 5/12/64, box 1, Memos to the president file; draft national security action memorandum (Johnson to Rusk and McNamara), undated, ibid.
10. Spiegel, *Other Arab-Israeli Conflict*, 132; Rostow memo, "How We Have Helped Israel," 5/19/66, box 7, Memos to the president file.
11. Bundy to Johnson, 5/6/64, box 1, ibid.
12. Nasser in Mohamed Hassanein Heikal, *The Cairo Documents* (Garden City, N.Y., 1973), 238–39.
13. Bundy to Johnson, 4/21/65, box 3, Memos to the president file.
14. Ball telephone conversation with Johnson, 2/7/65, box 6, Ball papers.
15. Spiegel, *Other Arab-Israeli Conflict*, 132–34.
16. Rostow to Johnson, 4/29/66, box 7, Memos to the president file.
17. Rostow memo, 5/19/66, ibid.
18. Memo of conversation, 8/2/66, box 20, NSC history file.
19. Rostow to Johnson, 6/10/66, box 8, ibid.
20. Johnson to Eshkol, 5/17/67 and 5/21/67, box 17, NSC history file.
21. Johnson to Nasser, 5/22/67, ibid.
22. Johnson to Kosygin, 5/19/67, ibid.
23. Johnson speech, 5/23/67, *Weekly Compilation of Presidential Documents*, 5/29/67.
24. Rusk to Johnson, 5/26/67, box 16, Memos to the president file; Rostow to Johnson, 5/25/67, box 17, NSC history file.
25. Rostow to Johnson, 5/25/67, ibid.
26. Rostow to Johnson, 5/17/67, ibid.
27. Rostow to Johnson, 5/23/67, ibid.
28. Rostow to Johnson, 5/26/67, ibid.
29. Rusk to Johnson, 5/26/67, ibid.
30. Minutes of NSC meeting, 5/24/67, box 1, NSC meetings file.
31. Rostow agenda notes for 5/26/67 meeting, box 1, Rostow file.
32. Memo of conversation, 5/26/67, box 17, NSC history file; DOS administrative history, vol. 1, pt. 4, sec. H, p. 57; Abba Eban, *An Autobiography* (New York, 1977), 354–59.
33. Notes from dinner, 5/26/67, President's daily diary file.

34. Eugene Rostow interview, 12/2/68.

35. Eban, *Autobiography*, 361; Rostow to Johnson, 5/27/67, box 17, NSC history file.

36. Memo of conversation, 5/57/67, ibid.

37. Rostow to Johnson, 5/28/67, ibid.

38. Nasser press conference transcript, 5/28/67, ibid.

39. Yost to DOS, 5/30/67; Rostow to Johnson, 5/29/67, ibid.

40. Rusk and McNamara to Johnson, 5/30/67, box 18, NSC history file.

41. Rostow to Johnson, 6/1/67, with Johnson note, ibid.

42. Wheeler to McNamara, 6/2/67, ibid.

43. Yost to DOS, 6/2/67, ibid.

44. Anderson to Johnson, 6/2/67, ibid.

45. Wheeler interview, 8/21/69.

46. Johnson to Eshkol, 6/3/67, box 18, NSC history file.

47. Memo of telephone conversation, 6/5/67; Saunders memo, 11/17/68 (relating events of 6/5/67); Christian statement, 6/5/67; ibid.; Rusk statement, 6/5/67, *Weekly Compilation*, 6/12/67.

48. Eshkol to Johnson, 6/5/67, box 18, NSC history file.

49. Rostow to Saunders, 11/17/68, ibid.

50. Johnson, *Vantage Point*, 297–98.

51. Rostow to Johnson, 6/5/67, box 18, NSC history file.

52. McCloskey statement, 6/5/67, *American Foreign Policy: Current Documents*, 506 n. 67.

53. Wattenberg to Johnson, 5/31/65, box 67, President's appointment file; Levinson and Wattenberg to Johnson, 6/7/67, box 18, NSC history file.

54. Eshkol message in Rostow to Johnson, 6/6/67, box 18, NSC history file; Rusk statement, 6/5/67, *Bulletin*, 6/26/67.

55. Johnson, *Vantage Point*, 299; DOS administrative history, vol. I, pt. 4, sec. H, p. 4.

56. Notes of NSC meeting, 6/7/67, box I, NSC meetings file.

57. CIA report, "Attack on U.S.S. *Liberty* Ordered by Dayan," 11/7/67, cited in Donald Neff, *Warriors for Jerusalem* (New York, 1984), 265 n.

58. Eban to Johnson, 6/8/67; Harman to Johnson, 6/8/67, box 18, NSC history file.

59. Notes of NSC meeting, 6/9/67, box 19, NSC history file.

60. Johnson to Kosygin, 6/8/67 (two messages), ibid.

61. DOS administrative history, vol. I, pt. 4, sec. H, p. 143.

62. Goldberg statement, 6/9/67, *Bulletin*, 6/26/67.

63. Johnson, *Vantage Point*, 302.

64. DOS to all diplomatic and consular posts, 6/9/67, box 19, NSC history file; CIA to White House, 6/8/67, box 18, NSC history file.

65. Johnson, *Vantage Point*, 302–3.

66. Rostow to Johnson, 6/7/67, box 18, NSC history file.

67. McPherson interview, 1/16/69.

68. McPherson to Johnson, 6/11/67, box 18, NSC history file.

69. Barbour to DOS, 6/13/67 and 6/15/67, ibid.

70. Johnson speech, 6/19/67, *Bulletin*, 7/10/67.

71. Rostow to Johnson, 6/15/67, box 18, NSC history file.

72. Rusk to Johnson, 6/17/67, ibid; Rostow to Johnson, 6/21/67, box 11, Rostow files.

73. DOS intelligence note, 6/22/67; Harriman in Rostow to Johnson, 6/24/67; box 229, Country file.

74. Rostow to Johnson, 6/14/67; Roche to Johnson, 6/19/67; ibid.

75. Memos of conversations, 1/23/67 and 1/25/67, box I, Declassified and sanitized documents from unprocessed files.

76. Notes for 6/23/67, President's daily diary file.

77. Johnson statement, 6/25/67, *Current Documents*, 432–33.

8. Vietnam in Context

1. McNamara to Johnson, 12/21/63, *Foreign Relations of the United States, 1961–1963*, vol. 4 (Washington, 1991), 732–35.
2. Mansfield to Johnson, 12/7/63, box 1, Memos to the president file.
3. George McT. Kahin, *Intervention* (New York, 1986), 201–2.
4. Mansfield to Johnson, 1/6/64 and 2/1/64, box 1, Memos to the president file.
5. Rusk to Johnson, 1/8/64, box 1, Vietnam file.
6. McNamara to Johnson, 1/7/64, ibid.
7. Bundy to Johnson, 1/6/64, *Foreign Relations of the United States, 1964–1968*, vol. 1 (Washington, 1992), 14–15.
8. Johnson to Khanh, 2/2/64, box 1, Vietnam file.
9. McNamara to Bundy (copy to Johnson), 3/2/64, box 2, Vietnam file; McNamara to Johnson, 3/16/64, box 1, NSC meetings file.
10. Notes of meeting, 3/17/64, box 1, NSC meetings file.
11. Johnson message to Congress, 5/18/64, box 4, Vietnam file.
12. Johnson, *Vantage Point*, 158–59.
13. Forrestal to Johnson, 5/29/64, box 5, Vietnam file.
14. Bundy to Johnson, 5/25/64, box 1, Memos to the president file.
15. Wheeler to McNamara, 7/27/64; McCone to Johnson, undated, box 1, NSC meetings file.
16. Notes of meeting, 8/4/64, 12:35 P.M., ibid.
17. George C. Herring, *America's Longest War* (New York, 1986), 121.
18. Notes of meeting, 8/4/64, 6:15 P.M., box 1, NSC meetings file.
19. Notes of meeting, 8/4/64, box 1, Bundy papers.
20. Greenfield to Rusk, 8/6/64, box 7, Vietnam file.
21. Notes of meeting, 8/4/64, box 1, Meeting notes file.
22. Taylor to DOS, 9/2/64, *Foreign Relations, 1964–1968*, vol. 1, 724–27.
23. Notes of meeting, 9/14/64, box 8, Vietnam file.
24. CIA report, 10/23/64, box 9, Vietnam file.
25. Notes of meeting, 12/1/64, box 1, Bundy papers.
26. Notes of meeting, 1/6/65, ibid.
27. Notes of meeting, 1/27/65, ibid.
28. Notes of meeting, 1/27/65, ibid.
29. Bundy to Johnson, 2/7/65 and 2/16/65, box 2, Memos to the president file.
30. Notes of meeting, 2/6/65, box 1, NSC meetings file.
31. Bundy to Johnson, 2/7/65, box 2, Memos to the president file.
32. Notes of meeting prior to NSC meeting, 2/8/65, box 1, NSC meetings file.
33. Notes of NSC meeting, 2/8/65, ibid.
34. Notes of meeting, 2/10/65, ibid.
35. Johnson address to Congress, 3/15/65, *Public Papers.*
36. Rostow to Rusk, 12/16/64, box 11, Vietnam file.
37. Herring, *America's Longest War*, 131.
38. Taylor to DOS, 4/14/65, box 3, Memos to the president file.
39. Rusk to Saigon, 4/22/65, box 41, NSC history file; Larry Berman, *Planning a Tragedy* (New York, 1982), 62–63.
40. Kahin, *Intervention*, 320.
41. Johnson message to Congress, 5/4/65, *Public Papers.*
42. Ibid.
43. McNamara to Johnson, 7/20/65, box 43, NSC history file.
44. Notes of meeting, 7/21/65, 10:40 A.M., box 1, Meeting notes file.
45. Notes of meeting, 7/21/65, 2:45 P.M., ibid.
46. Notes of meeting, 7/22/65, ibid.
47. Notes of meeting, 7/25/65, ibid.

48. Johnson address, 7/28/65, *Public Papers.*

49. Notes of NSC meeting, 7/27/65, box 1, Meeting notes file.

50. Notes of meeting with congressional leaders, 7/27/65, ibid.

51. Notes of meeting, 7/27/65, box 1, NSC meetings file.

52. Notes of meeting, 8/5/65, box 2, ibid.

53. Ibid.

54. Notes of meeting, 5/16/65, box 1, Meeting notes file.

55. Bundy to Johnson, 12/4/65, box 5, Memos to the president file.

56. Johnson, *Vantage Point,* 236–37.

57. Notes of meeting, 1/25/66, box 1, Meeting notes file.

58. Notes of meeting, 1/28/66, ibid.

59. Notes of meeting, 1/29/66, ibid.

60. Notes of meetings, 1/29/66 and 6/22/66, box 2, NSC meetings file.

61. Melvin Small, *Johnson, Nixon, and the Doves* (New Brunswick, N.J., 1988), 82–96.

62. Rostow to Johnson, 3/3/66, box 13, Walt Rostow papers.

63. Johnson state of the union address, 1/12/66, *Public Papers;* memo of conversation, 1/17/67, box 52, President's appointment file: diary backup.

64. Kathleen J. Turner, *Lyndon Johnson's Dual War* (Chicago, 1985), 116, 184.

65. Notes of meeting, 2/10/68, box 2, Tom Johnson file.

66. Notes of meetings, 3/20/68 and 3/22/68, box 2, Meeting notes file; Herbert Y. Schandler, *The Unmaking of a President* (Princeton, N.J., 1977), 227.

67. Notes of meeting, 9/12/67, box 2, Meeting notes file.

68. Notes of meeting, 11/2/67, ibid.

69. McNamara to Johnson, 11/1/67, box 3, Rostow files.

70. Notes of meeting, 3/4/68, box 2, Tom Johnson file.

71. Notes of meeting, 3/36/68, box 2, Meeting notes file; Bundy to Johnson, 3/21/68, Bundy folder, Office files of the president.

72. Notes of meeting, 3/36/68, box 2, Meeting notes file.

Select Bibliography

Acheson, Dean. *Present at the Creation*. New York, 1969.

Ambrose, Stephen E. *Eisenhower*. Vol. 2. New York, 1984.

Anderson, Patrick. *The Presidents' Men: White House Assistants of Franklin D. Roosevelt, Harry S. Truman, Dwight D. Eisenhower, John F. Kennedy, and Lyndon B. Johnson*. Garden City, N.Y., 1968.

Ball, George W. *The Past Has Another Pattern*. New York, 1982.

Bar-Zohar, Michael. *Embassies in Crisis: Diplomats and Demagogues Behind the Six-Day War*. Translated by Monroe Stearns. Englewood Cliffs, N.J., 1970.

Barnet, Richard J. *Intervention and Revolution: The United States in the Third World*. New York, 1968.

Barrett, David M. *Uncertain Warriors: Lyndon Johnson and His Vietnam Advisers*. Lawrence, Kans, 1993.

Berman, Larry. *Lyndon Johnson's War*. New York, 1989.

———. *Planning a Tragedy: The Americanization of the War in Vietnam*. New York, 1982.

Bill, James A. *The Eagle and the Lion: The Tragedy of American-Iranian Relations*. New Haven, Conn., 1988.

Bohlen, Charles E. *Witness to History*. New York, 1973.

Bornet, Vaughn Davis. *The Presidency of Lyndon B. Johnson*. Lawrence, Kans., 1983.

Bowles, Chester. *Promises to Keep*. New York, 1971.

Braestrup, Peter. *Big Story: How the American Press and Television Reported and Interpreted the Crisis of Tet 1968 in Vietnam and Washington*. Boulder, Colo., 1977.

Brands, H. W. "America Enters the Cyprus Tangle," *Middle Eastern Studies*, July 1987.

———. *Cold Warriors: Eisenhower's Generation and American Foreign Policy*. New York, 1988.

———. *India and the United States*. Boston, 1990.

———. "Johnson and Eisenhower: The President, the Former President and the War in Vietnam," *Presidential Studies Quarterly*, Summer 1985.

———. "The Limits of Manipulation: How the United States Didn't Topple Sukarno," *Journal of American History*, December 1989.

———. *The Specter of Neutralism: The United States and the Emergence of the Third World, 1947–1960*. New York, 1989.

Brinkley, Douglas. *Dean Acheson: The Cold War Years, 1953–1971*. New Haven, Conn., 1992.

Brown, Seyom. *The Faces of Power: Constancy and Change in United States Foreign Policy from Truman to Johnson*. New York, 1968.

Bundy, McGeorge. *Danger and Survival: Choices About the Bomb in the First Fifty Years*. New York, 1988.

Burke, S. M. *Mainsprings of Indian and Pakistani Foreign Policies*. Minneapolis, 1974.

Cabot, John Moors. *First Line of Defense: Forty Years' Experiences of a Career Diplomat*. Lanham, Md., 1979.

Califano, Joseph A., Jr. *The Triumph and Tragedy of Lyndon Johnson*. New York, 1991.

Caro, Robert. *The Years of Lyndon Johnson*. New York, 1982–.

Clifford, Clark, with Richard Holbrooke, *Counsel to the President*. New York, 1991.

Cohen, Warren. *Dean Rusk*. Totowa, N.J., 1980.

Conkin, Paul K. *Big Daddy from the Pedernales: Lyndon Baines Johnson*. Boston, 1986.

Cooper, Chester L. *The Lost Crusade: America in Vietnam*. New York, 1970.

Costigliola, Frank. *France and the United States*. New York, 1992.

Crozier, Brian. *De Gaulle*. New York, 1973.

Dallek, Robert. *Lone Star Rising: Lyndon Johnson and His Times, 1908–1960*. New York, 1991.

Donovan, Robert J. *Nemesis: Truman and Johnson in the Coils of War in Asia*. New York, 1984.

Draper, Theodore. *The Dominican Revolt*. New York, 1968.

Dugger, Ronnie. *The Politician: The Life and Times of Lyndon Johnson—The Drive for Power, from the Frontier to Master of the Senate*. New York, 1982.

Dulles, Foster Rhea. *American Policy toward Communist China, 1949–1969*. New York, 1972.

Eban, Abba. *An Autobiography*. New York, 1977.

Ehrlich, Thomas. *Cyprus*. New York, 1974.

Ennes, James M., Jr. *Assault on the* Liberty. New York, 1979.

Evans, Rowland, and Robert Novak, *Lyndon B. Johnson*. New York, 1966.

Farnsworth, David, and James McKenney. *U.S.–Panama Relations*. Boulder, Colo., 1983.

Fitzgerald, Frances. *Fire in the Lake: The Vietnamese and the Americans in Vietnam*. Boston, 1972.

Foreign Relations of the United States, 1961–1963. Vol. 4: *Vietnam, August–December* 1963. Washington, 1991.

Foreign Relations of the United States, 1964–1968. Vol. I: *Vietnam,* 1964. Washington, 1992.

Fulbright, J. William. *The Arrogance of Power.* New York, 1967.

———. *The Crippled Giant.* New York, 1972.

Gaddis, John L. *Strategies of Containment.* New York, 1982.

Galloway, John. *The Gulf of Tonkin Resolution.* Rutherford, N.J., 1970.

Gatzke, Hans W. *Germany and the United States.* Cambridge, Mass., 1980.

Gelb, Leslie H., and Richard K. Betts. *The Irony of Vietnam: The System Worked.* Washington, D.C., 1979.

Geyelin, Philip. *Lyndon B. Johnson and the World.* New York, 1966.

Gleijeses, Piero, *The Dominican Crisis.* Baltimore, 1978.

Goldman, Eric F. *The Tragedy of Lyndon Johnson.* New York, 1969.

Goodwin, Doris Kearns. *Lyndon Johnson and the American Dream.* New York, 1977.

Goodwin, Richard N. *Remembering America: A Voice from the Sixties.* Boston, 1988.

———. "The War Within," *New York Times Magazine,* August 21, 1988.

Goswami, P. K. *Ups and Downs of Indo-U.S. Relations.* Columbia, Mo., 1984.

Goulden, Joseph C. *Truth is the First Casualty: The Gulf of Tonkin Affair.* Chicago, 1969.

Graff, Henry F. *The Tuesday Cabinet.* Englewood Cliffs, N.J., 1970.

Green, Stephen. *Taking Sides: America's Secret Relations with a Militant Israel.* New York, 1984.

Halberstam, David. *The Best and the Brightest.* New York, 1972.

———. *The Making of a Quagmire.* New York, 1964.

Halperin, Maurice. *The Taming of Fidel Castro.* Berkeley, Calif., 1981.

Halperin, Morton H. *The Decision to Deploy the ABM.* Washington, 1973.

Harrison, James P. *The Endless War: Vietnam's Struggle for Independence.* New York, 1982.

Harrison, Selig S. *The Widening Gulf: Asian Nationalism and American Policy.* New York, 1978.

Hathaway, Robert M. *Great Britain and the United States.* Boston, 1990.

Heikal, Mohamed Hassanein. *The Cairo Documents: The Inside Story of Nasser and His Relationship with World Leaders, Rebels, and Statesmen.* Garden City, N.Y., 1973.

Henggeler, Paul R. *In His Steps: Lyndon Johnson and the Kennedy Mystique.* Chicago, 1991.

Herring, George C. *America's Longest War: The United States and Vietnam,* 1950–1975. New York, 1979.

Hilsman, Roger. *The Politics of Policy Making in Defense and Foreign Affairs.* New York, 1971.

Hoopes, Townsend. *The Limits of Intervention.* New York, 1969.

Humphrey, David C. "Tuesday Lunch at the Johnson White House," *Diplomatic History,* Winter 1984.

Isaacson, Walter, and Evan Thomas. *The Wise Men: Six Friends and the World They Made: Acheson, Bohlen, Harriman, Kennan, Lovett, McCloy.* New York, 1986.

Johnson, Lyndon Baines. *The Vantage Point.* New York, 1971.

Johnson, Richard A. *The Administration of U.S. Foreign Policy.* Austin, 1971.

Johnson, Sam Houston. *My Brother Lyndon.* New York, 1969.

Jones, Howard Palfrey. *Indonesia.* New York, 1971.

Jorden, William J. *Panama Odyssey*. Austin, 1984.

Kahin, George McT. *Intervention: How America Became Involved in Vietnam*. New York, 1986.

Kalb, Marvin, and Elie Abel. *Roots of Involvement: The U.S. in Asia, 1784–1971*. New York, 1971.

Kaplan, Lawrence S. *NATO and the United States*. Boston, 1988.

Karnow, Stanley. *Vietnam*. New York, 1983.

Kaufmann, William W. *The McNamara Strategy*. New York, 1964.

Kolko, Gabriel. *Anatomy of a War: Vietnam, the United States, and the Modern Historical Experience*. New York, 1985.

———. *The Roots of American Foreign Policy*. Boston, 1969.

Kolodziej, Edward A. *French International Policy under de Gaulle and Pompidou*. Ithaca, N.Y., 1974.

Kurzman, Dan. *Santo Domingo*. New York, 1965.

LaFeber, Walter. *Inevitable Revolutions: The United States in Central America*. New York, 1984.

———. *The Panama Canal*. New York, 1978.

Lefever, Ernest W. *Crisis in the Congo*. Washington, 1965.

Levinson, Jerome, and Juan de Onis. *The Alliance That Lost Its Way: A Critical Report on the Alliance for Progress*. Chicago, 1970.

Lewy, Guenter. *America in Vietnam*. New York, 1978.

Little, Douglas. "From Even-Handed to Empty-Handed." In Thomas G. Paterson, ed., *Kennedy's Quest for Victory*. New York, 1989.

Lodge, Henry Cabot. *The Storm Has Many Eyes*. New York, 1973.

Lowenthal, Abraham F. *The Dominican Intervention*. Cambridge, Mass., 1972.

Mandelbaum, Michael. *The Nuclear Question*. New York, 1979.

Martin, John Bartlow. *Overtaken by Events: The Dominican Crisis from the Fall of Trujillo to the Civil War*. New York, 1966.

McLellan, David S., and David C. Acheson, eds., *Among Friends: Personal Letters of Dean Acheson*. New York, 1980.

McMahon, Robert J. *Colonialism and the Cold War: The United States and the Struggle for Indonesian Independence, 1945–1949*. Ithaca, N.Y., 1981.

Mueller, John E. *War, Presidents and Public Opinion*. New York, 1973.

Muslin, Hyman L., and Thomas H. Jobe. *Lyndon Johnson: The Tragic Self*. New York, 1991.

Neff, Donald. *Warriors for Jerusalem: The Six Days That Changed the Middle East*. New York, 1984.

Newhouse, John. *Cold Dawn: The Story of SALT*. New York, 1973.

O'Balance, Edgar. *The Wars in Vietnam*. New York, 1983.

Oberdorfer, Don. *Tet*. Garden City, N.Y., 1971.

Packenham, Robert A. *Liberal America and the Third World*. Princeton, N.J., 1973.

Palmer, Norman D. *South Asia and United States Policy*. Boston, 1966.

Papandreou, Andreas. *Democracy at Gunpoint*. Garden City, N.Y., 1970.

Parker, Phyllis R. *Brazil and the Quiet Intervention, 1964*. Austin, 1979.

Podhoretz, Norman. *Why We Were in Vietnam*. New York, 1982.

Powers, Thomas. *The War at Home: Vietnam and the American People, 1964–1968*. New York, 1973.

Prados, John F. *Presidents' Secret Wars: CIA and Pentagon Covert Operations Since World War II*. New York, 1986.

Quandt, William B. *Decade of Decisions: American Policy Toward the Arab-Israeli Conflict,* 1967–1976. Berkeley, Calif., 1977.

Quester, George H. *The Politics of Nuclear Proliferation.* Baltimore, 1973.

Quill, J. Michael. *Lyndon Johnson and the Southern Military Tradition.* Washington, D.C., 1977.

Rabe, Stephen G. *The Road to* OPEC: *United States Relations with Venezuela,* 1919–1976. Austin, 1982.

Rabin, Yitzhak. *The Rabin Memoirs.* Boston, 1979.

Ranelagh, John. *The Agency: The Rise and Decline of the* CIA. New York, 1986.

Reedy, George E. *Lyndon B. Johnson.* New York, 1982.

Rostow, Walt W. *The Diffusion of Power: An Essay in Recent History.* New York, 1972.

Rubin, Barry. *Paved with Good Intentions: The American Experience and Iran.* New York, 1980.

———. *Secrets of State: The State Department and the Struggle Over U.S. Foreign Policy.* New York, 1985.

Rusk, Dean, as told to Richard Rusk. *As I Saw It.* New York, 1990.

Rudolph, Lloyd I., ed. *The Regional Imperative: The Administration of U.S. Foreign Policy Towards South Asian States Under Presidents Johnson and Nixon.* New Delhi, 1980.

Ryan, Paul R. *The Panama Canal Controversy.* Stanford, Calif., 1977.

Schandler, Herbert Y. *The Unmaking of a President: Lyndon Johnson and Vietnam.* Princeton, N.J., 1977.

Schlesinger, Arthur M., Jr. *The Bitter Heritage: Vietnam and American Democracy.* Boston, 1966.

———. *The Crisis of Confidence.* Boston, 1969.

Schoenbaum, David. *The United States and the State of Israel.* New York, 1993.

Schoenbaum, Thomas J. *Waging Peace and War: Dean Rusk in the Truman, Kennedy, and Johnson Years.* New York, 1988.

Shapley, Deborah. *Promise and Power: The Life and Times of Robert McNamara.* Boston, 1992.

Sidey, Hugh. *A Very Personal Presidency: Lyndon Johnson in the White House.* New York, 1968.

Skidmore, Thomas E. *The Politics of Military Rule in Brazil,* 1964–1985. New York, 1988.

Slater, Jerome. *Intervention and Negotiation: The United States and the Dominican Revolution.* New York, 1970.

Small, Melvin. *Johnson, Nixon, and the Doves.* New Brunswick, N.J., 1988.

Spector, Ronald H. *After Tet.* New York, 1992.

Spiegel, Steven L. *The Other Arab-Israeli Conflict: Making America's Middle East Policy, from Truman to Reagan.* Chicago, 1985.

Stavins, Ralph, Richard Barnet, and Marcus Raskin. *Washington Plans an Aggressive War.* New York, 1971.

Steel, Ronald. *Pax Americana.* New York, 1967.

Stern, Laurence. *The Wrong Horse: The Politics of Intervention and the Failure of American Diplomacy.* New York, 1977.

Stoessinger, John G. *Crusaders and Pragmatists: Movers of Modern American Foreign Policy.* New York, 1979.

Sullivan, Marianna P. *France's Vietnam Policy.* Westport, Conn., 1978.

Summers, Harry G., Jr. *On Strategy: A Critical Analysis of the Vietnam War.* Novato, Calif., 1982.

Szulc, Tad. *Dominican Diary.* New York, 1965.

Taylor, Maxwell D. *Swords and Plowshares.* New York, 1973.

Thompson, James C. *Rolling Thunder.* Chapel Hill, N.C., 1980.

Treverton, Gregory F. *The Dollar Drain and American Forces in Europe.* Athens, Ohio, 1978.

Trewhitt, Henry L. *McNamara.* New York, 1971.

Turner, Kathleen J. *Lyndon Johnson's Dual War: Vietnam and the Press.* Chicago, 1985.

VanDeMark, Brian. *Into the Quagmire: Lyndon Johnson and the Escalation of the Vietnam War.* New York, 1991.

Weintal, Edward, and Charles Bartlett, *Facing the Brink: An Intimate Study of Crisis Diplomacy.* New York, 1967.

Weintraub, Sidney, ed. *Economic Coercion and U.S. Foreign Policy.* Boulder, Colo., 1982.

Weissman, Stephen R. *American Foreign Policy in the Congo.* Ithaca, N.Y., 1974.

Westmoreland, William C. *A Soldier Reports.* Garden City, N.Y., 1976.

White, William S. *The Professional: Lyndon B. Johnson.* Boston, 1964.

Wilson, Harold, *A Personal Record.* Boston, 1971.

Windchy, Eugene C. *Tonkin Gulf.* Garden City, N.Y., 1971.

Wittner, Lawrence S. *American Intervention in Greece.* New York, 1982.

Index